Technical Change
in Asian Agriculture

# Technical Change in Asian Agriculture

*Editor*

R. T. Shand

AUSTRALIAN NATIONAL UNIVERSITY PRESS
CANBERRA
1973

First published in Australia 1973

Printed in Australia for the Australian National University
Press, Canberra

United Kingdom, Europe, Middle East, Africa and Caribbean: Angus
& Robertson (U.K.) Ltd, London

North, South, and Central America: International Scholarly Book
Services, Inc., Portland, Oregon

Southeast Asia: Angus & Robertson (S.E. Asia) Pty Ltd, Singapore

Japan: United Publishers Services Ltd, Tokyo



Registered in Australia for transmission by post as a book
National Library of Australia Card no.
and ISBN 0 7081 0702 8
Library of Congress Catalog Card no. 72-85677

# Foreword

THIS volume presents a timely assessment of what has been commonly referred to as the 'green revolution' in Asia's agriculture. It originated in a workshop seminar held in Canberra, January 1971 at the 28th International Congress of Orientalists. The aim of that seminar was to assemble and consider evidence of the progress and impact of the new agricultural technology, by bringing together scholars who were actively involved in field studies of the topic. Papers delivered at the seminar have been revised, further contributions have been solicited to widen the geographic coverage and scope of the inquiry, and the editor has written a final integrating chapter.

While such technical change is novel for some Asian countries it is by no means new for the region as a whole. Thus historical precedents are considered in an initial section of three papers. Iinuma examines the impact of foreign science and expertise introduced to Meiji Japan and explores their relation to, and eventual fusion with, advanced indigenous knowledge and practice. Hayami reviews the transmission of advanced technology in rice production from Japan to Taiwan and Korea. Carr and Myers examine the development of ponlai rice in Taiwan in detail and assess the social benefits of the research effort involved.

The core of the volume consists of the empirical analyses of recent technical change in foodgrain production in five countries: India, Indonesia, Philippines, Thailand and Burma. Each reports the findings of field studies in which the authors have been, and mostly still are, involved. Each considers the recent record of adoption of new high yielding varieties and of the use of other complementary factors in the input 'package' of the new technology. The scope of these studies varies quite widely but individually and collectively they prove the value of continuing field-level assessment of the progress of the new technology.

These studies are then followed by a number of challenging contributions concerned with actual, or still potential, secondary repercussions of the new technology, on which the individual authors are well qualified to write. Kaneda has been examining agro-industrial linkages over a considerable period in Pakistan. Whitcombe's critical comments of the long-term repercussions of major canal systems and their accompanying infrastructure come from a long-term study of the topic, the first results of which have recently been published. Castillo brings to bear a wealth of experience in the field of socio-economic research in rural development. Ladejinsky's unique expertise in land reform needs no introduction, while Dantwala's experience and erudition in writings on policy have long been recognised.

This volume effectively complements an earlier one, *Agricultural Development in Asia,* with the same editor. In it, contributors considered agricultural development experience on a country by country basis up to the time of the introduction of the new technology. That volume laid the basis for an understanding of past agricultural progress. This volume serves to update the first through a period of accelerated growth in agricultural production. In combination, the volumes provide rich insights into some of the most pressing and contentious questions of our time—the capacity of Asia to feed a fast growing population and to raise living standards at the same time. The editor's chapter is a particularly useful and cautionary reminder of the questions to be recognised by policy makers grappling with this greatest of Asian problems. Unless these are taken seriously, the rich promise of new agricultural technologies is likely to result in damaging social and political instability.

J. G. Crawford

# Notes on Contributors

JIRO IINUMA, Associate Professor, Research Institute for Humanistic Studies, Kyoto University

YUJIRO HAYAMI, Professor of Economics, Tokyo Metropolitan University; recently Visiting Professor, University of Minnesota

CAROLLE CARR, economist; formerly with the International Bank for Reconstruction and Development

RAMON H. MYERS, Professor of Economics, University of Miami, Coral Gables, Florida; formerly Department of Economics, Research School of Pacific Studies, The Australian National University

P. K. MUKHERJEE, formerly Chief, Programme Evaluation Organisation, Indian Planning Commission; in 1971 Visiting Fellow, The Australian National University; currently with the United Nations, New York

BRIAN LOCKWOOD, formerly Research Fellow, Research School of Pacific Studies, The Australian National University; now Senior Lecturer, University of the South Pacific, Fiji

SAJOGYO, Professor, Department of Social Economics, Bogor Agricultural University, Bogor, Indonesia

WILLIAM L. COLLIER, Agricultural Development Council and Visiting Professor, University of Bogor, Indonesia

RANDOLPH BARKER, formerly a member of the Faculty of Cornell University; since 1965, Visiting Professor at the University of the Philippines; currently Agricultural Economist, International Rice Research Institute

VIOLETA CORDOVA, graduate of the University of the Philippines, College of Agriculture; Research Assistant, International Rice Research Institute

DELANE E. WELSCH, Rockefeller Foundation and Visiting Professor, Department of Agricultural Economics, Kasetsart University, Thailand

SOPIN TONGPAN, Lecturer, Department of Agricultural Economics, Kasetsart University, Thailand

H. V. RICHTER, formerly with the *Economist* Intelligence Unit; now Research Fellow in the Department of Economics, Research School of Pacific Studies, The Australian National University

HIROMITSU KANEDA, Associate Professor of Economics, University of California, Davis; in 1968-9 served as Research Adviser at Pakistan Institute of Development Economics, Karachi

ELIZABETH WHITCOMBE, Senior Research Fellow, Department of Economics, Research School of Pacific Studies, The Australian National University

GELIA T. CASTILLO, Associate Professor of Rural Sociology, College of Agriculture, University of the Philippines, Laguna

ALICE M. DE GUZMAN, SOLEDAD L. PAHUD, LORNA PAJE, Research Assistants, College of Agriculture, University of the Philippines, Laguna

WOLF LADEJINSKY, Resident Representative, The International Bank for Reconstruction and Development, New Delhi

M. L. DANTWALA, Professor of Economics, University of Bombay, Bombay

R. T. SHAND, Senior Fellow, Department of Economics, Research School of Pacific Studies, The Australian National University

Acknowledgments

Chapter 9. To Professors Frank C. Child and W. Eric Gustafson who contributed useful comments to the improvement of this paper.

Chapter 13. To the Indian Economic Association to whom this paper was also delivered as a Presidential Address to the 53rd Annual Conference.

Chapter 14. To various contributing authors, members of the Department of Economics, Research School of Pacific Studies and particularly Mr E. K. Fisk, for much constructive and helpful comment on drafts.

All contributors wish to record their gratitude to Mrs Heather Harding, Secretary of the Department of Economics, Research School of Pacific Studies, Australian National University and to members of her staff (Mrs Harriss, Miss Hall, Mrs Inglis, Mrs Newman) not only for their patient assistance in the preparation and typing of the manuscript, but for helping to make the original workshop seminar at the 28th International Congress of Orientalists such an enjoyable experience. As in the earlier volume we are indebted to Mr H. Gunther and Mr M. Pancino of the Department of Human Geography, Australian National University, for maps.

*Canberra, 1973* R.T.S.

# Contents


Foreword, by Sir John Crawford v

Notes on Contributors vii

Acknowledgments ix

Tables and Figures xiii

Currencies, Statistical Units and Conversions xviii


## Historical Precedents


1 The Introduction of American and European Agricultural Science into Japan in the Meiji Era *Jiro Iinuma* 1

2 Development and Diffusion of High Yielding Rice Varieties in Japan, Korea and Taiwan, 1890-1940 *Yujiro Hayami* 9

3 The Agricultural Transformation of Taiwan: the Case of Ponlai Rice, 1922-42 *Carolle Carr & Ramon H. Myers* 28


## Current Performance


4 High Yielding Varieties Programme in India: an Assessment *P. K. Mukherjee & Brian Lockwood* 51

5 Adoption of New High Yielding Rice Varieties by Java's Farmers *Sajogyo & William L. Collier* 80

6 The Impact of New Technology on Rice Production: a Study of Change in Three Philippine Municipalities, 1966-9 *Randolph Barker & Violeta Cordova* 108

7 Background to the Introduction of High Yielding Varieties of Rice in Thailand *Delane E. Welsch & Sopin Tongpan* 124

8 The High Yield Program in Burma *H. V. Richter* 144

## Broader Implications

9 Mechanisation, Industrialisation, and Technical Change in Rural Pakistan *Hiromitsu Kaneda* 161

10 The New Agricultural Strategy in Uttar Pradesh, India, 1968-70: Technical Problems *Elizabeth Whitcombe* 183

11 The Green Revolution at the Village Level: a Philippine Case Study, 1963-70 *Gelia T. Castillo, Alice M. de Guzman, Soledad L. Pahud & Lorna Paje* 202

12 Agrarian Reform in Asia: the Green Revolution and its Reform Effects *Wolf Ladejinsky* 235

13 From Stagnation to Growth: Relative Roles of Technology, Economic Policy and Agrarian Institutions *M. L. Dantwala* 259

## Conclusion

14 An Interim Judgment *R. T. Shand* 282

Index 313

# Tables and Figures

## Tables


| | | Page |
|---|---|---|
| 1.1 | Rice yields obtained by leading farmers, Japan, *circa* 1889 | 3 |
| 1.2 | Family rank of students of the Komaba Agricultural School, Japan, 1877-81 | 4 |
| 1.3 | Agricultural schools in Japan, 1882 | 5 |
| 1.4 | The subject matter of student notebooks of the Fukuoka Agricultural School, *circa* 1886 | 6 |
| 2.1 | Ratio of area planted in improved seeds, fertiliser input and rice yield per hectare: Japan, Taiwan and Korea, selected years | 17 |
| 2.2 | Comparison of the costs of ponlai and chailai rice production per chia, Japan, 1926-7 | 21 |
| 2.3 | Irrigation and double cropping ratios in paddy field, Korea and Taiwan, selected years | 23 |
| 2.4 | Factor shares in the cost of rice production, Japan and Taiwan, 1925-7 | 25 |
| 3.1 | Indexes of foodgrain production, yield and cultivated area in Taiwan, 1900-40 | 28 |
| 3.2 | Foodgrain supply, rice production and rice exports to Japan, 1901-40 | 30 |
| 3.3 | Main types of rice produced in Taiwan, 1901-40 | 31 |
| 3.4 | Percentage of total rice area devoted to ponlai in seven prefectures of Taiwan, 1926-42 | 33 |
| 3.5 | Rate of adoption of ponlai for seven prefectures and for all Taiwan, 1926-42 | 35 |
| 3.6 | Ponlai seed research expenditures and net social returns, Taiwan, 1911-42 | 40 |
| 3.7 | Rate of return on hybrid rice research expenditures, Taiwan, 1942 | 41 |
| 3.8 | Labour and work-animal inputs on Taiwanese rice farms, 1936/7 | 43 |
| 3.9 | First rice crop labour and work-animal inputs per unit of land, Taiwan Survey, 1936/7 | 44 |
| 3.10 | Input costs for an average-sized farm, Taiwan Survey, 1936/7 | 47 |

3.11 Comparison of revenue and costs per unit of land from cultivating chailai and ponlai, Taiwan, second crop of 1925 and first crop of 1926 48

3.12 Comparison of 1937 yield and area values of different rice, Taiwan 49

4.1 High yielding variety seed use amongst farmers in Survey villages, India, 1967/8 to 1969/70 61

4.2 Percentage area under high yielding variety crops to total area under the selected crop, India, 1967-70 71

4.3 Average dose of nitrogen applied by farm size group of participants, India, 1967-9 71

4.4 Percentage of farmers adopting package of high yielding variety practices by farm size groups, India, 1967-9 72

5.1 Decline in growth of rice production, Java, 1953-67 80

5.2 Rice producing regions and AES randomly chosen *kabupaten*, Java, 1969-71 84

5.3 Average farm size of the AES sample farmers, Java, 1968/9 85

5.4 Java's operated farm size distribution, 1963 85

5.5 Personal traits of the sample farmers, Java 86

5.6 Adoption patterns for IR varieties, Java, 1968-71 88

5.7 Adoption patterns for national improved and local varieties, Java, 1968-71 88

5.8 Paddy farm size and planting of rice varieties, Java, 1968-70 89

5.9 Land tenure status and planting of rice varieties, Java, 1968-70 90

5.10 Farmer purchase and sale of rice and planting of rice varieties, Java, 1968-70 92

5.11 Off-farm income earning and planting of rice varieties, Java, wet season, 1968/9 92

5.12 Sample farmers' participation in Bimas credit programs and their planting of rice varieties, Java, 1968-70 94

5.13 Average pre-harvest labour use in paddy farming, by rice varieties, Java, wet season, 1969/70 95

5.14 Inputs of hired and family labour, used by the sample paddy farmers, Java, wet season, 1969/70 97

5.15 Farmers' average use of fertilisers and pesticides, Java, wet season, 1969/70 compared with wet season, 1970/1 98

5.16 Percentage increases in yields for five seasons, Java, 1968-71 100

5.17 Rice yields by varieties for five seasons, Java, 1969-71 102

5.18 Sample farmers' average yields for four seasons, Java, 1969-71 104

5.19 Average net benefits of paddy farm operation, by rice varieties, wet season, Java, 1970/1 105
6.1 Comparison of farm size, yields of rice, and double cropping in three municipalities of Laguna, 1954/5 and 1966/7 110
6.2 Percentage of area planted in high yielding, glutinous, and other local varieties in three municipalities of Laguna, wet seasons 1966-9 and dry seasons 1967-9 112
6.3 Returns and costs per hectare for rice farms in three municipalities of Laguna, wet seasons 1966 and 1969 113
6.4 Fertiliser users, nitrogen dosage and rice yields by location and adoption of high yielding varieties, Laguna, wet seasons 1966 and 1969 115
6.5 Percentages of users and average expenses per hectare for herbicides and insecticides, by location and adoption of high yielding varieties, Laguna, wet seasons 1966 and 1969 116
6.6 Labour used and percentage hired for specific practices on forty-two farms planted in high yielding varieties in 1969, Laguna, wet seasons 1966 and 1969 118
6.7 Money advanced by landlords for specific items, Laguna, 1965 and 1969 118
6.8 Output distribution, by factors, on forty-two farms shifting from local varieties in 1966 to high yielding varieties in 1969, Laguna, wet seasons 120
6.9 Output distribution, by recipients, on forty-two farms shifting from local varieties in 1966 to high yielding varieties in 1969, Laguna, wet seasons 121
7.1 Volume of rice exports from Thailand, 1857-1944 125
7.2 Volume and value of rice exports from Thailand, 1946-71 126
7.3 Volume of Thai rice exports, percentage of total by country of destination, 1957-69 127
7.4 Volume of Thai rice exports by grade, percentage of total, 1966-9 128
7.5 Rice area harvested, damaged area, production and yield, Thailand, 1907-67 130
7.6 Thai rice area planted and yield per planted hectare by region, 1950/1 to 1966/7 132
7.7 Areas planted and irrigated, method of planting and type of rice, by geographic region, Thailand, 1959-70 133
8.1 Land use, Burma, 1966/7 146
8.2 Population and rice production, Burma, 1965/6 147
8.3 Irrigated areas, by crops, Burma, 1970/1 147
8.4 Lower Burma sown areas, 1970/1 148

| | | |
|---|---|---|
| 8.5 | Upper Burma sown areas, 1970/1 | 149 |
| 8.6 | The states and Chin special division sown areas, Burma, 1970/1 | 150 |
| 8.7 | High yield rice production, Burma, 1967-71 | 153 |
| 8.8 | Distribution of high yield rices, Burma, 1969-71 | 154 |
| 8.9 | Area of field crops other than rice, Burma, 1966-71 | 157 |
| 8.10 | Upland crops: irrigation, fertiliser and yields, Burma, 1970/1 | 157 |
| 8.11 | Yields of certain crops, Burma, 1967/8 | 158 |
| 8.12 | Value of quality seed distributed, Burma, 1961-71 | 158 |
| 9.1 | Agricultural labour force, planted area and gross crop production, Pakistan, selected years 1951-69 | 169 |
| 9.2 | Irrigation water (field availability), by sources, Pakistan, 1960 and 1965 | 171 |
| 9.3 | Estimated areas in Pakistan covered by improved varieties, 1965/6 to 1969/70 | 173 |
| 9.4 | Area and output of wheat and rice, and fertiliser use in Pakistan, 1955/6 to 1969/70 | 173 |
| 11.1 | Proportions of farmers adopting specific rice practices over six cropping seasons, Philippines, 1964/5 to 1967 | 203 |
| 11.2 | Yield distribution, Philippines, 1965-7 | 206 |
| 11.3 | Housing materials and values, Philippines, 1963 and 1969 | 212 |
| 11.4 | Respondents' views on ways to improve farming, Philippines, 1963 and 1969 | 216 |
| 11.5 | Tenure status and adoption of recommended rice practices, Philippines, 1963 and 1967 | 220 |
| 11.6 | Sources of loan in cash and/or kind, Philippines, 1963 and 1967 | 221 |
| 11.7 | Sufficiency of income or savings for old age, Philippines, 1963 and 1969 | 222 |
| 13.1 | Prices of selected agricultural commodities, 1962-9 | 281 |


# Figures


| | | |
|---|---|---|
| 2.1 | Rice yields per hectare planted for Japan, Taiwan and Korea, five-year moving averages, 1895-1935 | 11 |
| 2.2 | Changes in the percentage of area planted in improved varieties to total area planted in rice, Japan, 1875-1963 | 15 |
| 2.3 | Fertiliser's share in agricultural output for Japan, Taiwan and Korea, five-year moving averages, 1895-1935 | 26 |
| 3.1 | Supply and demand for rice with different supply elasticities, Taiwan, 1922-42 | 37 |

4.1 High yielding variety adoption, India, 1967/8 to 1969/70 60
4.2 High yielding variety paddy adoption, by decile groups of operational holdings, India, 1967/8 to 1969/70 66
4.3 High yielding variety wheat adoption, by decile groups of operational holdings, India, 1967/8 to 1969/70 68
5.1 Locations of sample villages, Java 83
6.1 Location of sample *barrios* and the major sources of irrigation in Calamba, Cabuyao and Biñan, Laguna 109
6.2 Typical cropping pattern for rice in three municipalities of Laguna 111
6.3 Distribution of increase in profits on forty-two Laguna farms, wet seasons, 1966-9 121
10.1 The United Provinces of Agra and Oudh 184

*Figures drawn by the Cartographic Office, Department of Human Geography, Australian National University*

# Currencies, Statistical Units, and Conversions

*Currencies*

| *Country* | *Unit* | *Rate per US dollar (selling) as at mid-1971* |
|---|---|---|
| Burma | Kyat (K) | 4·76 |
| India | Rupee (Rs) | 7·50 |
| Indonesia | Rupiah (Rp) | 415* |
| Pakistan | Rupee (Rs) | 4·75 |
| Philippines | Peso (₱) | 6·43 |
| Taiwan | New Taiwan Dollar (NT$) | 40·10 |
| Thailand | Baht | 21·00 |

*Area*

| | | |
|---|---|---|
| 1 acre | = | 0·40 hectare |
| 1 hectare | = | 2·47 acres |
| 1 mile | = | 1·61 kilometres |
| 1 sq. mile | = | 640 acres |
| | = | 259 hectares |
| | = | 2·59 sq. kilometres |
| 1 sq. kilometre | = | 100 hectares |
| | = | 0·39 sq. mile |
| 1 rai (Thailand) | = | 0·40 acre |
| | = | 0·16 hectare |
| 1 acre | = | 2·53 rai |
| 1 hectare | = | 6·25 rai |
| 1 cho (Japan) | = | 2·45 acres |
| | = | 0·99 hectares |
| | = | 10 tan |
| 1 acre | = | 0·41 cho |
| 1 hectare | = | 1·01 cho |
| 1 chia (Taiwan) | = | 2·40 acres |
| | = | 0·97 hectare |
| 1 acre | = | 0·42 chia |
| 1 hectare | = | 1·03 chia |

* Import rate from August 1971

*Weights*

| | |
|---|---|
| 1 pound | = 0·45 kilogram |
| 1 kilogram | = 2·20 pounds |
| 1 long ton | = 2,240 pounds |
| | = 1·02 metric ton |
| 1 short ton | = 2,000 pounds |
| | = 0·91 metric ton |
| 1 metric ton | = 0·98 long ton |
| | = 1·10 short ton |
| | = 1,000 kilograms |
| | = 21·74 Spanish quintals |
| 1 Spanish quintal (Philippines) | = 46·01 kilograms |
| 1 quintal (India) | = 100 kilograms |

*Weight Equivalent of Volume Measures (approximate)*

| | |
|---|---|
| 1 bushel of paddy (US) | = 45 pounds |
| | = 20·4 kilograms |
| 1 bushel of paddy (Australian) | = 42 pounds |
| | = 19·1 kilograms |
| 48·99 bushels of paddy (US) | = 1 metric ton |
| 52·49 bushels of paddy (Australian) | = 1 metric ton |
| Burma: 1 basket of paddy | = 46 pounds |
| | = 20·9 kilograms |
| 48 baskets of paddy | = 1 metric ton |
| 1 basket of rice | = 76 pounds |
| | = 34·5 kilograms |
| Japan: 1 koku of rough rice (paddy) | = 106·3 kilograms |
| | = 234·4 pounds |
| 1 koku of brown (husked) rice | = 150 kilograms |
| | = 330·7 pounds |
| Philippines: 1 cavan of paddy (palay) | = 44 kilograms |
| | = 97 pounds |
| 22·7 cavans of paddy | = 1 metric ton |
| 1 cavan of milled rice | = 56 kilograms |
| | = 123·4 pounds |
| Philippines: 1 cavan of shelled rice | = 57 kilograms |
| | = 125·7 pounds |
| 17·5 cavans of shelled maize | = 1 metric ton |

*Weight Equivalent of Volume Measures (approximate) (cont.)*

| | |
|---|---|
| Thailand: 1 picul | = 60 kilograms |
| | = 132·3 pounds |
| 16·7 piculs | = 1 metric ton |
| India: 1 maund | = 37 kilograms |
| | = 82 pounds |

*Yields*

| | | |
|---|---|---|
| long tons per acre | × 2·51 | = metric tons per hectare |
| pounds per acre | × 1·12 | = kilograms per hectare |
| metric tons per hectare | × 0·40 | = long tons per acre |
| kilograms per hectare | × 0·89 | = pounds per acre |

1

# The Introduction of American and European Agricultural Science into Japan in the Meiji Era

*Jiro Iinuma*

THE Meiji era (1868-1912), especially the first half of it, represents both a gap and a period of continuity in the history of Japanese agriculture, as can be illustrated by the example of the plough.

The Japanese plough, which was used from ancient times to the Meiji era, appears to have been introduced to Japan from North China through Korea in the fifth century. Its chief characteristic is a long sole, after which it is called the long-sole plough. The long sole is suitable for shallow tillage, a technique that conserves soil moisture and was important for cultivators in North China, which is a dry zone. By contrast, the Japanese climate is humid and deep tillage is required to enable weeding of the fields. Owing to the unsuitability of the long-sole plough, the hoe was adapted and developed in its place.

In the Meiji era, when Japan started to study European and American agricultural science, the government established agricultural schools and experimental stations and employed European and American agricultural scholars. Some of them, Max Fesca especially, pointed out clearly for the first time the mistake of using the shallow-tillage plough in humid Japan. Fesca recommended the wider use of a no-sole plough which was employed at that time in a part of Japan (Fukuoka Prefecture).

The no-sole plough was the most primitive type of deep-tillage plough with the sole removed altogether, because it gets in the way of deep tillage. The depth of tillage using a no-sole plough was quite irregular and it took great skill to regulate it. As a consequence a short-sole plough was invented at the beginning of the 20th century. It was a kind of deep-tillage plough, but did not require the same operating skill as the no-sole plough and it rapidly became widespread throughout Japan at the latter's expense.

This plough accelerated drainage of wet paddy fields and thus it permitted an increase in the use of fertilisers and encouraged the breeding of new varieties of rice. Overall it provided the basis for a more productive system of rice cultivation, one which developed in the second half of the Meiji era. This was called the 'Meiji system' of rice cultivation.

It was the contact with European and American agricultural science that enabled the transition from shallow to deep tillage for the Japanese. Thus the short-sole plough bridges this gap in the history of Japanese agriculture; since it is at the same time a combination of the traditional long-sole and the no-sole ploughs, it also provides continuity. More generally too, in the whole method of rice cultivation, the Meiji era is both a gap and a period of continuity in the history of Japanese agriculture (Iinuma 1969b).

## THE DEVELOPMENT OF TRADITIONAL AGRICULTURE IN THE FIRST HALF OF THE MEIJI ERA

The small scale of farms (about 1 hectare on the average) and the intensive use of labour have been characteristic of Japanese agriculture since the 17th and 18th centuries. In feudal Japan, the development of agriculture was hampered by the various restrictions of the feudal lords. For example, the farmers could not determine the kind of crops to plant; they were restrained in their purchase and sale of land; and, because they had to give up the greater part of the crop as land tax to the feudal lords, they had inadequate capital for investments to improve agriculture. In spite of this, agriculture developed steadily over the period, especially after the 18th century, when development was accelerated by the expansion of commerce.

With the removal of these feudal restrictions after the Meiji restoration (1868), agriculture, and especially rice cultivation, developed more rapidly. But the way had already been paved for this in feudal times. Measures already taken included comparative experiments on varieties of rice, variety selection and exchange between different districts; land improvement and manuring; the practice of harvesting two crops a year; and the publication of agricultural books (Furushima 1949: Chapter 6 and Yasuda 1952: 1-5).

Naozo Nakamura (1819-1882), for example, collected many varieties of rice from almost all parts of Japan (of which in 1882 he recognised more than 700 as being especially excellent). He cultivated them in his own fields, selected the best, and undertook to spread them throughout Japan. Such outstanding farmers were able to obtain a much higher yield of rice than the average at that time (Table 1.1).

These leading farmers' techniques of rice cultivation were based not on science, but on their experiences alone, and were a mixture of

TABLE 1.1

RICE YIELDS OBTAINED BY LEADING FARMERS, JAPAN, *CIRCA* 1889

| *Prefecture* | *Name* | *Yield in about 1889 (koku per tan)* [a] | *Average Yield in Prefecture 1899 (koku per tan* [a]*)* |
|---|---|---|---|
| Fukushima | K. Okonogi | 2·4–2·5 | 1·14 |
| Ibaragi | S. An | 2·6 | 1·25 |
| Chiba | S. Nara | 8·2 | 2·89 |
| Niigata | T. Katagiri | 7·6 | 2·54 |
| Gifu | S. Hirayama | 5·2 | 1·49 |
| Aichi | K. Ezaki | 1·8–2·0 | |
| Aichi | K. Ota | 4·0 | |
| Osaka | H. Takaishi | 1·7–2·5 | |
| Shimane | S. Iwatsuki | 4·6 | 1·26 |
| Yamaguchi | K. Watanabe | 3·5 | |
| Kochi | K. Okazaki | 2·4–2·5 | |
| Saga | Ch. Tobimatsu | 6·8 | 2·28 |
| Kumamoto | S. Takeuchi | 3·0 | |

[a] 1 tan = 0·1 hectare (approximately) and 1 koku = 150 kilograms.

*Source:* Saito 1968: 290.

rational and irrational techniques. For example, Enri Hayashi (1831-1906), the most famous farmer at that time, invented new methods of rice cultivation, one being the cold-water method (putting rice seeds into water in the coldest season of the year) and the other the soil method (putting rice seeds into the soil), but both methods were absolutely irrational from a scientific view-point. This experience-based approach is naturally a limited one (Yasuda 1952: 9-13 and Saito 1968: 189-200).

## THE INTRODUCTION OF EUROPEAN AND AMERICAN AGRICULTURAL SCIENCE IN THE FIRST HALF OF THE MEIJI ERA

### *Role of Central Government*

After the Meiji restoration, the government made a concerted effort to introduce European and American agricultural science into Japan. It published translations of European and American agricultural books. It established a European and American agricultural implement station and nursery for new crops introduced from Europe and America at Tsukiji in Tokyo in 1872. This was removed to Shinjuku in Tokyo in the following year, and expanded into an experimental station. A branch was subsequently established at Mita in Tokyo in 1877, and another, an olive and grape garden, was opened in Kobe

TABLE 1.2

FAMILY RANK OF STUDENTS OF THE KOMABA AGRICULTURAL SCHOOL, JAPAN, 1877–8

| | *Family Rank* | *Preparatory Course* | *Agricultural Course* | *Veterinary Course* | *Practical Course* | *Total* |
|---|---|---|---|---|---|---|
| The first graduates | Samurai | 22 | 19 | 27 | 19 | 87 |
| (admitted in 1877) | Others | 5 | 1 | 2 | 9 | 17 |
| | Total | 27 | 20 | 29 | 28 | 104 |
| The second graduates | Samurai | 8 | 18 | 22 | 6 | 54 |
| (admitted in 1878) | Others | 1 | 1 | 1 | 0 | 3 |
| | Total | 9 | 19 | 23 | 6 | 57 |

*Source:* Iinuma 1969a: 5.

in Hyogo Prefecture in 1879. In 1875, a sheep ranch was established at Sanrijuka in Chiba Prefecture.

The most important measure of all was the establishment of agricultural schools. In 1872, Hokkaido-Kaitakushi (the Agency for Development of Hokkaido) established at Shiba in Tokyo a school for the development of Hokkaido, which in 1875, was removed to Sapporo, Hokkaido, and was named Sapporo Nogakko (Sapporo Agricultural School). In later years, this became the Agricultural Department of modern-day Hokkaido University. Hokkaido, the most northern part of Japan, was scarcely developed in feudal times and was properly developed only after 1868.

Komaba Nogakko (Komaba Agricultural School) had the largest influence on the agriculture of the main island of Honshu. It was established at Komaba in Tokyo by the government in 1877, and was expanded into the Agricultural Department of what is now Tokyo University.

At the Komaba School a two years' preparatory course and three years' regular course, and also a three years' practical course were offered. The regular course was divided into an agricultural course and a veterinary course to which an agricultural chemistry course was soon added. The preparatory course was opened to boys 13-15 years old (fifty three places), the agricultural course to boys 16-18 years old (thirty places), the veterinary course to boys 16-20 years old (thirty places), and the practical course to men 20-30 years old (fifty three places).

The regular course was opened to all men, but for the practical course, admission was limited to men who had their own fields and practised agriculture. Initially, however, almost all the students of Komaba School were Samurais (Table 1.2) and, being of Samurai outlook, they respected the study of theories, but held the practice of agriculture in contempt. This outlook controlled Japanese agricultural scholarship for a long time.

The first five teachers of Komaba School were all Englishmen and lectured on English agricultural science which bore no relation to Japanese agriculture of that time. Their lectures were therefore almost useless for Japanese farmers. They were not reappointed on the expiration of their contract in about 1881, and were replaced by German teachers. These German teachers concerned themselves mainly with Japanese soils and manures, and their lectures proved to be very useful. They studied for a long time in Japan (e.g. Oscar Kerner 1881-92, Max Fesca 1882-94) and were responsible for the establishment of agricultural chemistry in Japan. Under their influence Komaba School became the centre for the study of and education in Japanese agricultural science, and agricultural chemistry has played a central role in the development of Japanese agricultural science (Iinuma 1969a: 1-24 and Tsukuba 1959: 143-64).

TABLE 1.3

AGRICULTURAL SCHOOLS IN JAPAN, 1882

| *Name* | *Prefecture* | *Year Founded* | *Years of Education* | *Teachers* | *Students* |
|---|---|---|---|---|---|
| *Government* | | | | | |
| Niigata Kannojyo | Niigata | 1875 | 3 | 4 | 19 |
| Nogyo Denshujyo | Ishikawa | 1877 | 3 | 9 | 43 |
| Gifu Nogakko | Gifu | 1878 | 4 | 9 | 36 |
| Nogakko | Hiroshima | 1879 | 3 | 5 | 32 |
| Fukuoka Nogakko | Fukuoka | 1880 | 3 | 7 | 38 |
| Kooriyama Nogakko | Fukushima | 1880 | 3 | 3 | 22 |
| Nogakko | Tottori | 1881 | 2 | 1 | 19 |
| Nogyo Koshujyo | Yamanashi | 1882 | 3 | 3 | 15 |
| *Private* | | | | | |
| Gakunosha Nogakko | Tokyo | 1875 | 5 | 8 | 80 |
| Jyui-Gakko | Tokyo | 1881 | 3 | 8 | 13 |
| Oharino Nogakko | Akita | 1881 | 2 | 1 | 12 |

*Source:* Iinuma 1969a: 25.

TABLE 1.4

THE SUBJECT MATTER OF STUDENT NOTEBOOKS OF THE FUKUOKA AGRICULTURAL SCHOOL, *CIRCA* 1886

| *Science* | | *Agricultural Science* | |
|---|---|---|---|
| Physics | 1 | Cultivation | 8 |
| Classification of plants | 2 | Rotation of crops | 1 |
| Plant physiology | 1 | Soils | 4 |
| Classification of animals | 4 | Improvement of soils: agricultural implements | 1 |
| Animal physiology | 2 | Forestry | 1 |
| Geology | 1 | Livestock | 3 |
| Inorganic chemistry | 6 | Livestock pathology | 1 |
| Organic chemistry | 9 | Silkworm: bee | 1 |
| | | Poultry | 1 |
| | | Rural economics | 1 |
| | | Horticulture | 1 |
| Total | 26 | Total | 23 |

*Source:* Iinuma 1969a: 58.

### *Role of Local Authorities*

Imitating the central government, local authorities (prefectural, town and village) also established agricultural schools and experimental stations (Table 1.3). For example, Fukuoka Nogakko (Fukuoka Agricultural School) was established by the Fukuoka Prefectural Office in 1879. This school was open to men of 15-25 years (twenty places), and offered a three years' course. There were about ten teachers, all of whom were Japanese. One half of them lectured in science and the other half in agricultural science. Among them were two Komaba graduates, who played leading roles. According to the reminiscences of one of the first graduates of Fukuoka School, the teachers lectured on European and American agricultural science, so the lectures bore no relation to the agriculture of Fukuoka Prefecture.

The author recently discovered the notebooks of these students, apparently written in 1886. Twenty six volumes were on science and twenty-three on agricultural science (Table 1.4). The large number of notebooks concerned with chemistry in the science category seem to have been characteristic of the Komaba School (Iinuma 1969a: 25-61 and Susuda 1970: 15-27).

### Rono *Period*

As indicated above, these agricultural schools and experimental stations established by the government and local authorities were of small relevance to the existing Japanese agriculture and most were abolished in about 1887, as a consequence of the financial stringency of the time.

In their place, the government and local authorities employed *ronos* (meaning leading farmers) who acted as leaders in agriculture for about fifteen years (about 1880-95) throughout the country. The most famous *rono* was Enri Hayashi whose methods are described above. It was during the '*rono* period', that graduates of both the surviving national agricultural schools (Sapporo and Komaba) made concerted efforts to 'Japanise' the European and American agricultural science taught in these two schools; that is, to place the *ronos*' proven techniques of rice cultivation on a scientific base with the aid of the European and American teachings.

For example, Tokiyoshi Yokoi (1860-1927), a Samurai with no prior knowledge of agriculture, graduated from the Komaba School in 1880 and was employed at the Fukuoka School in 1882. He taught there until the closure of this school in 1887. From notebooks made by students of his lectures (*circa* 1886), it was clear that he had a regard for both the agricultural scholars' theories of feudal Japan and the leading farmers' techniques. Fukuoka Prefecture where he lived was agriculturally the most progressive part of Japan at that time. In the Fukuoka School, he examined these traditional theories and techniques in the light of European and American agricultural science. As a result of this, he succeeded in Japanising the foreign agricultural science. For example, he invented the '*ensuisen*' (the method of selection of rice seeds by salt water). This seems to have been derived from the traditional method of Fukuoka farmers. On the other hand, he did not blindly follow traditional theories and techniques. For example, he criticised clearly the methods of rice cultivation advocated by Hayashi of Fukuoka Prefecture. Afterwards, he became a professor of the Agricultural Department of Tokyo University (Iinuma 1969a: 603-945; Tsukuba 1959: 138-56 and Saito 1968: 260-85).

### *Meiji System*

The unity between the traditional practices and theories and foreign agricultural science became the basis for the new method of rice cultivation which evolved in the second half of the Meiji era and was called the 'Meiji system'. The spread of the Meiji system led to substantial development of rice cultivation and expansion of production.

In 1893, the national experimental station was established at Nishigahara in Tokyo on this new basis and subsequently government and local authorities established agricultural schools and experimental stations of a new type. Almost all of them still remain today. From that time, however, the *rono's* leadership rapidly lost its influence, and

Japanese agricultural scholars paid little attention either to the agricultural scholars' theories of feudal Japan or to the farmers' techniques which held sway in the *rono* period (Saito 1968: 287-96 and Tsukuba 1959: 153-64).

## CONCLUSION

In conclusion, we must emphasise the fact that European and American agricultural science introduced into Japan did not become useful until it had been integrated with traditional theories and techniques. The Meiji system was the outcome of this integration of practices.

2

# Development and Diffusion of High Yielding Rice Varieties in Japan, Korea and Taiwan, 1890-1940[1]

*Yujiro Hayami*

A dramatic rise in the potential of food cereal production in tropical and sub-tropical Asia since the mid-1960s has been heralded widely as a 'green revolution'. The core of the 'revolution' was the development and propagation of high yielding varieties of food cereals, rice and wheat, adaptable to local ecologies. The new varieties are designed to be high yielding in response to high levels of inputs. They are typically sturdy short-stemmed plant types capable of carrying the heavy burden of grains resulting from heavy doses of fertilisers with intensive controls of insects, pests and weeds and an adequate supply of water.

The varieties that possessed such characteristics had been available in temperate-zone developed countries, but their transfer to the tropics had been prohibited by different ecological conditions. The transfer was first made possible when locally adaptable varieties were developed according to the 'proto type' existing in different ecological zones, through co-ordinated adaptive research by international teams of scientists, particularly those in the International Centre for Corn and Wheat Improvement (CIMMYT) in Mexico and the International Rice Research Institute (IRRI) in the Philippines.[1]

The development and diffusion of fertiliser-responsive high yielding varieties through local adaptation, though of a recent origin in the tropics, has been a major contributing factor in the agricultural growth of Japan since the last quarter of the nineteenth century. Furthermore, this process was transferred to Japan's overseas colonies, Korea and Taiwan, during the inter-war period.

This essay aims to review this historical experience of transferring the high yielding variety technology in rice among different regions within Japan, and from Japan to Korea and Taiwan, in order to give a perspective on the current development in Asian agriculture.

1 This paper draws heavily on Hayami 1971, Hayami and Ruttan 1971, especially Chapters 7, 9 and 10, Hayami and Yamada 1968.

## SEQUENTIAL RISES IN RICE YIELDS IN JAPAN, KOREA AND TAIWAN

Before proceeding to analyse the process of diffusion of rice technology in Japan, Korea and Taiwan, it is appropriate to review the pattern of technical progress in rice production within these regions as reflected in yield trends for rice. It is also relevant to explore the motivation underlying the policies that led to these remarkable yield trends.

Distinct time sequences can be observed in the rises of rice yield per hectare in Japan, Korea and Taiwan (Figure 2.1).[2] In Japan, the first steep increases in national yield were mostly contributed by the western part of the country. Around 1905 the rate of increase in this part decelerated, but a fast rate in the eastern part maintained the national trend until the late 1910s. In the 1920s, rice yields in Taiwan and Korea rose as if to compensate for the stagnation in rice yield per hectare in Japan. Such synchronised sequences reflect not only regional differences in the diffusion of rice production technology, but also policies that were motivated by a desire to exploit agricultural production potential for industrial development.

Industrialisation and economic growth are conditioned by the availability of an agricultural surplus, especially in the form of food staples that can be utilised as basic wage goods for industrial workers. The food staples may of course be imported from abroad, but this implies a drain of foreign exchange needed for the import of capital goods and technical expertise which are also critical for the development process.

In Japan rice has been by far the most important wage good: it was the source of more than 60 per cent of total calorie intake of the population before World War II.[3] In the industrial development which took place during this period when labour intensive light industries predominated, especially in the manufacture of export goods, the supply of cheap rice to industrial workers was a critical factor in keeping wages low, returns to capital high and international competitive power of industry strong. Given the constraint of foreign exchange, this task of supplying cheap rice had to be accomplished without importation of rice from foreign countries on a large scale.

[2] Data used for Japan, Taiwan and Korea cover periods after the cadastral surveys were completed (respectively 1890 in Japan, 1906 in Taiwan, and 1918 in Korea). The reliability of data after the cadastral surveys has been subject to some criticism (see Nakamura 1966 and 1969). Questions raised by Nakamura regarding Japanese official statistics have been widely discussed by Japanese scholars, Hayami 1968, Hayami and Yamada 1969, Nakamura 1968 and other scholars, Rosovsky 1968, Clark 1967. Although official statistics for Taiwan and Korea also need further detailed examination they are used in this study.

[3] For the critical role of rice as a wage good in Japanese economic development, see Hayami 1972.

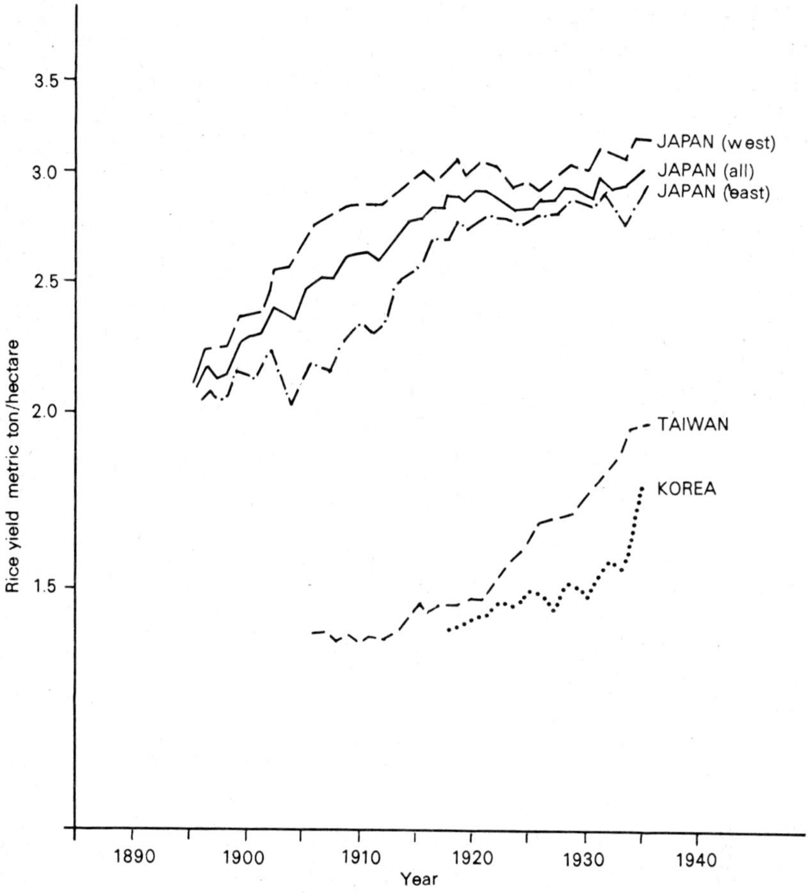


Figure 2.1 Rice yields per hectare planted for Japan, Taiwan and Korea, five-year moving average 1895-1935 (*Sources*: Hayami and Ruttan 1971: 201; Kayo 1958; Taiwan Government-General various years; Korea Government-General various issues).

Eastern prefectures: Aomori, Iwate, Miyagi, Akita, Yamagata, Fukushima, Ibaragi, Tochigi, Gunma, Chiba, Saitama, Tokyo, Kanagawa, Niigata, Nagano, Yamanashi, Shizuoka, Aichi.

Western prefectures: Toyama, Ishikawa, Fukui, Gifu, Mie, Shiga, Kyoto, Osaka, Hyogo, Nara, Wakayama, Tottori, Shimane, Okayama, Hiroshima, Yamaguchi, Tokushima, Kagawa, Ehime, Kochi, Fukuoka, Saga, Nagasaki, Kumamoto, Oita, Miyazaki, Kagoshima.

When modern economic growth began shortly after the Meiji restoration, Japan was already so densely populated that there was little room to expand the area of arable land (except on Hokkaido).

An increase in land productivity was the only possible way of increasing agricultural output and the surplus required to support industrial development. Initially, efforts were directed towards exploiting the technological potential in rice production domestically, starting with the more advanced districts of the west and then transferring the technology developed there to the east. Subsequently, when the potential was exhausted within Japan, policy switched to one of expanding rice production in Korea and Taiwan by a transfer of Japanese technology.[4]

These sequential rises in yield per hectare in Japan, Taiwan and Korea, enabled Japan to secure the supply of rice needed to meet the increasing demand of the developing industrial sectors through the first four decades of this present century and so avoided undue pressure on foreign exchange.

## DEVELOPMENT AND DIFFUSION OF RICE TECHNOLOGY WITHIN JAPAN

Shortly after the Meiji restoration government tried to modernise Japanese agriculture with the introduction of large-scale mechanised farming of the Anglo-American type. Western farm machineries, crops and livestock were imported and instructors were invited from Britain and the United States. This policy of direct 'technology borrowing' proved unsuccessful because of differences in both climatic and economic conditions. During the 1880s government quickly shifted to a strategy of agricultural development with the emphasis on raising yields of traditional food staples, and above all of rice.[5] Efforts were concentrated on the development of a labour-intensive and land-saving technology by screening and tailoring Japan's indigenous techniques with modern agricultural sciences from Germany (soil science and agricultural chemistry of the Liebig tradition).

In 1881 British agricultural instructors in the Komaba Agricultural School were replaced by German scientists. In 1885 the itinerant instructor system was established in which the instructors travelled throughout the country holding agricultural extension meetings. The government employed as instructors not only the graduates of the Komaba School but also veteran farmers (*rono*) in order to combine the best practical farming experience with the new scientific knowledge of the inexperienced college graduates. In contrast with the earlier emphasis on direct transplanting of western technology, the

[4] This process is described in Hayami and Ruttan 1970b.

[5] For a discussion of the earlier trials of technological importation and the subsequent policy reorientation, see Hayami and Ruttan 1971, Chapter 7, Ogura 1963 and Iinuma in this volume.

itinerant instruction system was designed to disseminate the best seed varieties already in use by Japanese farmers and the most productive cultural practices for Japan's traditional staple crops, rice and barley.

In order to provide better information for the itinerant lecturers, the Experiment Farm for Staple Cereals and Vegetables was set up at Nishigahara in 1886. In 1893 it was further strengthened and designated the National Agricultural Experiment Station with six branches over the nation. The itinerant instruction system was subsequently absorbed into the program of the National Agricultural Experiment Station.

The initial research conducted at the Experiment Farm for Staple Crops and Vegetables and at the National Agricultural Experiment Station was primarily at the applied end of the research spectrum. The major projects were simple field experiments which compared the various varieties of seeds or various husbandry techniques (e.g., check-row planting of rice seedlings versus irregular planting). Facilities, personnel and, above all, the state of knowledge did not permit research beyond simple comparative experiments.

These experiments nevertheless provided a basis for the rapid rise in agricultural productivity during the later years of the Meiji period. This was because of the existence of substantial indigenous technological potential which could be further tested, developed and refined at the new experiment stations and a strong propensity to innovation among farmers with whom the research workers interacted effectively.

During the 300 years of the Tokugawa period which preceded the Meiji restoration, farmers were subject to the strong constraints of feudalism. Personal behaviour and economic activity were highly structured within a hierarchical system of social organisation. Farmers were bound to their land, and were in general, not allowed to leave their village except for such pilgrimages as Ise-Mairi (to the Ise Grand Shrine). Neither were they free to choose what crops to plant or what varieties of seeds to sow. Barriers which divided the nation into feudal estates actively discouraged communication. Though feudal lords were anxious to raise agricultural productivity within their territories, in many cases they prohibited the export of the improved techniques from their territorial boundaries. Even within the territory of a lord, the diffusion was not quite free. It is recorded that a village called Maesawa in Toyama Prefecture placed a guard at its border to prevent a variety of rice seed selected in the village from being taken out.

With the reforms of the Meiji restoration such feudal restraints were removed. Farmers were free to choose their crops to plant, the seeds to sow and the techniques to practise. Nationwide communication was facilitated with the introduction of a modern postal service and railroads, and the cost of disseminating information concerning new technology was greatly reduced. The land tax reform which granted a fee simple title to the farmers and transformed a feudal share crop tax to a fixed rate cash tax increased the farmers' incentive to innovate.

The farmers, especially of the *gono* class (landlords who personally farm part of their holdings) vigorously responded to such new opportunities. They voluntarily formed agricultural societies called *nodankai* (agricultural discussion societies) or *hinshukokankai* (seed exchange societies) and searched for higher pay-off techniques. Such rice production practices as the use of salt water in seed selection, improved preparation and management of nursery beds and checkrow planting were discovered by farmers and propagated by the itinerant instructors, and sometimes enforced by the sabres of the police. The major improved varieties of seeds, up to the end of the 1920s, were also the result of selections by veteran farmers. For example, the shinriki variety, which was more widely diffused in the western half of Japan than any other single variety that has since been propagated, was selected in 1877 by Jujiro Maruo, a farmer in the Hyogo prefecture (the variety was called shinriki meaning the 'power of God' by the farmers, who were surprised by its high yield). Also, the kamenoo variety, which was propagated widely and contributed greatly to stabilising rice yields in northern Japan, was selected in 1893 by Kameji Abe, a farmer in the Yamagata prefecture.

The development and diffusion of these *rono* varieties were initiated within the western part of Japan, which includes the most advanced agricultural districts (Kinki and Northern Kyushu). This process was then transmitted to the relatively backward eastern part of Japan (Figure 2.2).[6] It must be emphasised that this was not a simple diffusion of seeds but rather the diffusion of the concepts and methods of selecting and improving seed varieties suited to local environments.

Experiment station research was successful in testing and refining the results of farmer innovation. The *rono* techniques were based on experience in the specific localities where they originated. They tended to be location specific and to require modification when transferred to other localities. Simple comparative tests effectively screened the *rono* techniques and varieties, and thereby greatly reduced the cost of supplying technical information to farmers. Slight modification or adaptations of indigenous techniques on the basis of experimental tests often gave them universal applicability.

In Japan the major motivation was, of necessity, to increase land productivity. The success of agricultural productivity growth in the Meiji period was achieved by a reorientation of agricultural develop-

6 For more detailed analysis, see Hayami and Yamada 1968.

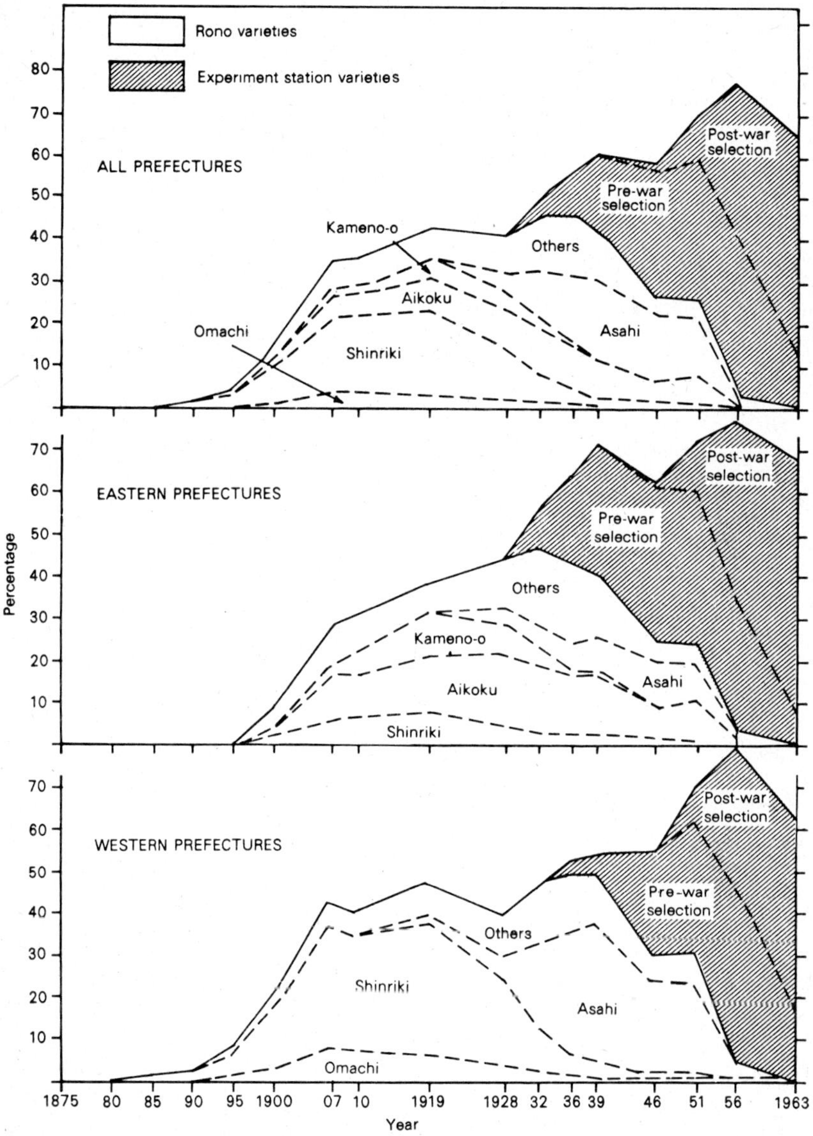


Figure 2.2 Changes in the percentage of area planted in improved varieties to total area planted in rice, Japan, 1875-1963. *Source:* Hayami and Yamada 1968: 152.

ment policy towards the development of a technology suited to the resource endowment situation in Japan.[7] Effective interaction between farmers and scientists was deliberately encouraged as part of the process of screening, adapting and diffusing the best indigenous technology.

Another important element in the effective response of technical change to Japanese resource endowments was improvement in the supply of inputs which substitute for land, that is, fertiliser. This depended on progress in the intersectoral distribution of labour which accompanied industrialisation and economic growth. Agricultural suppliers, particularly fertiliser firms, perceived the pressing demand from farmers for land substitutes and exploited the opportunities. Improved efficiency in transportation, especially the introduction of the steamship, greatly reduced the cost of shipping herring meals from Hokkaido. The search for a cheaper source of nitrogen brought about the enormous inflow of Manchurian soybean cake in the 1900s and 1910s.[8]

The increasing demand for fertiliser induced innovations in the fertiliser industry, which in turn reduced the cost of fertiliser to the farmer and induced innovations of the fertiliser-using type. The history of seed improvement in Japan is a history of developing varieties that were increasingly fertiliser responsive. Varieties selected by *rono* in response to the inflow of cheap Manchurian soybean cakes (such as shinriki and kamenoo) were characterised by high fertiliser responsiveness. They did not easily lodge and were less susceptible to disease at higher levels of nitrogen application. With the spread of such varieties accompanied by the increase in fertiliser input, first among western prefectures and then to and in eastern prefectures, the national average yield of rice increased rapidly from 1895 to 1920 (rows 1-3 in Table 2.1).

The development and diffusion of these fertiliser-responsive high yielding varieties were also based on the relatively well established water-control facilities in Japanese paddy fields. Even at the beginning of the Meiji restoration almost all paddy fields in Japan were irrigated, although the water supply was not necessarily sufficient and appropriate drainage was lacking in many cases.[9] These irrigation systems were built during the long peaceful feudal Tokugawa period primarily by

7 The development of technology consistent with resource endowments is examined in Hayami and Ruttan 1970a.

8 The decline in prices of fertilisers relative to those of agricultural products is shown in Hayami 1964.

9 The fact that no statistics have ever been collected on the irrigated area shows that in Japan 'paddy field' has been identified as 'irrigated field'. Construction of drainage facilities has been the primary objective of land-improvement projects.

Table 2.1

RATIO OF AREA PLANTED IN IMPROVED SEEDS, FERTILISER INPUT AND RICE YIELD PER HECTARE, JAPAN, TAIWAN AND KOREA, SELECTED YEARS

| | *1895* | *1905* | *1915* | *1920* | *1925* | *1930* | *1935* |
|---|---|---|---|---|---|---|---|
| Japan | | | | | | | |
| 1 Ratio of area planted in improved varieties to total paddy area planted in rice | 0·07 | 0·30 | 0·44 | 0·42 | 0·42 | 0·55 | 0·56 |
| 2 Fertiliser input per hectare of arable land (kilograms of $N + P_2O_5 + K_2O$) | 13 | 24 | 49 | 63 | 79 | 96 | 104 |
| 3 Rice yield per hectare planted in paddy rice (m. tons of brown rice) | 2·07 | 2·46 | 2·79 | 2·91 | 2·84 | 2·89 | 3·04 |
| Taiwan | | | | | | | |
| 4 Ratio of area planted in ponlai varieties to total paddy area planted in rice | | | | | 0·13 | 0·23 | 0·46 |
| 5 Fertiliser input per hectare of arable land (kilograms of $N + P_2O_5 + K_2O$) | | | | 12 | 20 | 33 | 55 |
| 6 Rice yield per hectare planted in paddy rice (m. tons of brown rice) | | | 1·47 | 1·47 | 1·63 | 1·75 | 1·97 |
| Korea | | | | | | | |
| 7 Ratio of area planted in Japanese varieties to total paddy area planted in rice | | | | 0·22 | 0·57 | 0·72 | 0·84 |
| 8 Fertiliser input per hectare of arable land (kilograms of $N + P_2O_5 + K_2O$) | | | | 1·3 | 3·4 | 12 | 28 |
| 9 Rice yield per hectare planted in paddy rice (m. tons of brown rice) | | | | 1·43 | 1·50 | 1·48 | 1·82 |

See estimation procedures and sources of data in Hayami 1971 and Hayami and Ruttan 1971: 202–3.

communal labour under the encouragement of feudal lords. Furthermore, during the Meiji period, landlords made substantial investments in land infrastructure, particularly in the improvements of drainage facilities. Government also assisted land-improvement projects.

Improvement in drainage conditions facilitated the introduction of horse ploughing. Deep ploughing by animal power and improved plough design contributed greatly to increases in the application of fertilisers and in the productivity of land.[10]

## THE TRANSFER OF RICE PRODUCTION TECHNOLOGY TO KOREA AND TAIWAN

By adequately testing and adopting veteran farmers' varieties and practices by rather simple experiments, Japan was able to exploit the substantial indigenous technological potentials in agriculture. Domestic rice production was able to supply about 95 per cent of domestic consumption during the period of the big gains in industrial development between the Russo-Japanese War (1904-5) and World War I. The technological potential was, however, exhausted, as exploitation was taking place in the absence of a supply of new potential. As already mentioned, the experiment stations in their early days contributed to agricultural productivity growth by exploiting indigenous potential rather than by creating new potential. The national experiment station did gradually shift emphasis towards more basic research, including original crop breeding projects at the Kinai Branch by cross breeding (1904), and at the Rikuu Branch, by pure line selection (1905). Results with major practical significance lagged, however, for more than two decades.[11]

In this situation the substantial expansion of demand arising from World War I caused a serious rice shortage culminating in the 1918 Kome Sodo (Rice Riot), which engulfed all major cities in Japan.

Japan was then faced with a choice between high rice prices, high costs of living and high wages on the one hand and a drain on foreign exchange from large-scale rice imports on the other. Both were unfavourable to industrial development. The reaction of the government

[10] This contribution is examined in this volume by Iinuma.

[11] The first major breakthrough in scientific rice breeding research was by Hiroshi Terao with the development of Rikuu no. 132 in the Rikuu Branch. But the appreciable contributions of organised research in seed improvement to the national average yields occurred only after the establishment of a nation-wide organised research system, Norinsho Shitei Hinshu Kairyoshiken Seido (System of Seed Improvement Experiments of Varities Assigned by the Ministry of Agriculture and Forestry) in 1926, and the creation and diffusion of Norin numbered varieties in the latter half of the 1930s as the research result of this system. Norin no. 1 was selected in 1931, but its large-scale propagation occurred after 1935.

was to increase rice imports from Taiwan and Korea by developing rice production in these colonies while suppressing domestic consumption of rice there by a squeeze on income by taxation and government monopoly sales. According to Tobata (1956: 597), until the Rice Riot 'development efforts in Taiwan were concentrated on sugar production and little was done in Korea. It was claimed that the development of rice production in those overseas territories should be suppressed since it was to foster the competitor against Japanese Agriculture.'

Under the program titled Sanmai Zoshoku Keikaku (Rice Production Development Program), the government invested in irrigation and water control and in research and extension, in order to develop and diffuse high yielding Japanese rice varieties adapted to the local ecologies of Korea and Taiwan. The expenditure for agricultural development by the Government-General of Korea before and after the start of the Rice Production Development Program (1920) clearly indicates this drastic policy reorientation. The total agricultural development expenditure jumped from a 1915-19 total of ¥3·5 million to a 1920-4 total of ¥18·6 million; expenditure for experiment stations from ¥1·1 million to ¥2·8 million; the expenditure for land-improvement projects including irrigation and drainage facilities from only ¥334,000 to as much as ¥12 million (Kobayakawa 1959: 117-19). The results of this policy reorientation were rapid increases in rice yields both in Taiwan and Korea, accompanied by stagnation in Japanese rice yields.[12]

*Taiwan*

The most spectacular success was attained in Taiwan with the development of the ponlai varieties. The ponlai varieties 'developed by cross-breeding of Japanese varieties or between Japanese and traditional Taiwan (chailai) varieties to have photo-sensitivities different from the original Japanese varieties'.[13] They are more fertiliser-responsive, and with adequate water control and cultural practices are higher yielding. They are also better suited to Japanese tastes than are the chailai varieties (indica).

It was not easy to adapt Japanese varieties to the tropical climate of Taiwan. Even before the policy reorientation which followed the Rice Riot, when the effort of the Government-General was directed to improving the chailai varieties to satisfy Taiwan's domestic demand,

[12] See Hayami and Ruttan 1970b.

[13] Iso 1964: 18. Dr Iso was director of the Agricultural Division, The Central Research Institute of the Government-General of Taiwan. He is known for his contribution to development of the ponlai varieties.

research had been conducted to adapt Japanese varieties to a tropical ecology, although on a small scale. A breakthrough came in the finding by Iso that the Japanese varieties could be grown successfully by reducing the sojourn of the rice seedling in the nursery bed to half of the period used in traditional practice.[14]

With this breakthrough and under the pressing demand of Japan, the Government-General shifted emphasis from improvement of the chailai varieties to development and propagation of the ponlai varieties. Areas planted with the ponlai varieties grew from 400 hectares in 1922, the first year this statistic was recorded, to 131,000 hectares in 1930, and 296,000 hectares (almost one half of the paddy field area planted) in 1935. This rapid diffusion was based on the high pay-off of the ponlai varieties. According to the rice production cost survey conducted by the Government-General in 1926-7, both profit (total revenue minus total cost) and farm family income (profit plus family labour wages) per chia (0·97 hectare) were very much higher with the ponlai varieties than with the chailai varieties (Table 2.2).

The cost comparisons in Table 2.2 show that differences in fertiliser expense, wages, and rent are particularly significant. Larger expenditures on fertiliser for the production of ponlai varieties clearly reflect their higher fertiliser responsiveness. Higher wage costs show that the ponlai varieties require more labour and better husbandry, including checkrow planting, deep ploughing, more intensive weeding and insect control. Higher rent for the ponlai varieties indicates that these varieties were grown in areas with better water control. It appears that the economic implications of the ponlai varieties for Taiwan in the 1920s were essentially equivalent to those of 'miracle rice' in Asia today (Barker and Quintana 1969).

With the rapid spread of the ponlai varieties, both fertiliser inputs and rice yields in Taiwan increased for 1920-35 in a way similar to the Japanese experience for 1895-1920 but at a much faster rate (compare rows 4-6 with rows 1-3 in Table 2.1).

### *Korea*

The growth spurt in Korean rice yields lagged for almost a decade behind Taiwan (Figure 2.1). Japan and Korea are situated in about the same latitudes, and the climate of Korea shows a much greater similarity to Japan's than to Taiwan's; indeed rice varieties of northern Japan can be directly transplanted to Korea. In 1920 the ratio of area planted in Japanese varieties to total paddy area was already 22 per

[14] From 50-60 days to 30-40 days for the first crop; from 30-40 days to 15-20 days for the second crop (Iso 1964: 76-7).

TABLE 2.2

COMPARISON OF THE COSTS OF PONLAI AND CHAILAI RICE PRODUCTION PER CHIA, JAPAN, 1926–7

| | 1926 Second Crop | | | 1927 First Crop | | |
|---|---|---|---|---|---|---|
| | *Ponlai* (*1*) | *Chailai* (*2*) | (*1*) − (*2*) | *Ponlai* (*3*) | *Chailai* (*4*) | (*3*) − (*4*) |
| | yen | | | | | |
| Seed | 8·12 | 3·97 | 4·15 | 8·31 | 4·17 | 4·14 |
| Fertiliser[a] | 51·26 | 25·50 | 25·76 | 63·58 | 26·90 | 36·68 |
| Wage | 112·38 | 93·86 | 18·52 | 126·20 | 103·83 | 22·37 |
| Family labour | 73·30 | 54·57 | 18·73 | 81·51 | 62·75 | 18·76 |
| Hired labour | 39·08 | 39·29 | −0·21 | 44·69 | 41·08 | 3·61 |
| Implements and building | 3·76 | 4·24 | −0·48 | 4·55 | 4·83 | −0·28 |
| Miscellaneous | 2·46 | 1·01 | 1·45 | 9·25 | 11·73 | −2·48 |
| Tax and rate | 1·41 | 1·66 | −0·25 | 1·81 | 1·57 | 0·24 |
| Rent | 147·20 | 121·70 | 25·50 | 176·40 | 133·42 | 42·98 |
| Total cost | 326·59 | 251·94 | 75·15 | 390·10 | 286·45 | 103·65 |
| Total revenue[b] | 382·04 | 285·31 | 96·73 | 466·76 | 285·26 | 181·50 |
| Profit[c] | 55·45 | 33·37 | 22·18 | 76·66 | −1·19 | 77·85 |
| Farm family income[d] | 128·75 | 87·94 | 40·81 | 158·17 | 61·56 | 96·61 |

[a] Include self-supplied fertilisers.

[b] Include the value of straw.

[c] Profit = total revenue − total cost.

[d] Farm family income = profit + wage for family labour.

*Sources:* Hayami 1971; Hayami and Ruttan 1971: 206; Taiwan Government-General, Bureau of Colonial Development 1928, No. 6: 11, 48–9, 62–3, 82–3, 112–13, 241, 249 and No. 9: 11, 13, 15, 17, 50–1, 64–5, 118–19, 152–3. Data are for tenant farmers.

cent, and this ratio continued to rise rapidly from 1920 to 1930. Yet little increase in rice yields or in fertiliser inputs can be observed for Korea during this period (rows 7-9 in Table 2.1).

An apparent contradiction exists: in spite of an earlier start in the diffusion of Japanese varieties, the level of fertiliser input per hectare was low and the yield 'take-off' lagged in relation to that in Taiwan. The key to this contradiction seems to be the differences in the level of irrigation and water control. Table 2.3 compares progress in paddy field irrigation in Taiwan and Korea. Since data on the area of irrigated paddy fields are not available for Taiwan, we calculated the ratio of irrigated paddy field area on the assumption that irrigation was developed only for rice and sugar production (which seems a rather reasonable approximation). The ratios thus calculated correspond well with the ratios of double cropped paddy area. In Taiwan, irrigation is required for double cropping rice.

From Table 2.3 it is apparent that irrigation construction in Korea lagged behind that in Taiwan. In terms of the ratio of irrigated paddy area to total paddy area, Korea in 1925 did not reach the level of Taiwan in 1915. Judging from the movements in double cropping ratios (which are not comparable with Taiwan in absolute level because of the different climate) it seems reasonable to assume that progress in water control in Korea was greatly accelerated during the 1925-35 period. This is compatible with the expenditure patterns for land improvement projects of the Government-General as described earlier. In the literature on Korean agriculture it is common to identify the lack of irrigation as the critical cause for low productivity. Tobata and Ohkawa (1935: 2-3), for example, wrote:

> The first technical condition of rice production is nothing but water control. But the paddy field in Korea is a so called 'rain-fed paddy field' . . . accordingly a marshy paddy field with drainage problems, which is considered of low quality in Japan is considered a good paddy field . . . Who would dare to apply fertilisers under such conditions?

It was thus natural that a high investment priority was placed on irrigation when the Rice Production Development Program was initiated in 1920.

While the similarity in climate enabled the Japanese rice varieties to be transplanted easily to Korea, the latter's precarious water supply, even in most parts of the so-called 'irrigated paddy field', resulted in the introduction of Japanese varieties in earlier days which were not highly fertiliser-responsive. Koremochi Kato, Director of the Agricultural Experiment Station in Korea remarked in 1926 (Nihon Nogyo Hattatsushi Chosakai 1956: 176-7):

TABLE 2.3

IRRIGATION AND DOUBLE CROPPING RATIOS IN PADDY FIELD, KOREA AND TAIWAN, SELECTED YEARS

| | | Year | | | | |
|---|---|---|---|---|---|---|
| | *Hectares* | *1915* | *1920* | *1925* | *1930* | *1935* |
| Taiwan | | | | | | |
| 1. Area of paddy field | 1000 | 343 | 367 | 374 | 396 | 479 |
| 2. Sugar cane area harvested | 1000 | 83 | 105 | 127 | 106 | 118 |
| 3. Irrigated arable land area | 1000 | 239 | 268 | 350 | 442 | 466 |
| 4. Double cropping paddy area | 1000 | | 246 | 266 | 292 | 313 |
| 5. Ratio of irrigated area | 3 ÷ (1 + 2) | 0·56 | 0·57 | 0·70 | 0·88 | 0·78 |
| 6. Ratio of double cropping area | 4 ÷ 1 | | 0·67 | 0·71 | 0·74 | 0·65 |
| Korea | | | | | | |
| 7. Area of paddy field | 1000 | 1168 | 1531 | 1551 | 1605 | 1668 |
| 8. Irrigated paddy field area | 1000 | | | 758 | 953 | 1152 |
| 9. Double cropping paddy area | 1000 | 160 | 240 | 266 | 353 | 429 |
| 10. Ratio of irrigated area | 8 ÷ 7 | | | 0·49 | 0·59 | 0·69 |
| 11. Ratio of double cropping area | 9 ÷ 7 | 0·14 | 0·16 | 0·17 | 0·22 | 0·26 |

*Sources:* Hayami 1971; Hayami and Ruttan 1971: 208; Taiwan Government-General various issues; Korea Government-General (a) and (b) various issues.

> It is natural that our experiment station since its establishment has worked to select those from many Japanese varieties, which have better results under low level of fertilisation. . . . But, as water control has been developing recently farmers have been increasing fertiliser application and have become dissatisfied with the results.

In response to the demand for fertiliser-responsive varieties, the South Korea Branch of the Agricultural Experiment Station in Iri was set up in 1930 with the primary purpose of developing fertiliser-responsive high yielding varieties. During the 1930s varieties with higher responsiveness such as ginbozu and rikku no. 132 rapidly replaced less responsive Japanese varieties such as tamanishiki and kokurato.[15]

In short, the development of high yielding varieties in Korea in comparison with the ponlai varieties in Taiwan lagged for almost a decade owing to the constraint of water control, which worked to depress the pay-off of investment in developing the high yielding varieties. The relevant question is then why irrigation was developed earlier in Taiwan. Many factors were involved:

(1) annexation of Korea occurred a decade later than for Taiwan so that investment in infrastructure in general was later in starting;[16]

(2) irrigation had been developed in Taiwan during the early days of colonisation to promote the production of sugar cane, and the facilities could be utilised for rice production;

(3) production of Korean rice (japonica) had been a direct menace to Japanese rice producers and was suppressed, while Taiwan chailai rice (indica) was not a direct competitor;

(4) the Taiwan Government-General enjoyed revenue surpluses during the half decade preceding 1910 (called the 'Golden Age' of the Taiwan Government-General's treasury) and could afford to invest in large-scale construction of physical infrastructure, including railways, ports and irrigation (Kawano 1941: 11; Yanaihara 1929: 91-117).

## TRANSFER OF PRODUCTION FUNCTION

The emergence and propagation of high yielding varieties in Taiwan and Korea during the 1920s and 1930s represent a transfer of

[15] The area planted in tamanishiki and kokurato was more than 30 per cent of the total area planted in rice and close to 50 per cent of the area planted in Japanese varieties in Korea during the late 1920s. According to the experiments conducted by the Agricultural Experiment Station of the Government-General of Korea for 1927-9, those two varieties both recorded lower yields in high fertilisation plots while such varieties as ginbozu recorded higher yields in the high fertilisation plots (Nihon Nogyo Hattatsushi Chosakai 1956: 177-8).

[16] Taiwan was conceded to Japan by China in 1895 as a result of the Sino-Japanese war. Korea became a protectorate of Japan in 1905, and a territory of Japan in 1910.

TABLE 2.4

FACTOR SHARES IN THE COST OF RICE PRODUCTION, JAPAN AND TAIWAN, 1925–7

| | *Japan* | *Taiwan* | | | |
|---|---|---|---|---|---|
| | | *Ponlai* | | *Chailai* | |
| | *1925–7 Average* | *1926 Second crop* | *1927 First crop* | *1926 Second crop* | *1927 First crop* |
| Fertiliser[a] | 15·9 | 15·7 | 16·3 | 10·0 | 9·4 |
| Wage | 35·8[b] | 34·4 | 32·2 | 37·2 | 36·2 |
| Rent | 41·5 | 45·1 | 45·2 | 48·3 | 46·6 |
| Others | 6·8 | 4·8 | 6·3 | 4·3 | 7·2 |
| Total cost | 100·0 | 100·0 | 100·0 | 100·0 | 100·0 |

[a] Include self-supplied fertiliser.

[b] Include livestock labour.

*Sources:* Hayami 1972b; Hayami and Ruttan 1971: 210; Japanese data calculated from Ishibashi 1961: 82–95 (data are for tenant farmers); Taiwanese data calculated from Table 2.6.

agricultural technology among different ecological regions through organised adaptive research and investment in the transformation of environmental conditions, i.e., by irrigation.

This process is illustrated in Table 2.4 which compares factor shares in the cost of rice production in Japan with those of the ponlai and chailai varieties in Taiwan. The factor shares in the case of the ponlai varieties are very similar to those in Japan. If we assume a production function of the Cobb-Douglas type and equilibrium under competitive factor markets, the factor shares represent production elasticities of the respective inputs. In the propagation of the ponlai varieties, Japan's rice production technology was assimilated by Taiwan.

Both in the cases of Japan and of the ponlai varieties in Taiwan, fertiliser's share is larger and land's share (rent) is smaller than in the case of the chailai varieties. This clearly reflects the fertiliser-using and land-saving character of the new technology embodied in fertiliser-responsive varieties. This bias in technical change was consistent with a land-saving demand of the Taiwan economy in which population pressure against land was raising the price of land relative to the prices of other factors.[17]

[17] During the period 1920-35 arable land area increased by 11 per cent and agricultural population by 21 per cent, resulting in a decline in the land/man ratio by 10 per cent in Taiwan. In Korea, during the same period, arable land area increased by 3 per cent and agricultural population by 11 per cent, resulting in a decline in the land/man ratio by 8 per cent. Although comparable data are unavailable for Taiwan and Korea, Japan recorded a persistent rise in the price of arable land relative to the prices of other inputs from 1880 to 1960 (Hayami and Ruttan 1970a).

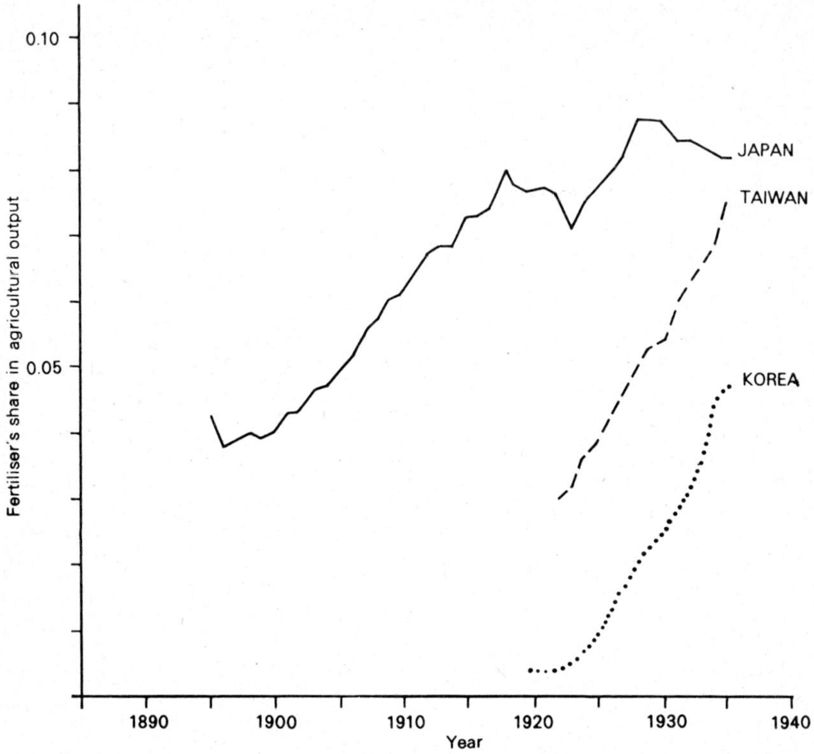


Figure 2.3 Fertiliser's share in agricultural output for Japan, Taiwan and Korea, five-year moving averages, 1895-1935. Fertiliser's share is total value of commercial fertiliser consumption divided by gross value of agricultural production (both in current prices). In the case of Taiwan, fertilisers applied to sugarcane and sugarcane production are deducted, respectively, from fertiliser consumption and agricultural production. *Sources*: Hayami 1971; Hayami and Ruttan 1971: 212).

Figure 2.3 shows the changes in fertiliser's share of total output in Japan, Taiwan and Korea.[18] Remarkable associations exist between the movements of fertiliser's share and of rice yields per hectare (Figure 2.1) in Japan, Taiwan and Korea. This correspondence suggests strongly that growth in rice yields in these three regions was a process of replacement of traditional varieties by fertiliser-responsive

[18] Owing to data limitations, self-supplied fertilisers such as manure and compost are not included in the fertiliser data. In Taiwan, because of its special nature, the sugarcane sector is excluded from the calculations (fertiliser input in sugarcane is deducted from total agricultural output). Sugarcane was produced either by plantations owned by sugar companies or by peasant farmers under the supervision and guidance of the sugar companies by contract. Fertiliser inputs in sugarcane were high even before the 1910s.

high yielding ones (e.g., chailai by ponlai). It indicates the process of assimilation of Japanese technology by Taiwan and Korea. In other words, it represents the transfer of Japan's agricultural production function to Taiwan and Korea.

## CONCLUSION

We have reviewed the process of technical change in rice production in Japan, Korea and Taiwan from the early period of Meiji to World War II. The core of technical progress was development and propagation of the high yielding varieties which respond to fertilisers when applied with an adequate supply of water, better land preparation and more intensive insect pest and weed controls. These varieties can be regarded as the forerunners of 'miracle rice' in Asia today. The process of diffusion was initiated in advanced agricultural districts in Western Japan and extended into Eastern Japan. During the interwar period the rice production technology was transferred to Korea and Taiwan with the objective of generating a growing food surplus in these colonies which could contribute to industrial growth in Japan.

The diffusion of high yielding varieties within Japan, and to Taiwan and Korea, involved local testing and adaptation, and breeding for specific environmental conditions. It also required the modification of environmental conditions by the construction of irrigation and drainage facilities. The process was not a simple transfer of known technology but rather the creation of new technology according to a 'proto type' existing in different ecological regions.

This appears to be the process which has led to the green revolution, and which will be required to exploit fully the new potential in food production in South and Southeast Asia. In order to sustain current development, countries in Asia will have to finance large-scale investments in adaptive agricultural research and in land and water infrastructure. Furthermore, if the green revolution is to lead to overall economic development, these nations will have to design a scheme to convert their current and potential food surpluses into industrial growth as was achieved with rice in Japan, Korea and Taiwan.

3

# The Agricultural Transformation of Taiwan: The Case of Ponlai Rice, 1922-42

*Carolle Carr and Ramon H. Myers*

JAPAN and Taiwan are the only two Asian countries to have experienced an agricultural revolution by the middle of the twentieth century. By an agricultural revolution we mean the adoption and use by family farms of new farm inputs, particulary new seeds and improved fertilisers, which bring about increasing returns to agriculture and create and sustain growth rates of agricultural output which substantially exceed the growth rate of population. During the period of Western colonial domination in Asia many countries successfully increased the supply of special crops, such as fibres, at very rapid growth rates, but foodgrains barely kept pace with population increase. It has been only in Japan and Taiwan that this rate of growth of agricultural output has been achieved primarily through supply increases in foodgrains.

Under Japanese rule in Taiwan a substantial increase was achieved in the supply of special products but it was in the production of foodgrains that performance was outstanding as the basic statistics of Table 3.1 show.

TABLE 3.1

INDEXES OF FOODGRAIN PRODUCTION, YIELD AND CULTIVATED AREA IN TAIWAN, 1900–40

| *Period* | *Cultivated Area* | *Yield* | *Production* |
|---|---|---|---|
| 1900–3 | 100 | 100 | 100 |
| 1919–21 | 150 | 132 | 198 |
| 1938–40 | 176 | 248 | 437 |

*Source:* calculated from data contained in Joint Commission on Rural Reconstruction 1966: 23–47. These foodgrains consisted of twenty food crops; excluded from the table are seven special crops, eight fibre crops, twenty five fruit crops, and thirty five vegetable crops, and various green manure crops. Production was calculated by aggregating output of twenty foodgrains and setting 1900–3 as the base year.

Foodgrain output rose more than fourfold, overall yield increased around two and one half times, while cultivated area increased only 76 per cent. Had yield remained constant over the period the cultivated area required to produce foodgrain supply of 1938-40 would have had to be nearly two and one half times the index figure for 1938-40. The large increase in yield and output makes this growth performance worthy of being termed an 'agricultural revolution'.

Rice comprised a key component of total foodgrain production. It has been Taiwan's most important crop in terms of land use: in 1901-3, roughly four fifths of total cultivated area was devoted to it and in 1939, the proportion was still about two thirds (Joint Commission on Rural Reconstruction, 1966: 11, 23). In the beginning of the period this crop alone made up nearly 50 per cent of foodgrain supply. As the price increased, rice came to be an important cash crop for farmers; sales to the cities increased, and exports to Japan rose dramatically. Table 3.2 shows the changing patterns of foodgrain supply, and the production and exports of rice to Japan for the 1901-40 period.

Rice declined as a share of total foodgrain supply over the period, and the available per capita rice supply declined after 1921-30. However rice exports to Japan increased, and they progressively accounted for a rising share of total rice production. Significantly, the supply of other foodgrains increased rapidly over the period; the output of sweet potatoes, a prime food for rural households, increased from 350,000 to 2,244,478 metric tons between 1901-3 and 1938-40. As rice steadily became more important as a cash crop for farms, the supply of other foodgrains continued to increase.[1]

The overall importance of rice in the foodgrain sector was illustrated in Table 3.2. The changing importance of the new rice seed (ponlai or horai as read in Japanese), imported from Honshu island of Japan, can be seen in Table 3.3 where its output is compared to native rice (chailai), glutinous and upland rice.

By the final decade of the period ponlai rice had become the most important source of rice production. In the post-war period ponlai's proportion increased to nearly 70 per cent of total rice production and by 1965 it occupied 68 per cent of total cultivated area under rice.

This essay focuses upon Japanese endeavours to increase rice production in Taiwan by improved seed selection and by securing the

[1] Unlike African countries where exports of cash crops have increased while domestic foodgrain markets have not developed during the past half century, a complex foodgrain market was already in existence in Taiwan before the Japanese occupation in 1895. As transportation improved and commercialisation increased, this market continued to evolve and handle the distribution of food to areas that had moved to greater specialisation because of their factor endowment and particular local comparative cost advantages.

TABLE 3.2

FOODGRAIN SUPPLY, RICE PRODUCTION, AND RICE EXPORTS TO JAPAN, 1901–40

| *Period* | *Production ('000 m. tons)* (1) *foodgrains* | (2) *rice* | (2)/(1) (*per cent*) | *Rice exports to Japan as percentage of* (2) ('000 *m. tons*) | *Rice supply after export to Japan* | *Available per capita rice*[a] (*kilograms*) |
|---|---|---|---|---|---|---|
| 1901–10 | 1231 | 572 | 46·5 | 15·1 | 485,399 | 141 |
| 1911–20 | 1514 | 671 | 44·3 | 17·8 | 551,431 | 153 |
| 1921–30 | 2053 | 880 | 42·8 | 30·8 | 608,316 | 144 |
| 1931–40 | 2951 | 1261 | 42·7 | 46·4 | 676,187 | 125 |

[a] This ratio was obtained by dividing average rice supplies, after subtracting exports to Japan (decennial periods), by average total island population for the same period.

*Sources:* data for foodgrains, total rice production, and population for after 1910 obtained from Joint Commission on Rural Reconstruction, Rural Economics Division 1966: 7 and 23–47. Foodgrains comprised twenty crops whose combined 1938–40 cultivated area accounted for 94·9 per cent of total cultivated land. By 1940 farmers had achieved a very high multiple cropping index and numerous fruit, vegetable, and special crops were being rotated with foodgrains. Data for rice export to Japan were obtained from the Bank of Taiwan 1958: 37. Population data for 1901–10 period (years 1905–10 only) obtained from Civil Affairs Bureau 1946: 76.

TABLE 3.3

MAIN TYPES OF RICE PRODUCED IN TAIWAN, 1901–40

| *Period* | *Rice Type (m. tons)* | | | *Share of Rice Output (per cent)* | | |
|---|---|---|---|---|---|---|
| | *Ponlai* | *Chailai* | *Other*[a] | *Ponlai* | *Chailai* | *Other*[a] |
| 1901–10 | — | 505,968 | 65,931 | 0 | 88·47 | 11·53 |
| 1911–20 | — | 583,280 | 87,251 | 0 | 86·99 | 13·01 |
| 1921–30 | 137,772 | 605,914 | 135,830 | 15·66 | 68·89 | 15·45 |
| 1931–40 | 583,999 | 509,304 | 167,884 | 46·31 | 40·38 | 13·31 |

[a] Other includes glutinous and upland rice.
*Source:* Taiwan Agricultural Statistics 1966: 23–8.

adoption of high yield varieties from Japan. The first section discusses seed improvement research and the adoption pattern of the new ponlai seed in Taiwan. The second section attempts to measure the social benefits to the Taiwan economy of adopting the seed. The final section examines the impact it had upon farm management throughout the island. The methodology utilised below relies upon historical data and provides a convenient means by which we can examine and assess an important historical precedent to the green revolution which began to sweep parts of Asia in 1967. The findings of our research strongly suggest that an agrarian policy which can produce a Taiwan-type agricultural revolution does not depend upon drastic restructuring of rural property and institutional foundations but mainly upon the skill, good luck, and capacity of governments to establish for themselves an efficient and viable working local administration and educational system to make new seeds profitable for farmers to adopt and use.

### RESEARCH AND ADOPTION OF IMPROVED SEED

By 1908 the Japanese colonial administration had established agricultural research stations in each of its administrative districts on the island (Myers and Ching 1964: 562-4). In each district officials took preliminary steps to create farmer associations comprised of landlords and wealthy farmers who paid dues to obtain new farming inputs and new farming knowledge from the association. Each association maintained contacts with district research stations. Many villages still remained outside the influence of this organisational network even by the mid-1920s, but by the 1930s most of the peasantry had been exposed to various potential means by which crop yields could be greatly increased. This infrastructure enabled the Japanese to distribute better native seed varieties after research stations had selected and standardised them. It also enabled the dissemination of the new ponlai variety from Japan.

In 1910, the administration took vigorous steps to increase rice production. Peasants at that time were using some 1100 rice seed varieties, so agricultural research stations began selecting the better seeds, standardising their use, and instructing the farmer associations to distribute them to the farmers and to encourage their use (Saiwasa 1923: 8-10). By the middle of the decade this policy had succeeded in reducing the number of seeds to 390, and by the early 1920s to 175 high yield seeds (Eigichi 1925). With administrative backing, farmers had increased total cultivated land between 1905 and 1920 by 24 per cent. As irrigation facilities expanded at the same time paddy land still occupied roughly half of total farm land. Rice prices trebled during this period, and with the increase in output led to an increase in the value of the 1924 rice crop of over 600 per cent of that in 1905 (Katsuji 1920). During the ten years from 1910 the demand for rice from Japan accelerated and prompted Japanese researchers to begin experimenting with rice seeds from Japan so that Taiwan could produce higher yield rice palatable to the Japanese consumer. At the same time farmers on Taiwan found it increasingly profitable to produce rice as a cash crop for the market.

Between 1911 and 1924 research stations experimented with the ponlai variety to adapt it to the soil and climate of Taiwan. Experiments ultimately showed that this seed ripened two to four weeks earlier than native varieties (Katsuji 1920), that more seedlings could be transplanted to a given unit of paddy land than the native variety seedlings, and that the ponlai stalk was shorter than the native variety and was thus better protected from the occasional high winds which buffeted the island. Its yield was also 10-30 per cent higher than those of native varieties. Finally, because of rising demand from Japan, the price of ponlai was higher than for native rice. As a cash crop, it began to compete with sugarcane.

Numerous problems had to be solved before this new rice seed could be distributed to farmers. First the peasant had to be instructed in proper germination and planting techniques. Ponlai had to be properly soaked, planted into seedling beds, and then transplanted at a prescribed time. Farmers had to fertilise the land more heavily when preparing the paddy fields. More water per unit of land was also necessary during the initial transplanting stage. After draining the fields, peasants had to weed their fields more carefully and intensively than they customarily did for native rice. The extra cost required to provide more fertiliser, and water and the additional labour required for plant care, made it troublesome and expensive for households to adopt the ponlai. However, as farmers learned that profits from its use

TABLE 3.4

PERCENTAGE OF TOTAL RICE AREA DEVOTED TO PONLAI IN SEVEN PREFECTURES OF TAIWAN, 1926–42

| *Year* | *Taipei* | *Hsinchu* | *Taichung* | *Tainan* | *Kaohsiung* | *Taitung* | *Hualien* |
|---|---|---|---|---|---|---|---|
| 1926 | 32 | 32 | 32 | 3 | 1 | 2 | 4 |
| 1927 | 25 | 28 | 19 | 4 | 1 | 1 | 5 |
| 1928 | 28 | 27 | 35 | 10 | 5 | 0 | 8 |
| 1929 | 25 | 21 | 28 | 9 | 1 | 0 | 9 |
| 1930 | 29 | 21 | 37 | 13 | 2 | 0 | 14 |
| 1931 | 26 | 26 | 38 | 14 | 4 | 0 | 14 |
| 1932 | 29 | 33 | 46 | 20 | 7 | 1 | 19 |
| 1933 | 36 | 41 | 51 | 24 | 14 | 8 | 31 |
| 1934 | 38 | 43 | 56 | 23 | 14 | 15 | 32 |
| 1935 | 43 | 49 | 63 | 29 | 27 | 24 | 36 |
| 1936 | 46 | 51 | 60 | 27 | 19 | 29 | 37 |
| 1937 | 50 | 59 | 61 | 26 | 26 | 37 | 42 |
| 1938 | 52 | 71 | 64 | 34 | 34 | 45 | 45 |
| 1939 | 53 | 56 | 60 | 35 | 33 | 49 | 49 |
| 1940 | 56 | 75 | 59 | 36 | 39 | 56 | 51 |
| 1941 | 59 | 78 | 63 | 43 | 46 | 59 | 59 |
| 1942 | 55 | 79 | 63 | 73 | 50 | 64 | 46 |

*Source:* Taiwan Sotokufu 1897–1942.

were usually two thirds higher than from native rice, and as the administration issued the new ponlai free of charge and freely issued instructions on how to use it, the rate of adoption by farmers grew progressively more rapid.

Between 1911 and 1923, during the experimental phase of accommodating the seed to the island's soil and climate, very little ponlai was planted. In 1917 the total area devoted to it was 76 hectares producing a yield of 1890 kilograms of rice. In 1923 farmer associations received these seeds to pass to their members. Officials at all levels encouraged the widespread adoption of the new seed. By 1925 the area under ponlai had increased to almost 26,000 hectares yielding 2130 kilograms/hectare. A year later roughly a third of the rice area in Taipei, Hsinchu, and Taichung districts was planted to the new seed. The shares of ponlai area in total cultivated rice area by district between 1926 and 1942 (Table 3.4) show that the southern and eastern districts (Tainan, Kaohsiung, Taitung and Hualien) lagged in adoption, but even in these over half of the rice area used ponlai by the early 1940s.

Farmers planted ponlai in the first crop season,[2] and then after observing the results gradually introduced the seed into the second crop cycle. They initially used it on inferior soils, but when yields proved to be higher than expected and the grain sold profitably, they quickly introduced it to the alluvial plain and paddy areas. In 1927, farmers slightly reduced their usage of ponlai; the proportion increased in 1928, only to be reduced again in 1929. We believe that the erratic adoption rate in the first four years was due primarily to the changing relative profitability of the variety caused by fluctuating sugarcane prices. As sugarcane and rice prices seesawed, farmers attempted to adjust plantings accordingly. Inevitably, many farmers incurred production costs for the variety which were higher than expected because of their unfamiliarity with its cultivation techniques. But, as they became more familiar with these, profits stabilised, and adoption increased rapidly thereafter. The fluctuating pattern in the use of the variety can be observed for particular areas (prefectures) where rice competed vigorously with sugarcane, and where its relative profitability was reduced by rising sugarcane prices, for example in the Tainan region (Table 3.5).

What factors enabled farmers in certain prefectures to adopt ponlai more rapidly? Information on a regional basis is too scanty to provide a conclusive answer. We tested the hypothesis that it might have spread faster in prefectures of high rice yield by performing a Spearman rank correlation of prefectural rice yields and adoption rates. No correlation was found between prefectural yield variations and adoption rates. More likely, local levels of profitability dictated the rate of adoption, though this statement begs the question because prefectural profitability depended upon a variety of different supply and demand conditions. On the supply side numerous factors such as seed availability, soil conditions, knowledge of market prices, and farmer managerial capabilities interacted to determine rice yields and farmer response to the changing market. It is however probably fair to conclude that there was little debate amongst farmers on the relative merits of ponlai and chailai rice as the greater yield and land value from growing the former became clearly apparent.

A further interesting question is whether the profitability of ponlai discouraged sugarcane production, i.e. whether the peasants switched from sugarcane to ponlai cultivation, thereby accelerating the diffusion rate of the variety. Peasants had traditionally varied their relative emphasis on rice and sugarcane depending upon relative prices. Data

[2] Farmers generally planted their first rice crop in February and harvested in August; their second rice crop in May and harvested in December, though there was variation by districts.

TABLE 3.5

RATE OF ADOPTION OF PONLAI IN SEVEN PREFECTURES AND FOR ALL TAIWAN, 1926–42

| *Year* | *All Island* | *Taipei* | *Hsinchu* | *Taichung* | *Tainan* | *Kaohsiung* | *Taitung* | *Hualien* |
|---|---|---|---|---|---|---|---|---|
| 1926 | 0·79 | — | — | — | — | — | — | — |
| 1927 | –0·25 | –0·32 | –0·18 | –0·59 | 0·33 | 0 | –0·5 | 0·25 |
| 1928 | 0·37 | 0·16 | –0·05 | 1·07 | 1·50 | 4·04 | –1·00 | 0·60 |
| 1929 | –0·28 | –0·15 | –0·30 | –0·30 | –0·11 | –0·80 | 0 | 0·14 |
| 1930 | 0·27 | 0·21 | 0 | 0·45 | 0·50 | 1·01 | 0 | 0·62 |
| 1931 | 0·07 | 0·14 | 0·29 | 0·04 | 0·09 | 1·00 | 0 | 0 |
| 1932 | 0·33 | 0·16 | 0·37 | 0·33 | 0·50 | 0·75 | 0 | 0·42 |
| 1933 | 0·28 | 0·33 | 0·36 | 0·20 | 0·25 | 1·00 | 1·00 | 0·80 |
| 1934 | 0·23 | 0·09 | 0·08 | 0·20 | –0·06 | 0 | 1·14 | 0·05 |
| 1935 | 0·19 | 0·21 | 0·24 | 0·28 | 0·33 | 1·08 | 1·15 | 0·18 |
| 1936 | –0·04 | 0·12 | 0·08 | 0·13 | –0·10 | –0·40 | 1·33 | 0·04 |
| 1937 | 0·14 | 0·16 | 0·32 | 0·04 | –0·05 | 0·47 | 1·38 | 0·22 |
| 1938 | 0·08 | 0·08 | 0·50 | 0·13 | 0·42 | 0·42 | 1·61 | 0·13 |
| 1939 | 0·04 | 0·04 | –0·71 | –0·17 | 0·05 | –0·05 | 1·80 | 0·16 |
| 1940 | 0·07 | 0·12 | 0·76 | –0·04 | 0·04 | 0·27 | 1·96 | 0·08 |
| 1941 | 0·16 | 0·12 | 0·16 | 0·46 | 0·30 | 0·29 | 2·24 | 0·32 |
| 1942 | 0·28 | –0·17 | 0·06 | 0 | 1·20 | 0·16 | 2·46 | –0·43 |
| Average | 0·16 | 0·08 | 0·12 | 0·14 | 0·32 | 0·58 | 0·91 | 0·22 |

*Sources:* formula for adoption rate from Griliches 1962: 327–30 and Babcock 1962: 336; calculations based on figures from Taiwan Sotokufu 1897–1942.

for the period of ponlai rice diffusion show that the share of land allotted to sugarcane declined slightly while that for rice remained roughly the same. While areas devoted to ponlai and sugarcane did fluctuate moderately from year to year, no large swings could be observed. Thus expansion of the new variety occurred primarily at the expense of native rice seeds. The main reason for the slight decline which did take place in share of land devoted to sugarcane was that farmers began cultivating fruits, fibres, and vegetables during the late 1920s and 1930s.

## SOCIAL BENEFITS OF ADOPTION

The Japanese invested considerable funds in agricultural research on Taiwan and it is of considerable relevance to estimate the rate of return realised on these, particularly in view of the current importance of research expenditures being made on seed and fertiliser. The procedure adopted below to estimate the realised social rate of return from Japanese experimental research to develop ponlai is naturally subject to a wide margin of error,[3] though the estimates derived below can be considered to be conservative.

First, the net social return from ponlai production was estimated by calculating the gross annual return for the 1926-42 period and deducting from it the additional annual costs of production. The annual or marginal costs, were obtained by multiplying the area under the variety by the 1926 base year costs for its production. The gross annual return was based upon an estimation of the increase in consumer surplus accruing to society from ponlai production. Net annual social returns were then cumulated for the 1926-42 period and expressed as an annual flow using two external interest rates, 5 and 10 per cent.

The second step was to estimate the cumulated past research expenditures for developing ponlai seed in Taiwan. These proved difficult to identify on the basis of available data. It was therefore decided to assume two alternative proportions, 3 and 10 per cent of total agricultural research expenditures for the 1911-42 period, as those which were devoted to ponlai seed research. To these two estimates were added special sums for rice improvement between 1923 and 1942. Both calculations probably overestimated greatly the funds used strictly for ponlai seed development, for this research activity was in fact an extremely small component of overall rice and other research station endeavours. By taking the two estimates of cumulated past research expenditures as a capital sum and dividing into the various

[3] The estimating procedure is based entirely upon the path-breaking work of Griliches 1958.

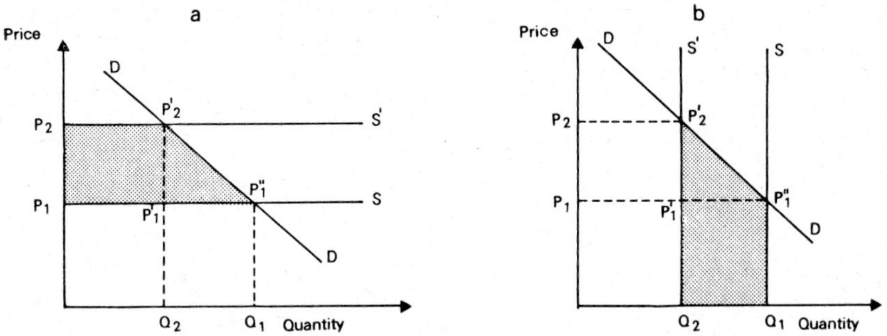


Figure 3.1 Supply and demand for rice with different supply elasticities, Taiwan, 1922-42.

estimated annual flows of net social returns, a range of realised rates of returns for ponlai research expenditures was obtained for the 1926-42 period.

The gain to society from the production of ponlai rice was measured by comparing the historical gross annual returns from its cultivation with hypothetical gross annual returns under the assumption of its absence over the period (Figure 3.1). It was assumed that, on average, ponlai rice yields were 15 per cent higher than those of the native varieties.[4] The annual gain to society does of course depend on assumptions regarding the price elasticities of demand for rice, but as will be seen below, different elasticities exert only a negligible effect on the results obtained.

The DD curve represents aggregate demand for rice and would

[4] The annual increased yield resulting from ponlai was calculated with the following formula:

$$\text{increased yield} = \frac{\text{ponlai yield} - \text{native rice yield}}{\text{native rice yield}}$$

For the 1926-42 period the regional increase in rice yield due to ponlai was calculated as follows:

| *Prefecture* | *Increased yield* |
|---|---|
| Taipei | 0·125 |
| Hsinchu | 0·027 |
| Taichung | 0·073 |
| Tainan | 0·396 |
| Kaohsiung | 0·201 |
| Taitung | 0·150 |
| Hualien | 0·200 |
| Average for 7 Prefectures | 0·167 |
| Average for Taiwan | 0·165 |

In the calculations which follow, a 15 per cent yield increase is assumed for the period.

*Source:* Taiwan Sotokufu 1897-1942.

include consumer demand in Japan as well. Assume different supply elasticities (Figures 3.1a and b) and a shift upward and to the left to represent the absence of supply if ponlai had not been produced. The loss to society due to the absence of ponlai appears as the area under the demand curve between the new and old supply curves. In Figure 1a, the area is $P_1$ $P_2$ $P'_2$ $P''_1$ and in Figure 3.1b is $Q_1$ $Q_2$ $P'_2$ $P''_1$. In both figures the loss to society can be interpreted as the increase in total cost to produce $Q_2$ and are the rectangles $P_1$ $P_2$ $P'_2$ $P'_1$ in Figure 3.1a and $Q_2$ $P'_1$ $P''_1$ $Q_1$ in Figure 3.1b plus the loss in consumer surplus caused by the rise in price or the triangles $P'_2$ $P''_1$ $P'_1$ in Figure 3.1a and $P_2$ $P''_1$ $P'_1$ in Figure 3.1b.

In order to measure loss or the areas within these identified boundaries the following variables are specified:

k = the percentage change in yield of 15 per cent
n = the price elasticity of demand for rice which we assumed to be 0·6 (FAO 1966)
P = the price of rice
Q = the annual rice supply

Now the loss in Figure 3.1b[5] can be measured:

(1) $\text{loss} = kP_1Q_1(1+\frac{1}{2}kn)$

In order to calculate the annual additional costs to produce ponlai, we used the following expression:

(2) marginal cost of producing ponlai $= A \times T$

where A denotes the area under ponlai seed and T the additional cost of ponlai production as of 1926. As the initial production costs are high and gradually fall, this assumption again biases the results downward. The cultivated area under ponlai seed was reported in annual crop statistics for Taiwan whereas the average ponlai cost for 1926 was obtained from a rural survey report (Sheng 1959: 46-7). After calculating the annual additional costs (Equation (2)) they were deducted from the annual gross social return to obtain the net social return.

[5] The measure of loss in Figure 3.1b is used because this gives a higher estimate of loss than in Figure 3.1a. By deliberately calculating a higher estimate of loss we thereby bias our final results downward (see Taiwan Sotokufu 1897-1942). In the case of measuring loss for Figure 3.1a, the expression becomes

$$\text{loss} = kP_1Q_1(1-\tfrac{1}{2}kn)$$

The ratio of loss in Figure 3.1b compared to Figure 3.1a is

$$(1+\tfrac{1}{2}kn)/(1-\tfrac{1}{2}kn)$$

which given the values assumed for k and n (0·15 and 0·6 respectively) works out to be approximately 3. Even if we alter the price elasticity of demand to 0.06, the ratio still remains 3. Therefore, inasmuch as the price elasticity of demand makes little in measuring loss, we assumed 0·6 and used the loss measure (Equation (1) for Figure 3.1b).

In Table 3.6 the two estimates for research expenditures and the estimates of net social returns which began to accrue in 1926 and end only in 1942 are presented. The annual net social returns and research expenditures in Table 3.6 were then capitalised at the two external interest rates of 5 and 10 per cent. If we let $R_t$ denote research expenditures during the period 1911-42 and $S_t$ stand for annual net social return for the period 1926-42, we can then use the following two expressions to derive the capitalised value for the past cumulative sums.

(3) Sums of past $R_t$ capitalised

$$=R_1(1+r)^n + R_2(1+r)^{n-1}\ldots + R_{n-1}(1+r)^2 + R_n(1+r)$$

(4) Sum of past $S_t$ capitalised

$$=B_1(1+r)^n\ldots + B_n(1+r)$$

The results obtained from (3) and (4) constitute the net cumulated past returns and the cumulated past research expenditures.

These results are presented in items 1 and 4 of Table 3.7. Past returns expressed as an annual flow were obtained by multiplying net cumulated past returns by the external annual interest rates of 5 and 10 per cent. These findings are presented as items 2 and 3 in Table 3.7. The internal rate of return from investment in ponlai research and dissemination was then obtained by dividing total net annual return by the cumulated past research and multiplying by 100. Note in Table 3.7 that the first estimate which assumed 3 per cent of total annual agricultural research expenditures for the 1911-42 period produced social returns of 352 and 311 per cent whereas the higher expenditure estimate of 10 per cent yielded 165 and 133 per cent. These findings compare favourably with rates of return from expenditures on developing hybrid corn in the United States and the new seeds in Mexico.[6]

[6] Estimates of rates of return to investment in agricultural research in the United States and Mexico are as follows:

| *Type of Study* | *Annual rate of return* (per cent) |
|---|---|
| *United States* | |
| Hybrid corn research as of 1955 | 700 |
| Hybrid sorghum research as of 1967 | 360 |
| Poultry research as of 1960 | 137 |
| Agricultural research and extension | 300 |
| *Mexico* | |
| Wheat research, 1943-63 | 750 |
| Corn research, 1943-63 | 300 |
| Total agricultural research, 1943-63 | 290 |

*Source:* Schultz 1968: 8.

Table 3.6

PONLAI SEED RESEARCH EXPENDITURES AND NET SOCIAL RETURNS, TAIWAN, 1911–42

| *Year* | *Total research expenditures*[a] *(yen)* | *Total research expenditures*[b] *(yen)* | *Net social returns (yen)* |
|---|---|---|---|
| 1911 | 3625 | 12,083 | |
| 1912 | 3594 | 11,980 | |
| 1913 | 3435 | 11,451 | |
| 1914 | 3497 | 11,657 | |
| 1915 | 9039 | 30,129 | |
| 1916 | 8802 | 29,339 | |
| 1917 | 9929 | 29,762 | |
| 1918 | 10,709 | 35,696 | |
| 1919 | 13,902 | 46,342 | |
| 1920 | 17,105 | 57,017 | |
| 1921 | — | — | |
| 1922 | — | — | |
| 1923 | 35,085 | 38,992 | |
| 1924 | 3356 | 11,186 | |
| 1925 | 35,825 | 45,091 | |
| 1926 | 37,179 | 47,053 | 1,674,688 |
| 1927 | 4413 | 14,711 | 1,394,279 |
| 1928 | 5154 | 17,179 | 1,394,279 |
| 1929 | 5279 | 17,597 | 1,189,855 |
| 1930 | 5033 | 16,778 | 1,287,890 |
| 1931 | 4810 | 47,814 | 902,834 |
| 1932 | 4313 | 45,953 | 1,896,923 |
| 1933 | 4387 | 44,386 | 2,138,024 |
| 1934 | 4551 | 44,957 | 3,599,159 |
| 1935 | 4893 | 22,689 | 4,431,057 |
| 1936 | 5100 | 23,500 | 5,224,344 |
| 1937 | 5631 | 25,898 | 5,386,443 |
| 1938 | 5927 | 26,051 | 6,321,789 |
| 1939 | 6143 | 20,476 | 6,336,143 |
| 1940 | 6444 | 27,548 | 5,716,247 |
| 1941 | 9946 | 33,153 | 7,042,999 |
| 1942 | 10,778 | 35,928 | 8,200,862 |

[a] Total research expenditures assuming 3 per cent of sums spent on all agricultural research plus expenditure for rice improvement programs starting in 1923.

[b] Total research expenditures assuming 10 per cent of sums spent on agricultural research plus expenditures for rice improvement programs starting in 1923.

TABLE 3.7

RATE OF RETURN ON HYBRID RICE RESEARCH EXPENDITURES, TAIWAN, 1942

| | $r$ = *5 per cent* | $r$ = *10 per cent* |
|---|---|---|
| (1) Net cumulated past returns (yen) | 67,365,200 | 70,573,088 |
| (2) Past returns expressed as annual flow (yen) | 3,368,260 | 7,057,309 |
| (3) Total net annual returns (yen) | 3,368,260 | 7,057,309 |
| (4) Cumulated past research expenditures (yen) | | |
| (a) 3 per cent total agricultural research plus rice improvement | 955,674 | 2,267,418 |
| (b) 10 per cent total agricultural research plus rice improvement | 2,042,788 | 5,311,465 |
| (5) Rate of return 100 × (3) / (4) | | |
| (a) 3 per cent total agricultural research plus rice improvement | 352 | 311 |
| (b) 10 per cent total agricultural research plus rice improvement | 165 | 133 |

## IMPACT OF NEW SEED ON FARM MANAGEMENT

What new managerial requirements were imposed upon rice cultivating farmers upon the adoption of a new innovation like the high yielding ponlai seed? To answer this question, a brief description is presented first of rice farming practices by Chinese farmers on Taiwan at the turn of the century which is then followed by an examination of inputs of labour and work animals used through the rice cultivation cycle and arranged by farm size. These preliminary comments outline and identify the kinds of costs which had to be borne by farms adopting ponlai and the extent of profitability required to justify wider diffusion of this new seed. We conclude by emphasising that importance of farmers acquiring new technical and managerial skills which can keep unit costs low and can justify the adoption of a new high yielding seed, a consideration which is of great significance for the current green revolution.

By 1900 Chinese farmers in Taiwan had already been cultivating rice for over two centuries. The slow accumulation of farming experience transmitted to each generation by father to son represented a fund of knowledge which slowly expanded, was adapted to local conditions, and was based upon practical experience and first hand observation. Peasants selected the riper seeds from the harvest, placed them in bamboo baskets, and stored them until planting in seedling beds.[7] To prepare for planting, peasants soaked the seeds in warm

[7] The discussion which follows was derived wholly from an economic survey of rural Taiwan conducted by the Japanese between 1898 and 1903 and reported in Rinji Taiwan Kyukanchosa dainibu 1905: 18-27.

water for a period depending upon the locale: in northern Taiwan four days, in the south for one night only. After drying the seeds outdoors, they were placed in straw to conserve heat, until a small sprout began to emerge.

Meanwhile, the peasant selected a suitable site, protected from the wind, near the farmstead for the seedling bed. After preparing, fertilising, and irrigating the soil, the seeds were broadcast over the bed. The peasants irrigated every evening and drained the water from the seedling beds each day and scattered straw and ashes on them. After the seedling had grown beyond an inch, human compost in water was spread on the seedling bed. Constant irrigation and fertilisation continued for forty to fifty days before the first harvest and thirty to forty days before the second rice planting period.

At the same time farmers began preparing the fields for the seedlings. They ploughed the soil to a depth of 6 to 7 inches, and about two weeks before transplanting, they irrigated, weeded and tilled the fields. Just before transplanting, water was allowed into the paddies. Each peasant selected a certain paddy area, collected a quantity of stalks from the seedling beds, and planted them about 9 inches apart in a series of lines across the field. The higher the quality of the soil the greater the density of stalks planted. After planting, more water was let into the fields. Fifteen days after transplanting, the peasants undertook the first weeding after which the fields were again flooded. Ten days later the weeding and irrigation cycle was repeated. After a further ten days, the final weeding was completed. Two major weedings in the manner just described were conducted for the second rice crop.

At harvest time, the men cut the rice stalks with sickles about 5 to 6 inches above the ground. Others followed behind to collect the rice stalks and carry them to an area where peasants flailed the grain to separate the rice from the stalks. The rice was then winnowed, dried and stored. A portion of this would be reserved for the following season's planting while the remainder was held for consumption or marketing.

Farming experience had taught the peasants that a fixed combination of labour and work animals was required to accomplish certain work tasks satisfactorily. This combination did not vary greatly even on large farms for certain tasks were indivisible such as soil preparation, weeding, irrigation and harvesting. This relationship can be observed in the first part of Table 3.8 that was based upon a 1936-7 survey of 174 farms in which labour and work-animal inputs per unit of land were measured by farm size groups.

TABLE 3.8

LABOUR AND WORK-ANIMAL INPUTS ON TAIWANESE RICE FARMS, 1936/7

A. *Total family farm inputs per unit of land by farm size*

| *Farm size range (chia)* [a] | *Input* | *Household* | | *Hired* | | *Total* | | *Ratio of total work-animal to labour inputs (chia)* | *Index of farm size* |
|---|---|---|---|---|---|---|---|---|---|
| | | *(days)* | *(per cent)* | *(days)* | *(per cent)* | *(days)* | *(per cent)* | | |
| 1–1·9 | Labour | 325·23 | 81·95 | 71·61 | 18·05 | 396·84 | 100 | 0·087 | 100 |
| | Work-animal | 31·79 | 92·57 | 2·55 | 7·43 | 34·34 | 100 | | |
| 2–2·9 | Labour | 288·37 | 81·75 | 64·38 | 18·25 | 352·75 | 100 | 0·084 | 144 |
| | Work-animal | 27·96 | 94·56 | 1·61 | 5·44 | 29·57 | 100 | | |
| 3–5·0 | Labour | 220·31 | 75·84 | 70·17 | 24·16 | 290·48 | 100 | 0·092 | 214 |
| | Work-animal | 24·39 | 91·04 | 2·40 | 8·96 | 26·79 | 100 | | |

B. *Family farm inputs per unit of land for rice production by farm size*

| *Farm size range (chia)* | *Average farm size (chia)* | *Total labour for first and second rice crops (man days)* | *Total labour inputs for farm (man days)* | *Per cent of labour inputs in rice production* | *Work-animal inputs in rice production (man days)* | *Work-animal inputs in farming (man days)* | *Per cent of work-animal inputs in rice production* |
|---|---|---|---|---|---|---|---|
| 1–1·9 | 1·963 | 205·87 | 396·84 | 52 | 33·61 | 34·34 | 98 |
| 2–2·9 | 2·823 | 201·73 | 352·75 | 57 | 30·03 | 30·03 | 100 |
| 3–5·0 | 4·197 | 179·06 | 290·48 | 62 | 26·93 | 26·93 | 100 |

[a] 1 chia = 0·97 hectare.

*Source:* Taiwan sotokufu shokusan kyoku 1940: 1–3.

TABLE 3.9

FIRST RICE CROP LABOUR AND WORK-ANIMAL INPUTS PER UNIT OF LAND, TAIWAN SURVEY, 1936/7

| *Farm Size (chia)* | *Input* | *Preparing Seed Bed* | *Field Irrigation* | *Transplanting* | *Fertilisation* | *Field Weeding* | *Water Control* |
|---|---|---|---|---|---|---|---|
| 1–1·9 | Labour days | 6·28 | 22·65 | 11·12 | 7·06 | 29·52 | 9·59 |
| | (per cent) | (5·55) | (20·01) | (9·82) | (6·24) | (26·08) | (8·47) |
| | Animal days | 0·76 | 18·15 | 0·09 | 0·87 | 0·35 | — |
| | (per cent) | (3·71) | (88·67) | (0·44) | (4·25) | (1·71) | — |
| 2–2·9 | Labour days | 4·52 | 18·74 | 11·95 | 6·29 | 27·22 | 7·27 |
| | (per cent) | (4·27) | (17·69) | (11·28) | (5·94) | (25·70) | (6·86) |
| | Animal days | 0·41 | 14·92 | 0·13 | 1·01 | 0·01 | 0·21 |
| | (per cent) | (2·42) | (88·07) | (0·18) | (5·96) | (0·06) | (1·24) |
| 3–5·0 | Labour days | 4·69 | 18·73 | 11·54 | 6·27 | 23·79 | 6·41 |
| | (per cent) | (4·89) | (19·54) | (12·04) | (6·54) | (24·82) | (6·69) |
| | Animal days | 0·68 | 12·93 | 0·17 | 0·70 | 0·01 | 0·24 |
| | (per cent) | (4·37) | (83·04) | (1·09) | (4·50) | (0·06) | (1·54) |

| *Farm Size (chia)* | *Input* | *Pest Control* | *Natural Calamity Control* | *Harvesting* | *Total* | *Index* | *Ratio of Animal Inputs to Labour Inputs* |
|---|---|---|---|---|---|---|---|
| 1–1·9 | Labour days | 1·05 | 0·12 | 25·81 | 113·20 | 100 | 0·181 |
| | (per cent) | (0·93) | (0·10) | (22·80) | (100·0) | | |
| | Animal days | — | — | 0·25 | 20·47 | 100 | |
| | (per cent) | — | — | (1·22) | (100·0) | | |
| 2–2·9 | Labour days | 1·55 | 0·16 | 28·23 | 105·93 | 94 | 0·160 |
| | (per cent) | (1·46) | (0·15) | (26·65) | (100·0) | | |
| | Animal days | — | — | 0·35 | 16·94 | 83 | |
| | (per cent) | — | — | (2·07) | (100·0) | | |
| 3–5·0 | Labour days | 0·90 | 0·40 | 23·12 | 95·85 | 85 | 0·162 |
| | (per cent) | (0·94) | (0·42) | (24·12) | (100·0) | | |
| | Animal days | — | 0·02 | 0·82 | 15·57 | 76 | |
| | (per cent) | — | (0·13) | (5·27) | (100·0) | | |

*Source:* Noka rodo chosa: suito jisaku noka 1940: 4–6.

This table shows that over the range observed, the largest size farms hired considerably more labour (per chia) than the smallest. This probably arose because family labour resources were insufficient to accomplish the various tasks involved in rice cultivation. It should also be noted however that, with respect to all farm activities and rice cultivation, the intensity of labour and work-animal inputs per unit of land varied inversely with farm size. Farmers either could not or chose not to maintain the same intensity of power inputs per unit of land for farms larger than 2 chia (roughly 2 hectares) in size. These larger farms cultivated crops other than rice, but the data show they actually committed a higher proportion of total labour and work-animal inputs to rice production. Thus larger farms (at least up to 5 chia) appeared to specialise more of their resources in rice. These different patterns of labour and work-animal input use are extremely important when considering the costs associated with different phases.

The pattern of labour and work-animal input allocation to the various tasks of the rice cultivation cycle, calculated from the 1936-7 survey data, are shown in Table 3.9. Of the nine sequential work tasks, irrigating, weeding and harvesting required the most labour and work-animal inputs. They accounted for over two thirds of the total labour inputs and nearly nine tenths of the work-animal inputs required. Seed bed preparation and transplanting absorbed a further 15 per cent of total labour inputs. Irrigating, weeding, and harvesting accounted for most of the labour and work-animal input irrespective of farm size, which supports the argument of the indivisibility of joint input use. The fact that large farms placed as much emphasis upon these three work tasks as smaller farms suggests that the same meticulous care given on small farms was necessary to produce high yields. This largely eliminated the possibilities of achieving economies of scale in production with the traditional technology. Input costs for an average sized farm (Table 3.10) showed the critical role of manual labour in the production process and the concentration of costs in the tasks of irrigation, weeding and harvesting.

Although the cultivation of ponlai required additional labour and fertiliser inputs per unit of land which raised costs of production above those for chailai rice, the higher yield obtained from the former and the higher rice prices induced by the expanding Japanese demand more than compensated for these and enabled the farmer to earn higher profits per unit of land. An island-wide survey of seventy farms growing primarily rice during the second crop of 1925 and eighty farms growing primarily rice during the first crop of 1926 provided data on revenue and costs per chia (roughly equivalent to 1 hectare). These farms grew both chailai and ponlai and were classified on an ownership basis into owner-cultivated and tenant farms (Table 3.11).

TABLE 3.10

INPUT COSTS FOR AN AVERAGE-SIZED FARM, TAIWAN SURVEY, 1936/7

| *Input* | *Preparing Seed Bed* | *Field Irrigation* | *Trans-planting* | *Field Fertiliser* | *Field Weeding* | *Water Control* | *Pest Control* | *Natural Calamity Control* | *Harvesting* | *Total* |
|---|---|---|---|---|---|---|---|---|---|---|
| Labour cost per day (yen) | 6·06 | 23·45 | 15·69 | 7·06 | 29·79 | 8·85 | 1·32 | 0·26 | 36·52 | 129·00 |
| Labour-animal cost per day (yen) | 0·62 | 13·80 | 0·09 | 1·16 | 0·09 | — | — | — | 0·21 | 15·97 |
| Total cost (yen) | 6·68 | 37·20 | 15·78 | 8·22 | 29·88 | 8·85 | 1·32 | 0·26 | 36·78 | 144·97 |
| (per cent) | (4) | (26) | (11) | (5) | (20) | (6) | (2) | (1) | (25) | (100) |

*Source:* Noka rodo chosa: suito jisaku noka 1940: 10–11.

Table 3.11

COMPARISON OF REVENUE AND COSTS PER UNIT OF LAND FROM CULTIVATING CHAILAI AND PONLAI, TAIWAN. SECOND CROP OF 1925 AND FIRST CROP OF 1926

| | | yen/chia | | | | | | | | |
|---|---|---|---|---|---|---|---|---|---|---|
| *Type of Rice* | *Cultivator* | *Total Revenue* | *Fertiliser, Seed, etc.* | *Labour* | *Tools* | *Taxes* | *Interest Charges*[a] | *Tenant Rent* | *Total* | *Net Profit* |
| A. *Second crop of 1925* | | | | | | | | | | |
| Chailai | Owner-cultivator | 307·74 | 33·26 | 88·45 | 3·97 | 12·83 | 95·31 | — | 233·82 | 70·92 |
| | Tenant | 330·36 | 33·11 | 91·24 | 4·81 | 2·01 | — | 140·64 | 271·81 | 58·55 |
| Ponlai | Owner-cultivator | 447·66 | 91·30 | 113·19 | 4·68 | 14·62 | 118·24 | — | 342·03 | 105·63 |
| | Tenant | 396·89 | 72·63 | 115·30 | 4·96 | 1·91 | — | 187·87 | 382·67 | 14·22 |
| B. *First crop of 1926* | | | | | | | | | | |
| Chailai | Owner-cultivator | 366·23 | 56·35 | 110·28 | 4·99 | 14·94 | 114·27 | — | 300·83 | 65·40 |
| | Tenant | 351·71 | 48·81 | 103·09 | 4·76 | 1·75 | — | 152·65 | 311·06 | 40·65 |
| Ponlai | Owner-cultivator | 482·07 | 98·53 | 119·37 | 5·22 | 18·31 | 133·82 | — | 375·25 | 106·82 |
| | Tenant | 491·98 | 90·69 | 131·03 | 5·03 | 1·21 | — | 187·75 | 415·71 | 76·27 |

[a] for all capital including land.

*Source:* Taiwan sotokufu shokusankyoku 1927: 1.

TABLE 3.12

COMPARISON OF 1937 YIELD AND AREA VALUES OF DIFFERENT RICE, TAIWAN

| | Type of Rice | | | | | |
|---|---|---|---|---|---|---|
| | *Ponlai* | | *Chailai* | | *Upland* | |
| *Harvest* | *Yield Value* (*yen per chia*) | *Land Value* (*yen per koku*) | *Yield Value* (*yen per chia*) | *Land Value* (*yen per koku*) | *Yield Value* (*yen per chia*) | *Land Value* (*yen per koku*) |
| First | 24·3 | 387·7 | 20·7 | 273·3 | 19·6 | 149·2 |
| Second | 23·5 | 339·6 | 20·3 | 250·7 | 18·5 | 147·0 |
| Total | 47·8 | 727·3 | 41·0 | 524·0 | 38·1 | 296·2 |

*Source:* Taiwan sotokufu shokusankyoku 1938: 46.

The largest cost components in producing ponlai for both owner-farmers and tenants were seed and fertiliser purchases and expenditures on labour. For tenants growing ponlai during the second harvest of 1925, the gains obtained were far below the profits obtained for the first harvest period in 1926. This was principally because there was no marked increase in revenue to match the higher costs. Revenue was only 20 per cent higher than from chailai for that crop season. In 1926 by contrast it was 40 per cent higher. Owner-cultivators received substantially higher revenue from ponlai in both crop seasons (46 and 32 per cent respectively).

A similar study of seventy nine rice growing farms during the first harvest period in 1927 revealed lower profits from ponlai as compared to the 1925/6 harvest periods. However subsequently as farmers became more adept with the techniques of cultivating ponlai, and its price continued to rise, profits continued to grow and greatly exceeded those obtained from chailai.

Survey evidence in 1937 strongly suggests that the difference in yields between chailai and ponlai had persisted since the mid-twenties and that owing to expanding demand for the latter rice from Japan, farmers obtained greater revenue and realised greater land values by growing it than from chailai and upland rice (Table 3.12).

It took some time in the late 1920s before farmers achieved optimum yield from ponlai. To achieve this, farmers had to learn to soak the seeds for the proper period, plant at the stipulated time, apply more than the customary fertiliser dosage, weed and irrigate more intensively after transplanting, and either make a careful selection of seed at harvest time for the next cycle of planting or purchase new seeds in the market. It is not clear whether tenants or owner farmers adopted ponlai more rapidly and which group thereby obtained higher returns to land and labour. Answers to these questions would require further analysis of the variables influencing profits on farms where different rice seeds were used.

Three reasons explain the success of the green crop revolution in Taiwan as illustrated by rice production during this historical period. First, the Japanese colonial administration, unlike European colonial administrations, became firmly committed to increasing output of food-grains and industrial crops. Rural Taiwan simply became an extension of the Japanese countryside, and was perhaps even more favoured in terms of government investment to create an infrastructure of research agencies, farmer associations, irrigation projects, a transportation system, and a program of primary education for farmers. Second, the Chinese farmer by 1900 was a skilled, energetic, and profit-motivated individual who could and would respond to new profit opportunities. In the case of rice cultivation, the demonstration that a certain new rice seed produced higher yields and more income than traditional rice, quickly led to the mastery of the necessary skills for its adoption. Biological advances could thus be simply integrated into the farm work schedule and mode of family farm management without major modifications to or departures from the existing or traditional patterns. Furthermore experience with technological improvements in the production of other foodstuffs such as the sweet potato, probably helped to accelerate the response of Taiwanese farmers to new opportunities for rice as a cash crop. Finally, the additional prospect of steadily rising prices for rice must have reduced the element of profit uncertainty on adoption and must have provided an added incentive for a speedy response from farmers. Thus this study shows that the combination of an innovation which is not demanding financially upon farmers and a market which promises attractive and relatively (market) secure profits constitute a situation which carries a high probability of a speedy adoption and diffusion rate for the innovation.

4

# High Yielding Varieties Programme in India: An Assessment

*P. K. Mukherjee and Brian Lockwood*

THE High Yielding Varieties Programme of 1966-7 has attempted to introduce modern agricultural methods based on the use of purchased inputs such as certified seed of high yielding, cereal varieties and chemical fertiliser, into a basically subsistence peasant agricultural system. Peasant farmers in many parts of India, producing cereal crops under a variety of agro-economic conditions, have recently made major changes in their methods of production, most of which have increased the dependence of the farm on the outside, both as a source of inputs and as a market for an expanded output. Two aspects of this change in the relationship between the farm and the Indian economy and society have received widespread comment. The first is the impact of the program on the Indian farmer, the extent to which he has been willing to participate by adopting the new varieties and other inputs. The second is the possible repercussions of the new agricultural technology on the distribution of income between different farmers, between farmers and rural landless, and between different regions. The first aspect will be discussed in this paper in some detail while the second will be touched upon only lightly.

We propose to treat the subject in three sections. In the first we outline briefly the main approaches to rural development which preceded the program in India and then describe the substance of the program itself. In section two, we present data collected by the Programme Evaluation Organisation (PEO) of the Indian Planning Commission (1967-70) to measure: the extent of farmer adoption of high yielding paddy and wheat varieties from 1967/8 to 1969/70; the distribution of adopting farmers between those with small farms and those with large farms; the extent of crop area commitment to the new varieties; and the extent to which adopting farmers have applied chemical fertiliser to their new varieties. Seed adoption patterns will be examined in all major paddy and wheat States while area commitment and fertiliser use will be limited to Tamil Nadu for paddy and Punjab for wheat. In the final section, we discuss, at a more general level, the main policy issues related to the new agricultural technology.

## THE HIGH YIELDING VARIETIES PROGRAMME AND ITS BACKGROUND

Agriculture in India, as in many underdeveloped countries, has been for centuries a way of life rather than a commercial vocation. For generations farmers have followed the agricultural practices of their fathers and have limited their inputs mainly to those available in the village economy. A large proportion of farm output is consumed on the farm and exchanged for various services, labour and inputs from the village. Farmers have limited contact with the wider economy, and few opportunities of moving into non-agricultural employment. As population increased there developed a desperate and bitter struggle to control land, the main factor of production and source of livelihood. In times of land surplus, a system of inheritance developed which was responsible for increasing subdivision and fragmentation of farms, and for accumulation of land by the stronger farmers and dispossession of the weakest. In each village now there are typically a few individuals who control extensive areas of land, many who own small landholdings, many others working land as tenants or share croppers, and still others who no longer have even this tenure and hold on land and are dependent entirely on payments for labour.

Land reform legislation has changed this situation only marginally in recent times. On the whole Indian agriculture presents a picture of low productivity, with the gains being shared very unevenly. This has been well documented, particularly in reports and programs designed to raise rural production and incomes. Probably the first comprehensive study of rural sector was that of the 1929 Royal Commission on Agriculture. On its recommendation Commissions on Irrigation and on the Problems of Famine were initiated during the early 1930s. A number of irrigation, land reform, agricultural research and rural welfare programs resulted from these after 1935 when popular ministries were established upon the introduction of provincial autonomy.

During the 1930s the population growth rate outstripped the growth rate of foodgrain production, and the separation of Burma, a rice surplus area, further aggravated the food situation. World War II and the 1942 Bengal famine demonstrated again the seriousness of the food situation in India, and in 1942 a Food Department was created to control foodgrain prices, to co-ordinate civil and military purchases, and to administer a supply and distribution system.

Several committees were appointed between 1943 and 1951 to deal with the production and distribution of foodgrains. A Foodgrains Policy Committee was appointed in July 1943 which, among other things, recommended the launching of a Grow More Food Campaign through an extension of area, an increase in irrigation potential, and the use of manures, fertilisers and improved seeds. Rationing and food control measures were introduced in big towns and famine-stricken areas in 1943.

A Second Foodgrains Policy Committee was appointed in 1947 after the partition. Referring to the Grow More Food Campaign, this committee said that the measures which were undertaken were doubtless in the right direction, but the objectives were too diversified, the efforts were inadequate and in most areas necessary drive and vigour were lacking. It recommended that a drive for food self-sufficiency be launched through works schemes such as the digging of wells and tanks, tubewells and input supply schemes for fertilisers, manure and seeds.

Another committee was appointed in 1952 to assess the campaign. This committee questioned the rationality of the food self-sufficiency objective, and criticised the lack of co-ordination between different programs, the inadequate and poorly trained staff, and resource inadequacy. On the other hand, it recognised that there had been some positive gains, particularly in the development of minor irrigation, and an increase in the production of foodgrains and commercial crops like cotton and jute. The 1952 committee recommended the formulation of a comprehensive agricultural plan to raise the level of rural living, and an important component of this was the creation of an extension service.

In the wake of the excitement and high expectations of political independence and responsible government, production of foodgrains and other agricultural raw materials in India continued to decline. In 1950/1, the output of cereals amounted to 41·7 million (metric) tons as against 46 million in 1949/50 and 43·3 million in 1948/9.

With this background a major concentration on the rural problem, particularly the foodgrain situation, was to be expected from the First Five Year Plan (1951-6). Its programs were similar to those instituted during the thirties—investment in irrigation, land reform and a massive attack on rural poverty, illiteracy and disease, through various programs of rural community development. Of these programs only the irrigation projects had a direct effect on agricultural production by providing irrigation for about 2·8 million hectares by 1956, and a further 3·6 million hectares by 1961. But of more importance in raising foodgrain production were a general expansion in crop area (by 12 per cent) and a period of favourable monsoons. By 1956 the Plan targets for foodgrains output had been exceeded and the government felt sufficiently satisfied with the rural sector to concentrate the Second Plan (1956-61) on industrial development. It was assumed that the Indian farmer could, and would, continue to raise the output of his farm if he were not exploited at every turn by landlords, money lenders and merchants. Hence the rural programs directed at improving the general social and economic environment of farmers—community development, land reform, credit co-operatives—were expanded, and there was little attempt, in the light of little apparent need, to improve farm technology.

This peripheral approach to agricultural development did not succeed in terms of increased production. In 1959 a Ford Foundation Team issued a report entitled *India's Food Crisis and Steps to Meet It* which recommended a direct attack on the production problems of Indian farmers. While the old rural programs continued under the Third Plan (1961-6), a new program was initiated on a trial basis in selected districts with high production potential. The Intensive Agricultural District Programme attempted to introduce to farmers the most effective package of practices and inputs for their individual farm conditions and to improve the input supply system, extension services and district administrations so that farmers could derive the anticipated benefit from the recommended package. It was assumed that significant increases in farm production could be obtained from the existing technology if the farmers would learn to apply it in the most efficient way. However progress was slow and unspectacular and it became evident that both the existing technology and the available supply of skilled manpower placed severe limits on the program's utility. For example the Intensive Agricultural District Programme emphasised the use of chemical fertiliser, but the available crop varieties were relatively unresponsive to fertiliser. While yields could be raised by 10 or even 20 per cent, many farmers considered this an inadequate return for the extra effort, costs and risk involved.

While the results were disappointing, the main reasons for this were becoming clear and led to greater interest in finding and developing foodgrain varieties which would be yield responsive to the package of inputs, particularly fertiliser. In 1963 the Indian Agricultural Research Institute in New Delhi imported a selection of dwarf wheats from the International Wheat Research Centre in Mexico for its own breeding program and began a series of trials in various locations in the wheat belt. By 1960 at least one useful paddy variety had resulted from a joint Indian Council of Agricultural Research/FAO Rice Hybridisation Project and this was introduced successfully in Tamil Nadu (ADT-27). Two years later several semi-dwarf indica and japonica rice varieties from Taiwan were tested in India, and the International Rice Research Institute started its work on dwarf indica rice varieties in the Philippines. Thus, as the technical limitations of the Intensive Agricultural District Programme were recognised, a solution was being found both in India and abroad. By 1965 a number of fertiliser-responsive varieties of wheat and rice and hybrids of maize, sorghum and millet, were available and some, particularly the Mexican dwarf wheats, already had been tested extensively in India with extremely encouraging results.

The next step was therefore to put the new varieties to work. The need to do this quickly and extensively was forced on the government by a continuously worsening foodgrain situation, culminating in particularly poor monsoons in 1965 and 1966 and conditions of severe foodgrain shortage and heavy imports. In August 1965, the Union Department of Agriculture made its first general announcement of a New Strategy of agricultural development which was:

1 to apply scientific techniques and knowledge of agricultural production at all stages, particularly in the fields; and
2 to select a few areas with assured rainfall and irrigation for concentrated application of a package of practices based on improved varieties of seeds responsive to heavy doses of fertilisers.

The Intensive Agricultural District Programme concept of a package of practices applied to specific areas of high production potential was retained as the basic structure, but the high yielding varieties were introduced as the key input in a new package.

By accepting the New Strategy the government was committed to its own package of complementing and supporting programs which included the following:

1 to make available the required types and quantities of inputs, particularly chemical fertilisers, and to allocate sufficient foreign exchange for this purpose;
2 to encourage investment in fertiliser factories and manufacture of other inputs in India;
3 to reorganise agricultural research into a co-ordinated program of all-India schemes aimed to raise India's agricultural productivity;
4 to intensify agricultural extension services and activities in selected program areas;
5 to provide adequate credit to farmers willing to grow the new varieties and to adopt the appropriate farm practices; and

6 to implement a cereal price policy which would be producer oriented, that is, would provide an economic incentive to farmers to invest in the fertiliser, seed and other inputs required for the new agricultural technology.

Total area under foodgrain crops in 1966/7 was about 121 million hectares of which about 24 million were irrigated and about 21 million were double cropped. The New Strategy was to reach farmers in selected districts through two related field programs—the High Yielding Varieties Programme and a Multiple Cropping Programme. The specific overall objective of the two programs, as written into the Fourth Plan (1966-71, later changed to 1969-74), was to attain foodgrain self-sufficiency by the end of the Plan period. The two field programs were expected to provide an additional 26 million tons of foodgrains by 1974.

Planning for the New Strategy, which began in 1965, consisted first of setting the overall production objectives (targets) and then translating these into area and input requirements. This exercise was conducted in New Delhi, although New Delhi was not to implement the program at the field level. The annual national targets were divided between States (which would be responsible for the field programs) after Centre-State consultations, and the State Departments of Agriculture subdivided their targets between districts with assured water (rainfall or irrigation), a large area under the relevant crop, a sound co-operative structure, and adequate administrative, technical and extension resources.

Two important aspects of the High Yielding Varieties Programme were not effectively planned and carried out. The first was the selection and testing of varieties in the areas where they were to be introduced, and the second was the contact with the farmers who were expected to accept the package of inputs more or less on faith. Few farmers accept advice of this nature on faith, particularly if it is official. The only high yielding varieties to have been subjected to extensive field trials in the areas selected for the program were the Mexican wheats. By 1967 the essential agronomic facts were known and communicated to farmers only in the wheat program areas. The various dwarf paddy varieties and maize, sorghum and millet hybrids had been tested at research institutions but not tried in many of the districts selected for the High Yielding Varieties Programme, and their basic agronomy was not known under field conditions.

Consequently, of the five crop programs implemented in 1966, only that for wheat had received anything like adequate preparation, and since then only the wheat program has been an unmitigated success. All the paddy varieties introduced from Taiwan failed to some degree, particularly during the *kharif* (monsoon) season. Indeed, the only early success was with the Indian-developed ADT-27 in Tamil Nadu. The paddy program received a boost in 1967/8 with the introduction of dwarf indicas (IR8 and IR5) from the Philippines but again these were distributed without adequate field trials and they have not been uniformly successful. Two new Indian dwarf indicas released in 1968/9, and a number under trial in the research institutes hold promise for the future, but the main contribution of the high yielding paddy program to date appears to lie in the increasing stress being laid on the *rabi* season—that is, in growing high yielding paddy as a second season crop, thereby expanding the area under multiple cropping.

In supporting the New Strategy the government has taken certain positive steps. Varietal and other production-oriented research has been supported through a number of all-India co-ordinated schemes. Main provisions of the Seed Act passed in 1966 have recently been implemented, giving hope for the development of a modern seed industry with high product standards. Seed Corporations, with the power to control quality through certification, have been set up at the Centre and in most States. The demand for fertiliser has been met by imports and expanded domestic production. Shortages and badly timed deliveries of fertilisers which characterised the first two years of the High Yielding Varieties Programme have been largely overcome. This has been no mean achievement as total fertiliser consumption in India has risen from 0·3 million tons of nutrients in 1960/1 to about 2·6 million tons in 1969/70.

Irrigation has also received considerable attention, particularly in the encouragement of private (farm) investment in tubewells and pumpsets. Concomitant with this there has been a complete reorientation of the rural electrification program, directed at energising private irrigation facilities. In 1969, on the recommendation of the All-India Rural Credit Review Committee, a Rural Electrification Corporation with an outlay of Rs1500 million was established to assist State Electricity Boards finance electricity supplies to rural areas for irrigation purposes. It is estimated that one and a quarter million privately owned pumpsets and tubewells will be energised during the Fourth Plan, again a substantial achievement.

The main program of rural credit in India, which pre-dates the First Plan, was mainly an attempt to reduce the power of money lenders for welfare considerations. The co-operate credit structure has not performed well in most States and has not provided adequate credit for productive purposes for the High Yielding Varieties Programme participants. Recently, the government has encouraged banks, both commercial and public to invest in rural development programs such as the financing of tubewells, and is in the process of redirecting short-term government and co-operative credit to small farmers. Finally, the government is encouraging a selective mechanisation program through research, tractor and power tiller production, and State Agro-Industry Corporations. Whatever the immediate production gains of the New Strategy a firm basis is being laid now for a large and important commercial farming sector in India.

In both the *kharif* (monsoon) and *rabi* (post-monsoon) seasons of the three years 1967/8 to 1969/70 the Programme Evaluation Organisation (PEO) of the Planning Commission has carried out a systematic study of the High Yielding Varieties Programme. The formulation of policy in the States, the program's administrative machinery, the supply of inputs and credit, and the response of farmers have been studied for all five crop programs in all States where the particular crops are important. The results of this study have been published in evaluation reports issued after the completion of each season. In 1969 Sir John Crawford initiated a joint research project between the Australian National University and the Programme Evaluation Organisation in which the latter's survey data would be subjected to a more detailed analysis than is possible in the course of preparing program evaluation reports.

As this overall study is at an early stage at the time of preparing this paper, we are able to present data on only a limited aspect of the High Yielding Varieties Programme, namely on the spread of high yielding variety seed use in the villages where the paddy and wheat programs were studied.[1]

## ADOPTION OF HIGH YIELDING VARIETIES: DATA

Farm participation in the high yielding wheat program was studied in ninety villages in six States. The high yielding paddy program during the *kharif* season was studied in 138 villages (twelve States) and during the *rabi* season, in sixty nine villages (six States). At the beginning of each village survey, lists were drawn up of all farm households (including tenants) and two items of information were obtained for each household: the size of the operational farm holding (owned land plus land net of leasing), and whether or not high yielding variety seed of the relevant crop had been sown during the current and previous seasons. It is this information which is presented below: the number of farmers in the survey villages who used high yielding variety seed in the years and season specified. Our data on the areas

[1] Since the time of writing a comprehensive report on High Yielding Varieties Programme participation has been published (Lockwood, Mukherjee and Shand 1971).

sown by the adopting farmers in these villages has yet to be tabulated fully, but some figures are available for wheat in the Punjab and *kharif* and *rabi* paddy in Tamil Nadu (Madras). For the three crops in these two States we are able to give preliminary estimates of the quantities of chemical fertiliser applied to the high yielding crops also.

*Adoption of High Yielding Wheat and Paddy Seed by States*

Figure 4.1 and Table 4.1 show the percentage of farmers in the survey villages of each State who used seed of high yielding paddy or wheat in 1967/8 (the second year of the High Yielding Varieties Programme) to 1969/70.

At first sight the general picture of high yielding variety adoption is confusing in its diversity: there were considerable differences between crops and States in the levels of farmer participation in any one year and in the rate and direction of change in participation over time. Although not shown here, there was considerable variation also between different districts in most States and, of course, between the villages studied in each district. However, taking all villages in all States together for each crop there has been a steady increase over the four years in the proportion of farmers using high yielding variety seed. This upward trend is strongest in the wheat villages where, in 1969/70, almost 70 per cent of the farmers sowed dwarf varieties. It is more modest but still significant in the *rabi* paddy villages with 39 per cent farm participation in 1969/70, and in the *kharif* paddy villages with 28 per cent farm participation. These varieties were unknown in the villages before 1966 and were not introduced in some villages until 1967.

*Wheat*: the water-fertiliser requirements of dwarf wheat varieties limit their effective use to irrigated areas, that is, only to approximately 50 per cent of the Indian wheatlands. The availability of irrigation water is therefore an important cause of variation between villages, districts and States in the extent of dwarf wheat adoption. But to a considerable extent this factor is greatly modified in our data as the villages studied were selected because they had irrigation facilities which were adequate or at least better than average for their districts. Nevertheless the rank order of the States in Figure 4.1 based on the proportion of farmers using dwarf wheat seed in 1969/70 corresponds closely to a ranking of States by the proportion of wheat crop irrigated. While all farmers in the selected Punjab villages, and most in the Uttar Pradesh villages, could grow irrigated wheat, the apparent stagnation in Rajasthan at around 35 per cent of the farmers could represent an irrigation ceiling and the decline in participation in Bihar from 53 per cent of farms in 1968/9 to 34 per cent in 1969/70

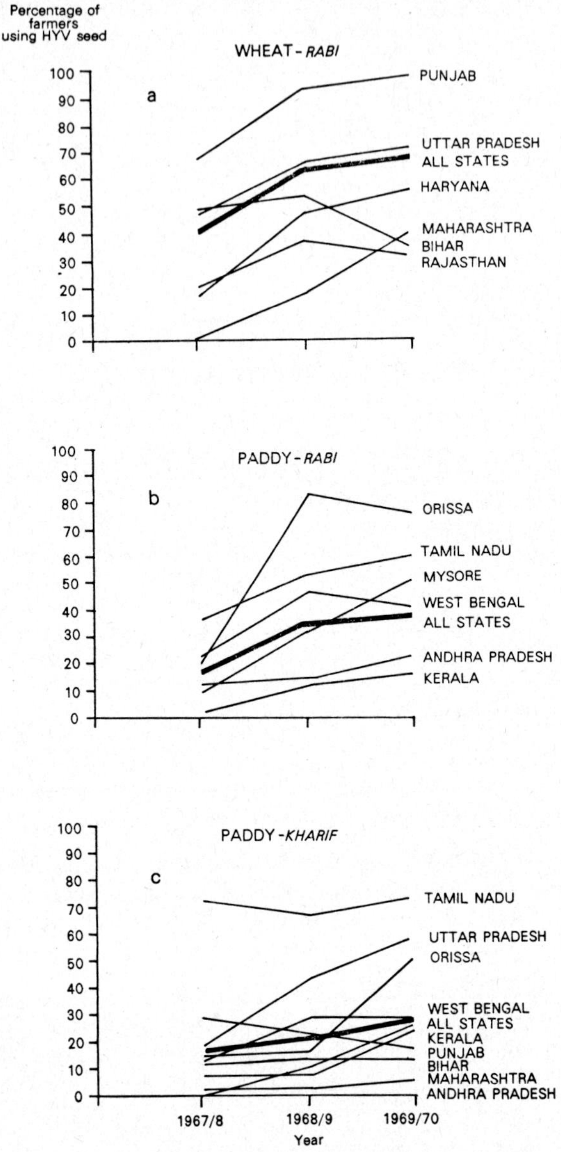


Figure 4.1 High yielding variety adoption, India, 1967/8 to 1969/70.

TABLE 4.1

HIGH YIELDING VARIETY SEED USE AMONGST FARMERS IN SURVEY VILLAGES, INDIA, 1967–8 TO 1969–70

| *Crop* | *State* | *1967/8* | | | *1968/9* | | | *1969/70* | | | *Change from 1967/8 to 1968/9 in percentage of farmers using HYVs* | *Change from 1967/8 to 1969/70 in percentage of farmers using HYVs* |
|---|---|---|---|---|---|---|---|---|---|---|---|---|
| | | *No. of villages* | *Total farmers* | *Percentage of farmers using HYV seed* | *No. of villages* | *Total farmers* | *Percentage of farmers using HYV seed* | *No. of villages* | *Total farmers* | *Percentage of farmers using HYV seed* | | |
| Paddy–*kharif* | Andhra Pradesh | 17 | 4981 | 2 | 17 | 4982 | 3 | 17 | 4977 | 5 | + 1 | + 3 |
| | Bihar | 12 | 1273 | 29 | 12 | 1286 | 23 | 12 | 1286 | 19 | − 6 | −10 |
| | Kerala | 3 | 1052 | 0 | 3 | 1058 | 10 | 3 | 1058 | 27 | +10 | +27 |
| | Tamil Nadu | 14 | 1868 | 72 | 14 | 1876 | 67 | 14 | 1870 | 73 | − 5 | + 1 |
| | Maharashtra | 12 | 2603 | 11 | 12 | 2603 | 14 | 12 | 2598 | 14 | + 3 | + 3 |
| | Orissa | 12 | 1763 | 15 | 12 | 1765 | 18 | 12 | 1762 | 51 | + 3 | +36 |
| | Punjab | 6 | 469 | 7 | 6 | 469 | 8 | 6 | 459 | 24 | + 1 | +17 |
| | Uttar Pradesh | 12 | 781 | 18 | 12 | 781 | 42 | 12 | 780 | 55 | +24 | +37 |
| | West Bengal | 18 | 1469 | 14 | 18 | 1469 | 29 | 18 | 1467 | 28 | +15 | +14 |
| | *Aggregate*[a] | 115 | 16,883 | 18 | 115 | 16,913 | 22 | 115 | 16,879 | 28 | + 4 | +10 |
| Paddy–*rabi* | Andhra Pradesh | 14 | 2440 | 11 | 14 | 2441 | 15 | 14 | 2418 | 24 | + 4 | +13 |
| | Kerala | 12 | 1624 | 2 | 12 | 1629 | 12 | 12 | 1629 | 17 | +10 | +15 |
| | Tamil Nadu | 11 | 1685 | 35 | 11 | 1685 | 53 | 11 | 1657 | 60 | +18 | +25 |
| | Mysore | 6 | 367 | 10 | 6 | 368 | 32 | 6 | 367 | 50 | +22 | +40 |
| | Orissa | 6 | 874 | 20 | 6 | 874 | 82 | 6 | 873 | 75 | +62 | +55 |
| | West Bengal | 15 | 1446 | 21 | 15 | 1446 | 46 | 15 | 1445 | 41 | +25 | +20 |
| | *Aggregate* | 64 | 8436 | 17 | 64 | 8443 | 35 | 64 | 8389 | 39 | +18 | +22 |
| Wheat–*rabi* | Bihar | 6 | 397 | 48 | 6 | 396 | 53 | 6 | 395 | 34 | + 5 | −14 |
| | Haryana | 12 | 1364 | 16 | 12 | 1364 | 48 | 12 | 1364 | 56 | +32 | +40 |
| | Maharashtra | 6 | 1166 | 1 | 6 | 1170 | 18 | 6 | 1171 | 41 | +17 | +40 |
| | Punjab | 24 | 2278 | 66 | 24 | 2287 | 94 | 24 | 2271 | 98 | +28 | +32 |
| | Rajasthan | 6 | 388 | 20 | 6 | 389 | 37 | 6 | 373 | 31 | +17 | +11 |
| | Uttar Pradesh | 36 | 3603 | 46 | 36 | 3606 | 67 | 36 | 3598 | 72 | +21 | +26 |
| | *Aggregate* | 90 | 9196 | 40 | 90 | 9212 | 63 | 90 | 9172 | 69 | +23 | +29 |

[a] The aggregate for all-India includes data for Jammu and Kashmir, Assam and Mysore also.

could be partly a withdrawal after trying dwarf wheat on rainfed or inadequately irrigated land. In the cases of Rajasthan and Bihar there are undoubtedly other important factors such as poor supplies and poor distribution of inputs (including fertilisers), weak co-operatives and marketing facilities.

The dwarf wheat program has been accepted most readily by the farmers of the Punjab. By 1969/70, 98 per cent of the farmers in the twenty four villages (in four districts) were growing dwarf wheat and only 4 per cent of the total wheat area of these villages remained under local varieties. Government estimates for Punjab State for 1969/70 show 1·4 million out of a total of 2·2 million hectares under wheat were sown to dwarf varieties. The 1·4 million hectares represents almost 90 per cent of the irrigated wheat area.

In Uttar Pradesh 70 per cent of the farmers in the thirty five villages surveyed (from districts in western, central and eastern Uttar Pradesh representing a wide range of agro-climatic conditions) grew dwarf wheat in 1969/70. Government estimates for the State indicate that about 85 per cent of the irrigated wheat land was sown to dwarf varieties. In Haryana only 56 per cent of the farmers in the twelve villages surveyed (two districts) grew dwarf wheat, while government estimates show 56 per cent of the irrigated wheatland of the State under dwarf varieties.

The proportion of farmers using dwarf wheat in Rajasthan and Bihar declined in 1969/70, although government estimates show a steady rise in dwarf wheat sowings to 30 per cent of the irrigated wheat area in Bihar, and about 10 per cent in Rajasthan. In these States, and Haryana also, the official figures of irrigated area appear to be gross over-estimates, particularly in terms of effectiveness or dependability (quantity and timing) of water. However, in these States there have been additional problems in implementing the High Yielding Varieties Programme, problems of input supply and marketing which, coupled with the uncertainty of water and relatively poor programs of minor irrigation and rural electrification, have tended to make the production of dwarf wheat a more risky venture for farmers in many districts. It should be said that in all these States, from Rajasthan to Bihar, the dwarf wheat varieties have performed well when given the necessary package of inputs and practices.

*Paddy*: the program for paddy was implemented in 1966/7 before the semi-dwarf varieties introduced from Taiwan had been tested under Indian farming conditions. The varieties were recommended for use in the main paddy season, the monsoon season, and were grown by farmers on their best paddy land, that is, the lowland fields which were easily flooded by the heavy monsoon rains and irrigated from canals and tanks. When IR8 and IR5, the dwarf indica varieties developed at the International Rice Research Institute in the Philippines, were introduced in 1967/8, they were also recommended for lowland fields during the *kharif* season. The main emphasis of the paddy program was on the *kharif* crop. Adoption experience suggests that had the new varieties been tested extensively prior to their introduction, the emphasis may well have been different.

Farmers found that certain varieties introduced in 1966 were quite unsuited to local conditions in either growing season, on waterlogged or even well drained fields. The variety Tainan 3 for example, introduced to Kerala in 1966, proved to be completely unsuited to conditions there. Taichung native 1, the main variety introduced to Andhra Pradesh, West Bengal and Bihar, gave yields well below expectations, mainly because farmers (and the Department of Agriculture) did not know how to use it. The same fault of introducing a variety without careful prior trials locally was responsible for the failure of IR5 and IR8 in many areas in which they were encouraged for the *kharif* season.

Farmers found that both the Taiwan and Philippine varieties give higher yields and are less subject to pests and diseases if grown during the *rabi* season (where irrigation water is adequate), and that they do much better on upland fields during the *kharif*. Our data on farm participation in the use of high yielding seed show a much greater response during *rabi* than during *kharif* in several important paddy States. In Andhra Pradesh, for example, the *kharif* program has stagnated at an extremely low level but there has been a steady increase in the number of farmers growing IR8 as a *rabi* crop. In Tamil Nadu, although ADT-27 has been grown successfully and widely as a *kharif* crop by 70-80 per cent of the farmers in the survey villages, there has been a rapid increase in the use of this variety during the *rabi* season, a rise from 35 per cent of the farmers in the villages surveyed in 1967/8 to 65 per cent in 1969/70. In Orissa and West Bengal, even though there was a decline in farm participation during the 1969/70 *rabi*, a larger proportion of farmers grew IR8 as a *rabi* crop in 1969/70 than as a *kharif* crop. The general picture is similar on the western side of the peninsula in Mysore.

Wherever waterlogging and flooding are common, and this is a characteristic of most important paddy areas during the monsoon season, the high yielding varieties have been found to be unsuitable, and farmers have either given them up or shifted them to upland fields; but when irrigation water is available and dependable, an increasing number of farmers are using them as a second (*rabi*) crop. In districts where waterlogging and flooding are not major problems during the monsoon, e.g. eastern Uttar Pradesh, Orissa and Kerala, farmers are growing dwarf paddy successfully as a first (*kharif*) crop.

Within the paddy States there was considerable variation in adoption between districts. For example, in Andhra Pradesh the adoption of IR8 as a *kharif* and *rabi* crop has been more widespread in West Godavari District than in Nellore and Nizamabad Districts. In 1969/70 in West Godavari 20 per cent of the farmers in the villages surveyed used dwarf paddy as a *kharif* crop and 56 per cent as a *rabi* crop. In Nellore and Krishna districts fewer than 10 per cent of the farmers used dwarf varieties in either season. In Bihar 64 per cent of the farmers surveyed used dwarf varieties during the *kharif* in Shahabad district in 1968/9 and 1969/70, while in Gaya district, the percentage fell from 32 per cent in 1967/8 to 4 per cent in 1969/70. There was wide variation among the four paddy districts studied in Tamil Nadu, within each season and between seasons.

For paddy there was as much variation between different districts within a State as between States and most States had at least one district where the high yielding varieties have been well accepted. Such performances require that assessments of the paddy program should proceed by district rather than by State. In the paddy States in particular an important factor contributing to this variation appears to be the suitability of the particular variety released and its characteristics. In Tamil Nadu a number of tall local indicas have been classified officially as high yielding varieties. While these are improved local varieties they are not as responsive to fertiliser as IR8 for example and strictly speaking are not part of the new technology. Adoption of one such variety in Coimbatore District, Tamil Nadu explains a large part of the increased farm participation in the High Yielding Varieties Programme during *rabi*. Official distortion of the program by the inclusion of improved local varieties occurs in other States also.

### *Farm Size and Program Participation*

It is assumed that within a village the area operated by a farmer is an approximate measure of his socio-economic status vis-à-vis the other farmers in the village. By ranking all farmers in a village by size of operational holding and interpreting the order obtained in this wider context, a measure of relative socio-economic status is obtained. This allows comparison of the behaviour of different status groups within a village, and also similar status groups between villages and districts where agro-economic conditions may vary considerably. While actual rather than relative land holdings may be adequate for comparing farmers within a village, it is inadequate as a basis for comparing farms, in, say, Tamil Nadu where the largest farm may be only

5 hectares (20 per cent irrigated by tanks) with farms in Haryana where the largest could be 25-30 hectares (all irrigated by canals).

Farmers, in each village separately, were ranked by size of operational holding and the list was divided into ten equal groups. In Figure 4.2, number 1, the first interval on the horizontal axis, indicates the group of farmers which fall in the decile containing the smallest farms, and number 10 comprises the farmers in the decile of largest farms. The corresponding decile groups for all villages surveyed for the relevant crop in a State have been aggregated, and show the percentage of farmers in each decile group who used high yielding seed in 1967/8, 1968/9 and 1969/70.

The relationship between incidence of high yielding variety seed use and relative farm size for the entire sample of villages is shown in Figures 4.2a (*kharif* paddy) and b (*rabi* paddy) and 4.3a (wheat). In each case there was a strong positive linear association between the two variables with the incidence of high yielding seed use increasing from the smallest to the largest farms.[2] Apart from variation in the general level of adoption between the three crops, there appears to be no significant difference in the nature of the relationship as indicated by the slope and shape of the curves. Clearly on this aggregate level, smaller farmers have been slower or less inclined to adopt the high yielding varieties than their larger neighbours. The data suggest that the probability that a farmer in decile 10 would be found using the seed was about three times greater than that for a decile 1 farmer.

This is of course a highly generalised picture. There was considerable variation between States, and often between districts within States (Figures 4.2c-q and 4.3b-g).

*Wheat States*: in the Punjab, the use of high yielding seed has spread fairly evenly to farmers in all decile groups, and as the general level of participation increased over time, the inter-decile difference in percentage of adopting farmers declined. In 1969/70 there was virtually 100 per cent participation in all deciles. Lack of reliable irrigation could prevent this situation being reached in the Haryana and Rajasthan villages studied, particularly where reliability is a function of tubewell investment which is limited largely to the larger farms. An adoption ceiling seems to have been reached in 1968/9 in Uttar Pradesh and Haryana (and Rajasthan) and further increases may depend largely on improvements in irrigation facilities. In these two States the differences between the small and large farm groups in terms of adoption were substantial though not as striking as in Maharashtra, Bihar and Rajasthan.

[2] There was a discernibly higher level of adoption amongst the largest farmers (Group 10) than might be expected with a straight line relationship.

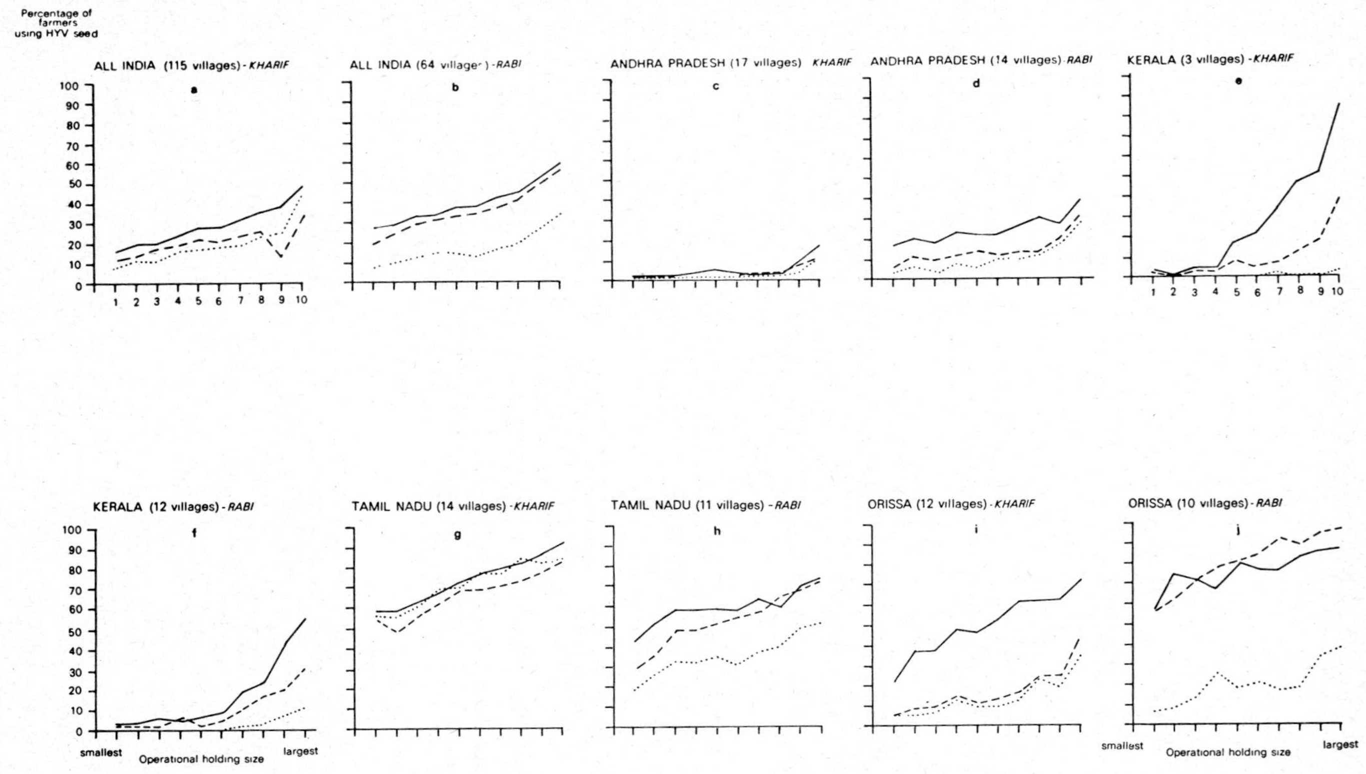
Percentage of farmers using HYV seed
ALL INDIA (115 villages) - KHARIF
a
ALL INDIA (64 village·) - RABI
b
ANDHRA PRADESH (17 villages) KHARIF
c
ANDHRA PRADESH (14 villages) RABI
d
KERALA (3 villages) - KHARIF
e
KERALA (12 villages) - RABI
f
TAMIL NADU (14 villages) - KHARIF
g
TAMIL NADU (11 villages) - RABI
h
ORISSA (12 villages) - KHARIF
i
ORISSA (10 villages) - RABI
j
100
90
80
70
60
50
40
30
20
10
0
1 2 3 4 5 6 7 8 9 10
smallest
Operational holding size
largest

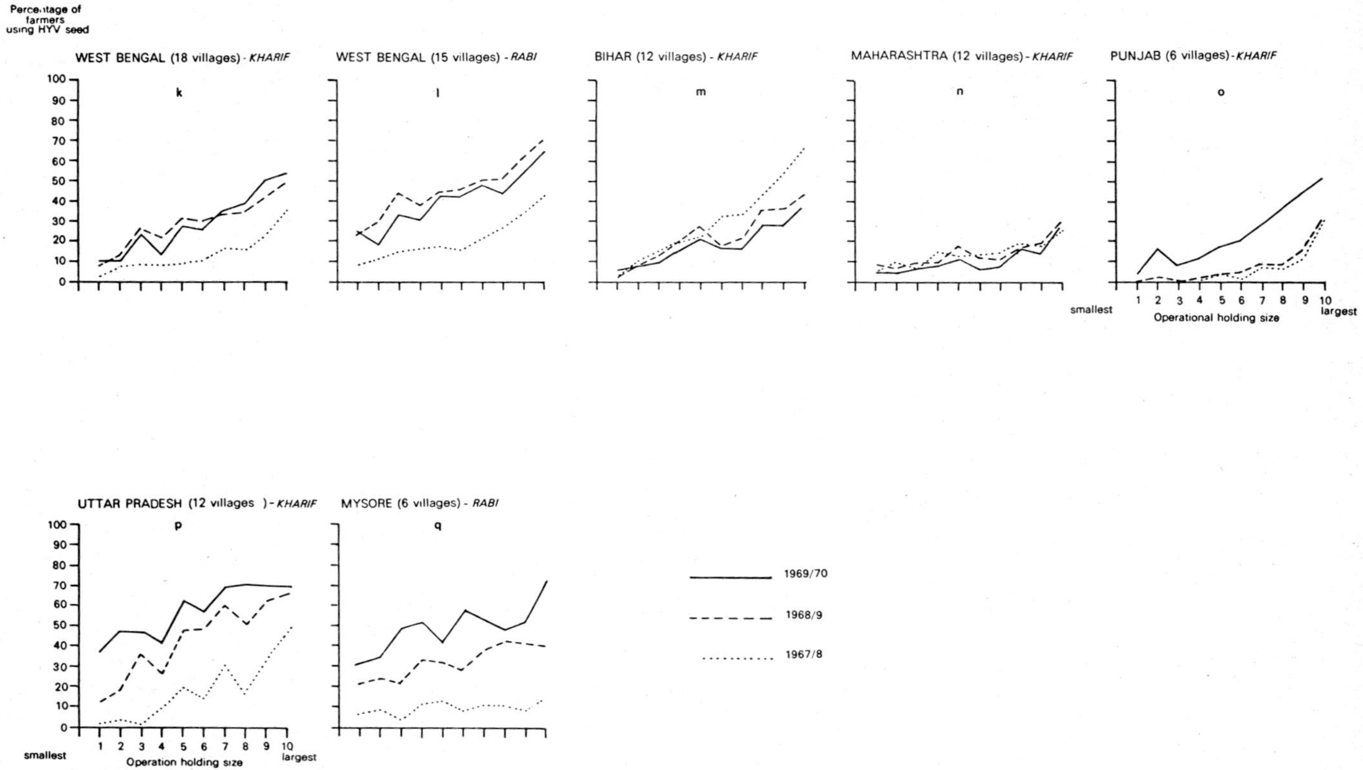


Figure 4.2 High yielding variety paddy adoption by decile groups of operational holdings, India, 1967/8 to 1969/70.

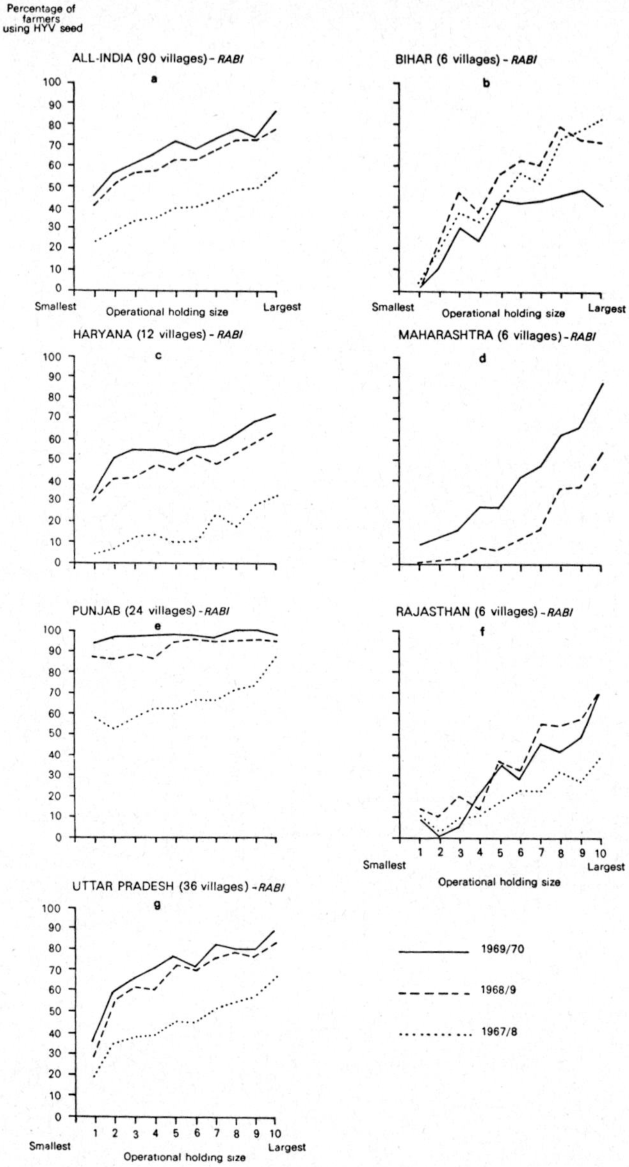


Figure 4.3 High yielding variety wheat adoption by decile groups of operational holdings, India, 1967/8 to 1969/70.

Wheat is not an important crop in Maharashtra. It is grown as a second (*rabi*) crop only when irrigation water is available, and it has to compete for this water with other crops such as sugarcane and millet. But our data show that many of the farmers who do grow some wheat have adopted the dwarf varieties. The steepness of the adoption curves probably indicates a strong relationship between farm size, irrigation, and the inclusion of wheat of any variety as a second crop.

In Gaya district, Bihar, in 1967/8 and 1968/9, the relationship between farm size and dwarf wheat use was strong, ranging from fewer than 5 per cent adopters amongst the smallest to 70-80 per cent of the largest farmers. This was very similar to the situation found in Rajasthan. In both these States there are major problems with irrigation and power and poorly organised supplies of other inputs, all of which are conditions which are generally unfavourable to small farmers. It was notable that, in Gaya district in 1969/70, there was a substantial decline in the proportion of adopting farmers in the medium to largest size groups (6 to 10), with the decrease greatest amongst the largest farmers. There was also a decline in high yielding seed use amongst the bigger farmers in Rajasthan. The reasons for these withdrawals are as yet not known.

*Paddy States*: the shape and slope of the adoption curves for paddy by States mostly show conformity with those characteristics of the all-State crop curve, being linear with a slope denoting a considerable difference between the rate of adoption of the smallest and largest farm size groups. This was apparent irrespective of season.

Changes in the general level of adoption over time have generally slowed to a parallel shift in the curves. The exceptions to this pattern over time were first Bihar where a decline in participation after 1967/8 took place amongst the bigger farmers. In Kerala and to a lesser extent Andhra Pradesh (*kharif*) the reverse was true: while little change occurred in adoption amongst smaller farmers, there were increases amongst bigger farmers, thus producing a steeper curve over time.

In Amritsar District of the Punjab and Gaya District of Bihar both the paddy and wheat programs were studied by the Programme Evaluation Organisation. In each district the farmers had the same access to the inputs and other facilities of the two programs and in each the level of farmer participation was significantly lower in the paddy program. Two points are worth stressing in relation to high yielding paddy. First, the limited extent of participation which had taken place in most States by 1969/70 was concentrated primarily amongst the bigger farms. Second, the program had made little impact in Andhra Pradesh where it was expected to be highly successful.

The data presented here provide a useful measure of the extent to which farmers have been willing to try the high yielding seed. This inevitably raises the question of the extent to which they have committed their land and other resources to the new varieties—the proportion of paddy land under high yielding paddy, the proportion of total fertiliser put on the high yielding crop and the actual dose.

As indicated earlier, this analysis is an early report of work in progress. Later reports will analyse information on such questions as area commitment, input use, costs of production, yields and on-farm investment. This will be carried out on a district basis where possible in order to present a detailed picture for specific areas of the changes which have resulted from the High Yielding Varieties Programme.

For example, the farmers who grew dwarf wheat in the Punjab villages studied in 1967/8 sowed about 58 per cent of their total wheat area to dwarf varieties (Table 4.2). In 1969/70 the average area commitment was 96 per cent. In Tamil Nadu the area commitment during *kharif* was 61 per cent of the total paddy area of adopting farmers and this remained at about the same level (58 per cent) in 1969/70. The area commitment to dwarf varieties during the *rabi* season was considerably higher, although it had also apparently reached a ceiling around 75-80 per cent. This is not surprising in view of the fact that most paddy farmers prefer to grow local varieties for their own consumption and therefore the almost complete switch to dwarf (wheat) varieties found in the Punjab cannot be expected in the paddy areas. The maximum level of adoption may well have been reached in the Tamil Nadu villages. It is notable that amongst both Punjab and Tamil Nadu adopters there was an almost equal proportion of crop area committed to high yielding varieties by all three groups of farmers.

The use of chemical fertiliser on the dwarf wheat crops appears on average to be well below the doses recommended by the State Departments of Agriculture (Table 4.3). In the Punjab 618 kilograms/hectare of nitrogen is recommended for wheat. The average dose applied in 1968/9 was only 464 kilograms and it fell to 353 kilograms in 1969/70. At the same time there was an evening out in the average doses applied by farmers in three broad farm size groups. In Tamil Nadu the recommendation stands at 371 kilograms of nitrogen per hectare. In 1967 the average application was only 153 kilograms. The following year it rose to 314 and in 1969 it fell to 237 kilograms. The average applications of farmers in three groups fluctuated in such a way as to suggest fairly careful experimentation which led to a fairly uniform application in the range 222-49 kilograms/hectare in 1969/70.

One further aspect of the move towards a modern agriculture is illustrated by the number of farmers adopting the complete package

TABLE 4.2

PERCENTAGE AREA UNDER HIGH YIELDING VARIETIES CROPS TO TOTAL AREA UNDER THE SELECTED CROP, INDIA, 1967–70

| *State/Crop* | *Season/Year* | | *Size Group* Below 2 hectares | 2–8 hectares | 8 hectares and above | All |
|---|---|---|---|---|---|---|
| Punjab: | *rabi* | 1967/8 | 67·7 | 55·8 | 60·8 | 57·9 |
| Wheat | *rabi* | 1968/9 | 68·5 | 74·5 | 79·3 | 76·8 |
| | *rabi* | 1969/70 | 95·6 | 96·2 | 96·5 | 96·3 |
| Tamil Nadu: | *kharif* | 1967 | 65·2 | 55·4 | 77·8 | 60·9 |
| Paddy | *kharif* | 1968 | 67·9 | 51·5 | 31·5 | 48·1 |
| | *kharif* | 1969 | 57·5 | 54·3 | 68·8 | 58·4 |
| | *rabi* | 1968/9 | 81·8 | 89·1 | 26·6 | 79·1 |
| | *rabi* | 1969/70 | 74·9 | 78·6 | 67·9 | 75·6 |

TABLE 4.3

AVERAGE DOSE OF NITROGEN APPLIED BY FARM SIZE GROUP OF PARTICIPANTS, INDIA, 1967–9

| *State/Crop* | *Season/Year* | | *Size Group* Below 2 hectares | 2–8 hectares | 8 hectares and above | All | *Average recommended dosage (kilograms/hectare)* |
|---|---|---|---|---|---|---|---|
| Punjab: | *rabi* | 1967/8 | 141 | 422 | 541 | 464 | 618 |
| Wheat | *rabi* | 1968/9 | 304 | 341 | 366 | 353 | |
| Tamil Nadu: | *kharif* | 1967 | 156 | 138 | 252 | 153 | |
| Paddy | *kharif* | 1968 | 240 | 447 | 262 | 314 | 371 |
| | *rabi* | 1968/9 | 250 | 225 | 242 | 237 | |

of inputs and practices recommended for the high yielding varieties (Table 4.4). Thirty four per cent of the dwarf wheat growers in the Punjab had adopted the four main items of the package—treated seed, fertiliser, plant protection and intercultural operations in 1967/8. The following year only 21 per cent of the farmers did so. In Tamil Nadu, the package was adopted by few in 1967 (14 per cent) but there was a small increase during the following year (19 per cent). In Tamil Nadu there was a very considerable increase in the extent to which the package was adopted as farm size increased.

TABLE 4.4

PERCENTAGE OF FARMERS ADOPTING PACKAGE OF HIGH YIELDING VARIETY PRACTICES BY FARM SIZE GROUPS, INDIA, 1967–9

| *State/Crop* | *Season/Year* | | *Size Group* | | | |
|---|---|---|---|---|---|---|
| | | | *Below 2 hectares* | *2–8 hectares* | *8 hectares and above* | *All* |
| Punjab: | *rabi* | 1967/8 | 40·0 | 29·1 | 48·0 | 33·6 |
| Wheat | *rabi* | 1968/9 | — | 22·9 | 18·2 | 20·6 |
| Tamil Nadu: | *kharif* | 1967 | 8·3 | 26·1 | 100·0 | 14·3 |
| Paddy | *kharif* | 1968 | 14·5 | 22·7 | 37·5 | 18·5 |
| | *rabi* | 1968/9 | 15·5 | 25·0 | 50·0 | 19·0 |

It is probable that on all four indicators of High Yielding Varieties Programme participation—percentage of farmers using high yielding variety seed, area commitment, quantity of fertiliser used, and adoption of the package of practices, the farmers of the Punjab and Tamil Nadu lead those of other States for wheat and paddy respectively. If this is so then a number of alternative explanations are possible. Farmers may be extremely cautious in following the recommended package; they have found it uneconomic, or perhaps they obtain acceptable yields without adopting fully the recommended technology.

## POLICY ISSUES

This analysis of a limited aspect of the High Yielding Varieties Programme—the use of dwarf wheat and paddy seed in the villages selected for study by the Programme Evaluation Organisation—indicates that an increasing number of farmers are adopting the main inputs of the New Strategy. It is reasonable to assume that the use of high yielding seed means, for most farmers, the use of chemical fertiliser and other inputs, and some change from the old cropping pattern and farm practices. We have seen that there has been an increase in the use of dwarf paddy in the second (*rabi*) season in a number of important paddy districts and to some extent this represents a new emphasis on the *rabi* and an increase in double cropping. In the wheat areas dwarf varieties have been adopted rapidly by a very high proportion of farmers with irrigation during the *rabi*. In many districts there has been a significant increase in the use of purchased inputs—particularly seed, fertiliser and water. The rapid spread of dwarf wheat has had a considerable impact on wheat production; the more halting and much slower spread of dwarf paddy has increased paddy output in some villages and districts but so far has not had a discernible impact on total paddy production. The term 'green revolution', which has been bandied about widely in the press, may have some meaning in the Punjab and parts of Uttar Pradesh and Haryana, and even in some parts of Tamil Nadu, but it is certainly a term ill chosen to describe the changes which have taken place in farm technology so far in other parts of India. It is clear however that in the irrigated areas a great many farmers are experimenting with the new inputs provided by the New Strategy; they are running their own cost/benefit studies and where the results are favourable, are adopting their own packages of new inputs.

While the results of the field programs, High Yielding Varieties and Multiple Cropping, have varied widely from place to place, the major supporting programs of the government, varietal research, the supply of fertiliser and other inputs, the provision of rural credit for inputs and investment in irrigation, and the support of wheat prices, have been pursued vigorously. They have been kept under constant review and this has resulted in a more efficient administration of input production and distribution, an expanded program of minor irrigation and rural electrification, improved credit facilities for small farmers, and a continuous program of varietal and agronomic research. The New Strategy has been in operation for only five years but this has been long enough for the emergence of some serious problems.

### *Millet, Sorghum and Maize Programs*

While data on the performance of high yielding varieties for millet, sorghum and maize have not been presented here, evidence is available to show that these programs have been far from successful in most of the districts surveyed by the Evaluation Organisation. In most of these districts there has been a considerable decline in the numbers of farmers using hybrid varieties of these crops; a general picture which is similar to the one shown for wheat and paddy in Bihar. In 1969/70, 63 per cent of the farmers who had grown hybrid maize earlier had dropped the variety altogether. For sorghum 56 per cent had reverted to local varieties and for millet about 50 per cent. In addition a significant number of those who continued to sow hybrid varieties had reduced the area under these varieties. The reasons given by the farmers for their reversion to local varieties differ from place to place, and between crops, but the most common were lack of irrigation water, attacks of pests and diseases, high costs of production and problems of finding a ready and remunerative market. In some areas, for example in the Punjab, Haryana and western Uttar Pradesh a reduction in hybrid sorghum acreage during the *rabi* has occurred with improvements in irrigation as the farmers find it more profitable to increase their wheat acreages. In the Punjab, paddy and sugarcane have replaced millet and maize during the *kharif*.

*Levels of Input Use*

Even where a large proportion of farmers have adopted the high yielding varieties, and where they comprise a large proportion of the total irrigated crop, farmers have not in general applied the recommended package of other inputs. Even in the Punjab the average dose of fertiliser used in 1969/70 was only 57 per cent of that recommended by the research institutions, and extension agencies. But this may well be because, through their own trials, the farmers formed their own views on the most economic level of inputs to use. Levels recommended by the extension people are based on yield trials at research institutions and these are not exposed to the test of the marketplace. Further, the recommendations are often not based on local trials so there is no reason to expect that they will be appropriate even as yield maximisers. Farmers' responses suggest that in economic terms the recommendations are often unrealistic.

While farmers have been adjusting input levels of high yielding variety seed and fertiliser, particularly in the light of water availability, on the finer points of pest and disease control and the use of soil analysis to assist in determining appropriate fertiliser and water requirements, both they and officials have yet to record any progress in most districts.

*Credit*

Participation in the High Yielding Varieties Programme involves farmers in a switch from a basically self-contained and non-monetary agronomy to one dependent on inputs purchased from outside. This change in the relationship between the farm and the market has been recognised by the program planners who have assumed that the participating farmers will not be able to purchase the main inputs from their own cash resources, and will need substantial credit. Consequently an important feature of the field program has been the provision of credit for production purposes through the co-operative structure and the Departments of Agriculture. Longer-term credit has also been provided for investment in irrigation, electrification, land levelling and other capital ventures.

In many States the co-operative structure has not been able to provide credit to those farmers who need it most. Since 1965 the credit system has undergone a number of modifications, first to try to improve the co-operative system, and more recently to draw the banks into the sphere of rural lending, particularly for capital investment. One major problem has been that institutional credit, particularly co-operative and bank credit, has been more readily available to the medium and large farmers, many of whom are now showing that they do not need it, than to smaller owners and tenant farmers who do need it. The uneven spread of credit may partially explain the slope of the adoption curves shown above in Figures 4.2 and 4.3. Data from the 1969/70 Programme Evaluation Organisation survey show that the larger wheat and paddy farmers have begun to finance their current inputs and capital investment substantially from their own cash resources and, have started to extend credit to neighbouring small and medium farmers. To some extent, the prevailing institutional credit structure could be helping to subsidise a new group of farmer money lenders.

Official emphasis is now given to channelling institutional short-term production credit to medium and small farmers. Small farmers' Development Agencies have been set up on an experimental basis in forty four districts (1970) to handle this and other programs designed to give greater development assistance to small farmers. Large farmers are going to have to rely increasingly on their own resources for production credit.

*Irrigation*

Many of India's large irrigation schemes do not provide the type of irrigation service required for high yielding variety production or systematic multiple cropping. Most are primarily power stations. Water is released for irrigation during the *kharif* season mainly as a protective measure against poor rainfall and in most years only a limited area receives water during the *rabi*. Field channels are poorly designed, often makeshift or are absent altogether, and water passes by the gravity flow system from higher to lower fields. Farmers have no control over the quantity or timing of such water as both depend on technical decisions made without much reference to agricultural needs, and on the location of fields in relation to the canals. Generally there is no provision for drainage and the lower fields become waterlogged and flooded. Timing and quantity of water could possibly be improved with closer co-operation between the irrigation engineers and agricultural development officials, and there has been some improvement in recent years. But with continuation of flow irrigation, flooding and waterlogging persist, and generate conditions in which pests and diseases flourish more readily, and fertiliser applied to upper fields washes to lower fields. No major program of land levelling and reshaping in the command areas of major irrigation projects is planned for the foreseeable future. In many areas therefore the quality of the irrigation service is an important, and probably the most important, constraint on the High Yielding Varieties and Multiple Cropping Programmes.

These problems have led to increased government emphasis on underground water use through energised pumps and tubewells, even in the command areas of major irrigation schemes. These have the advantage that farmers can control the timing and quantity of water applied to their fields. In some areas pumping has lowered the water table and reduced the incidence of waterlogging.

With this shift in emphasis to medium and small irrigation schemes the government has encouraged private investment by making available to farmers liberal credit facilities for sinking tubewells and connecting electricity. Punjab farmers have led the way in this as in the adoption of dwarf wheat, but there is now a great deal of private investment on minor irrigation in Uttar Pradesh, Haryana and some other areas. The country's underground water resources are currently being explored and mapped.

### *Rural Institutions*

Many observers of the contemporary Indian scene have emphasised the need for creating appropriate farmers' institutions to help in accelerating and stabilising the adoption process of new technology. It is well known, for example, that the co-operative agency is weak in many areas and that there is a need to provide a suitable and stable base from which such a movement can develop spontaneously and progress. In this matter, Indian farmers could well emulate the experience of their counterparts in Japan, Taiwan and South Korea. It could be encouraged by the extension agency which can identify the progressive farmers and those who are willing to follow their lead, and give them institutional recognition and support.

### *Labour vs Capital in Indian Agriculture*

The use of the new high yielding varieties requires the application of modern inputs in the form of pure seed, fertiliser, plant protection chemicals and water, which cannot be produced directly by the farmers. It is a land-saving, capital-intensive technology. Many consider that this is not an appropriate policy for India unless the technology can be simultaneously labour-absorbing. Ishikawa (1969) has argued that for many countries of the ECAFE region the appropriate technology is an intermediate one relying on pure-line selection of indigenous improved plant varieties, irrigated by minor and medium schemes constructed mainly by labour inputs, fertilised by farm produced natural manures, and worked with improved traditional implements. The same view has been expressed by Johnston and Cownie

(1969) who fear that the seed-fertiliser technology will encourage premature tractor mechanisation.

An intermediate technology of the type suggested by Ishikawa has been tried in India. This is essentially what was involved in the Intensive Agricultural District Programme (IADP), and it was the inability of this approach to accelerate rural development sufficiently to meet India's increasing foodgrain shortage which led to the formulation of the High Yielding Varieties Programme. In reaching the decision to include the New Strategy in the Fourth Plan the government placed a high priority on the matter of foodgrain self-sufficiency even though this necessitated heavy capital investment and use of foreign exchange. It was a major policy decision on the allocation of these two scarce and costly resources. To question this, one would need to question the objective of foodgrain self-sufficiency itself.

In the few districts where the High Yielding Varieties Programme has been widely adopted by farmers, there has been appreciable investment in farm machinery—tractors, thrashers, seed drills and tubewells fitted with electric or diesel pumps, in addition to the production inputs such as seed and fertiliser. Investment in minor irrigation is easily explained and in many areas is a necessary pre-condition of high yielding varieties adoption. Investment in tractors and other implements, all of which are labour-saving and have high capital costs, is the result of a number of factors. Probably the most important is the relative shortage, high cost and undependability of labour in the very busy period covering the harvesting of the *kharif* crops and the planting of the *rabi* crops. That is, it is an important factor where the farmer has adopted a double cropping pattern. A farmer who depends on labour and bullock power to prepare his fields for *rabi* sowing often has to postpone the thrashing and storing of his *kharif* crops for several weeks during which time they could be damaged by rain and pests. There can be little argument about this on grounds of technical efficiency and the reduction of risk.

It is not known whether or not this type of mechanisation, when it accompanies a shift from single to double cropping, significantly reduces the level of labour use on the farm over the entire year. It certainly reduces the demand for labour during certain busy periods, but it could increase the use of labour during the new second season. Empirical evidence on the effect of selected mechanisation on medium and large farms in India has not been collected systematically and until this is done this important question remains unanswered.

It should be noted, however, that the rapid mechanisation of the Punjab has encouraged the growth of a rural-based industry concerned with servicing machines and producing ploughs, harrows, trailers, etc. This high growth sector has attracted and absorbed a significant quantity of labour which earlier had depended on farm work for its livelihood.

### *Welfare Considerations*

The High Yielding Varieties Programme and other rural development programs in India have been introduced into a setting in which production resources, particularly land, are distributed very unevenly between those dependent on agriculture for a livelihood. Ladejinsky (1969a and b, 1970) and others have argued forcefully that the introduction of a capital-using agricultural technology in this situation will lead to further polarisation of wealth and to increasing social and political unrest. One cannot refute this if it is assumed that nothing else changes. While our data show that the new technology is by no means limited to the large farmers, favoured as they are in the distribution of inputs, credit and extension services, their share in total village output will inevitably increase. This will be so even if all farmers adopt the package of inputs equally and double their output and incomes. It cannot be avoided without a complete equalisation of land holdings. It can be softened by making inputs more readily available to small farmers and tenants so that they can raise the productivity of the land they do hold, and by giving the latter greater security of tenure. Possibly the only feasible way by which this kind of polarisation can be reduced is through a progressive land or farm income tax, and on this the government has as yet made no effective move.

Another aspect of the new technology which bears on the problem of income distribution is the increasing share of farm income (in High Yielding Varieties Programme areas) accruing to capital, that is, to the farmer rather than to his hired labour. It is not known whether the actual amount paid to labour increases, decreases, or remains static, but it is inevitable that labour's share will decline. If labour has no alternative employment this could indeed cause serious unrest. If this is to be avoided without watering down the ingredients of the new technology the pattern of broad rural development seen in the Punjab, in which labour is increasingly attracted away from the farms, is needed in other program areas.

As part of its overall strategy of rural development, the government has made some effort to reduce the income polarising effects of the high yielding varieties and related programs. In dryland areas where the High Yielding Varieties Programme has not penetrated, attempts are being made to introduce supplementary incomes through dairying and other livestock programs, pasture improvement, etc. In irrigated areas small farmers and tenants are being given special treatment in the distribution of credit and inputs. To a large extent these are regarded as welfare measures but at the same time they aim also at increasing productivity of land and labour. For obvious reasons these programs for small farmers and rural labourers are regarded as important by the government but they are not being allowed to interfere with the main objective of foodgrain self-sufficiency and the spread of the new technology.

Land reform remains a government objective but it is being seen less as a necessary precondition for agricultural development and more as a welfare or income redistributing measure. On the whole it remains a dead letter although our argument above, and recent events in Bengal, suggest that it may become increasingly urgent as the new technology spreads. The present uneven distribution of land ownership does not seem to be an important limitation or constraint on the High Yielding Varieties Programme.

## CONCLUSIONS

The high yielding varieties and related programs of the New Strategy have succeeded in introducing modern capital-using farm methods in some areas and already appear to have contributed substantially to the achievement of the government objective of foodgrain self-sufficiency by 1974. On the whole, the program has been forcefully implemented by the Departments of Agriculture of most States and problems of input shortages, which characterised the first year or two, have been largely overcome. The government has decided to pursue the programs of the New Strategy vigorously and to treat income distribution and welfare problems which arise through separate production oriented programs.

It is argued that the paddy and hybrid maize, sorghum and millet programs were introduced before sufficient knowledge of the characteristics of the varieties were known, or before suitable varieties were available for widespread distribution. The instability of the paddy program and the poor response to the maize and sorghum programs attest to this. But the wheat program was a major success and the varietal breeding work underway in India holds promise of an increasingly successful paddy program in the near future. Farmers have demonstrated their willingness to try the new technology and to adopt it where it proves profitable and feasible. To date however, there has been no green revolution, except perhaps in the Punjab, but there have been pocket developments in many well irrigated districts, disturbing the Gandhian concept of rural self-sufficiency and serenity.

# 5

# Adoption of New High Yielding Rice Varieties by Java's Farmers

*Sajogyo and William L. Collier*

IF we take three recent points in time and compare statistics of Java's population and rice production (Table 5.1) we find the growth of the latter has fallen well behind that of population. Between 1953 and 1967 total output of milled rice rose by 13 per cent and total population by 30 per cent. Wet rice area harvested showed little change over the period and yield differences between 1953 and 1967 accounted for the limited increase in output.

TABLE 5.1

DECLINE IN GROWTH OF RICE PRODUCTION, JAVA, 1953–67

| | *1953* | *1960* | *1967* |
|---|---|---|---|
| Total population (million) | 53·4 | 62·9 | 70·4 |
| Total rice production (million tons milled) | 4·42 | 5·05 | 4·99 |
| Rice area harvested | 4·02 | 4·32 | 4·02 |
| Rice yields (tons milled rice per hectare) | 1·14 | 1·21 | 1·29 |

*Source:* Ministry of Agriculture, Indonesia.

The Paddy Centra-Scheme (1959-62) was the first effort by the government to organise on a large scale the supply of inputs, especially fertilisers, to paddy planting peasants in Java. Fertiliser supply was raised from zero to more than 70,000 (nutrient) tons in 1962, by which time it was outpacing fertiliser use on the estate crops. In spite of budget problems, efforts were continued in 1964-7 to supply fertilisers to rice-growing peasants together with new National Improved varieties (bengawan, sigadis, remadja and shinta),[1] with the initial help of university students. At the time 'Bimas'[2] was coined as the new name for projects of such concentrated efforts.

1 Produced by research at the Central Agricultural Research Station in Bogor. Bengawan was released in the 1940s; the other three in the late 1950s and early 1960s.

2 Bimas = Bimbingan Masal or Mass Guidance or Extension.

Further developments saw the release of the new IR high yielding varieties (IR8 and IR5) in 1967 and their encouragement with the help of foreign companies from 1968 to 1971.[3] In 1968-70 annual fertiliser use for foodcrops reached levels two to three times higher than in the preceding period, and accounted for some 82 per cent of total fertiliser use in Indonesia (over 200,000 tons).

Aggregate statistics from the Bimas Committee of the Ministry of Agriculture for four planting seasons, 1969 dry season to 1970/1 wet season provided estimates of coverage of the newer varieties in Bimas program areas of 190,000 hectares (1969 dry season), rising to 287,200 hectares in the following wet season but falling back in the next dry season (1970) to less than one third of the preceding dry season. In 1970, the foreign companies began phasing out their activities. The 1970/1 wet season (under the all-government National Bimas program) area surpassed the earlier wet season level (306,000 hectares). In the last mentioned season, if areas under the Inmas program (supply of inputs without credit) were included, IR variety coverage had reached 10 per cent of total rice hectarage in Indonesia. Within Java, East Java showed a relatively larger impact (some 25 per cent of total hectarage) than Central Java (12 per cent) and West Java (10 per cent).

## THE AES STUDY OF PADDY FARMING, 1969-72

How have Java's peasants responded to the newer rice varieties that have shown much higher potential yields but have demanded higher levels of inputs, including cash? To study farmers' problems at the village level and those of the many different agents of change who have tried to influence farmers, the Agro-Economic Survey[4] (AES) has, since 1969, interviewed twice a year a sample of 600 farmers living in twenty villages on Java and 500 farmers in seventeen villages on the islands of Sumatra, Celebes and Bali. The study areas chosen were in the best irrigated paddy regions, and had marketing facilities, including machine processing, which were relatively conducive to better contacts with town and city market centres. Java was seen as a series of major rice producing areas: the northern lowland coastal plains with recent modern irrigation works, (two sample areas in West Java, two in Central Java); the mountain area (one sample area in

[3] The foreign companies involved in these programs were Ciba, Hoechst, Mitsubishi, and several others.

[4] The Agro-Economic Survey was created in 1965 as an ad hoc inter-ministerial organisation by a decree of the Vice-Premier of Economic Development. Some assistance has been provided by the Ford Foundation and, since 1968, also by the Agricultural Development Council, Inc.

West Java); river basin areas (one in Central Java, two in East Java[5]); and two plains areas on the southern coast (one in Central Java and one in East Java, with the latter separated by limestone hills from the sea). Two villages were selected in each area. Figure 5.1 shows the locations of these villages on Java. Table 5.2 lists the sample areas, all the villages or *kabupatens* included in each region and the randomly chosen *kabupatens.*

With six villages in West Java, eight in Central Java and six in East Java chosen for the study, comparisons made here are still on such a province-grouping basis that care must be taken in making generalised conclusions. The thirty sample farmers in each village were chosen randomly from three strata: five from a list of 'large farmers' (as interpreted in the village), fifteen from the list of Bimas participants (wet season 1968/9—the time of the first interview) and ten from a list of non-participants. If either of the latter two categories did not exist in a village the sample of twenty five 'other farmers' was taken randomly from the existing strata.

### *Types of Farmers Represented by AES Sample*

*Farm size*: if only two groupings are distinguished ('large farmers' and 'other farmers'), farm size in terms of total land operated and wet paddy field operated in the 1968/9 wet season can be shown as in Table 5.3.

By comparing the average farm sizes in Table 5.3 with the distribution in 1963 Agricultural Census (Table 5.4) one can judge how representative the sample farmers are.[6] In relation to the averages of total land operated, the AES sample farmers appear to be in the top 33-43 per cent (in classes of more than 0.5 hectares) of the total farm population in each of the three provinces. The 'large strata' sample farmers' averages are in Census classes 4 and 5 (Table 5.4). If we look at individual sample farmers, several fit into class 6. The averages in the 'others strata' or smaller sample farmers are mostly in class 3 (from 0·5 to 0·99 hectares). In the AES sample the 'large' farmers utilised or operated 78 per cent of the farm in wet paddy on average while the 'smaller' farmers operated a lower 66 per cent. There was a range of from 1·40 to 3·50 hectares under actual wet paddy on 'large' farms and from 0·47 to 0·66 hectares on 'other' farms.

[5] One of these two is also a river delta or coastal plain type.

[6] It will be noted that the class of 'less than 0·1 hectare' farmers is included in Table 5.4. These farmers were not interviewed in the 1963 Census, but their numbers were known.

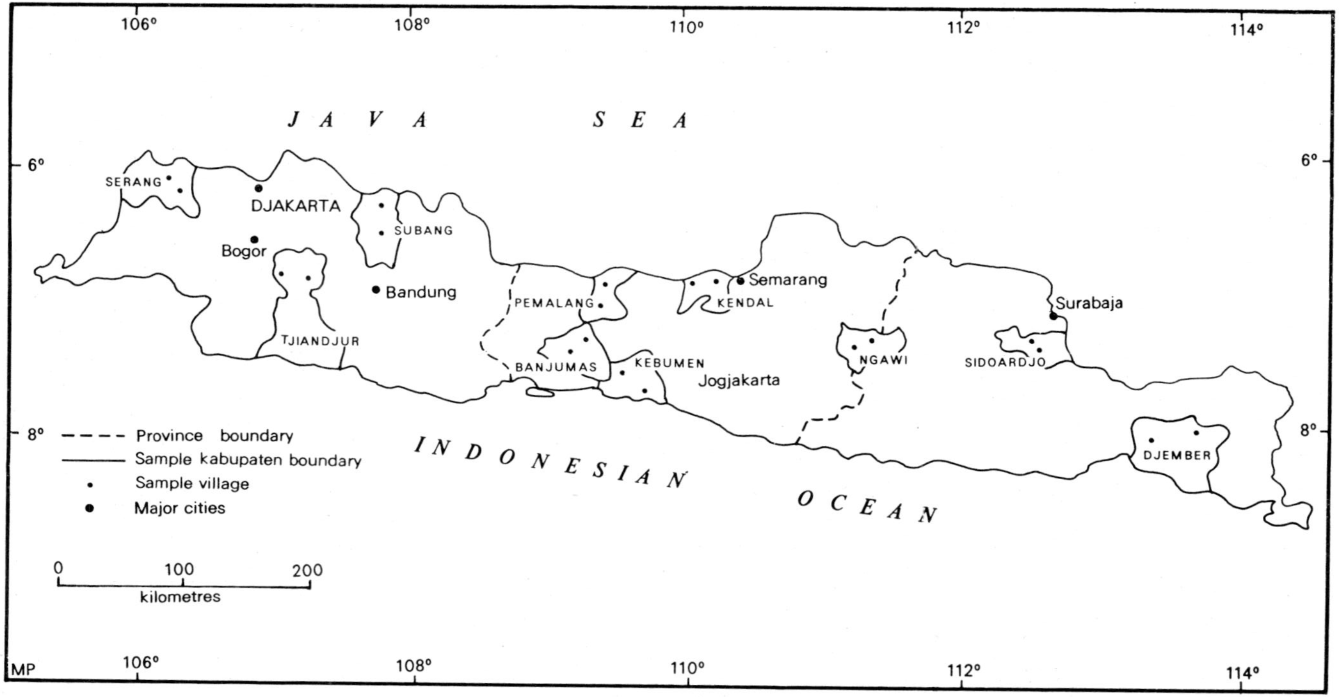


Figure 5.1 Locations of sample villages, Java.

TABLE 5.2

RICE PRODUCING REGIONS AND AES RANDOMLY CHOSEN KABUPATEN, JAVA, 1969–71

| *Province* | *Region* | Kabupatens *included in each region* | *Selected* kabupaten |
|---|---|---|---|
| Sumatra Utara | Lowland coastal plain, east coastal | Langkat, Deli Serdang, Simalungun, Asahan, Labuhan Batu | Deli Serdang and Simalungun |
| Sumatra | Lowland coastal | Padang Pariaman, South Coast | Padang Pariaman |
| Barat | Mountains | Pasaman, Lp. Kota, Tanah Datar, Solok, Sw-Sidjundjung | Tanah Datar |
| Lampung | Non-transmigration | Lampung Selatan | Lampung Selatan |
| | Transmigration (resettlement schemes) | Lampung Tengah | Lampung Tengah |
| Djawa Barat | North coast, lowland plain, western section | Serang, Tanggerang, Bogor, Lebak, Pandeglang | Serang |
| | North coast, lowland plain, eastern section | Bekasi, Krawang, Indramaju, Subang, Tjirebon, Kuningan, Madjalengka | Subang |
| | Uplands | Sukabumi, Tjiandjur, Garut, Tasik, Tjiamis, Sumedang, Bandung | Tjiandjur |
| Djawa Tengah | North coast, lowland plain | Brebes, Tegal, Pemalang, Pekalongan, Batang | Pemalang |
| | North coast, lowland plain | Kendal, Semarang, Demak, Kudus | Kendal |
| | South coast, lowland plain | Kebumen, Purworedjo, Kulon Progo, Tjilatjap | Kebumen |
| | Seraju River | Wonosobo, Bandjarnegara, Purbolinggo, Banjumas | Banjumas |
| Djawa Timur | Bengawan Solo River basin | Bodjonegoro, Tuban, Lamongan, Madiun, Magetan, Ngawi, Ponorogo, Patjitan | Ngawi |
| | Brantas River | Surabaja, Sidoardjo, Modjokerto, Djombang, Kediri, Ngandjuk, Blitar, Tulung Agung, Trenggalek, Malang | Sidoardjo |
| Djawa Timur | Outside of the main river basin areas, lowlands | Pasuruan, Probolinggo, Lumadjang, Bondowoso, Pamanukan, Djember, Banjuwangi | Djember |

*CARRIED OVER*

| *Province* | *Region* | Kabupatens *included in each region* | *Selected* kabupaten |
|---|---|---|---|
| Bali | Lowland plain | Gianjar | Gianjar |
| Sulawesi Selatan | Southwest coast, lowland plain | Gowa, Takalar, Maros | Maros |
| | Northwest coast, lowland plain | Palmao, Pinrang, Sidenreng, Roppang, Soppeng | Pinrang |
| | Eastern coast, lowland plain | Luwu, Madjo, Bone | Bone |

TABLE 5.3

AVERAGE FARM SIZE OF THE AES SAMPLE FARMERS, JAVA, 1968/9

| *Farm Size Strata* | *Total Land Operated (hectares)* | | | | *Wet Paddy Field Operated (hectares)* | | | |
|---|---|---|---|---|---|---|---|---|
| | *West* | *Central* | *East* | *All Java* | *West* | *Central* | *East* | *All Java* |
| Large | 3·33 | 1·83 | 3·86 | 2·73 | 1·75 | 1·40 | 3·50 | 2·13 |
| (n) | (30) | (40) | (30) | (100) | (30) | (40) | (30) | (100) |
| Other | 1·08 | 0·60 | 0·99 | 0·86 | 0·62 | 0·47 | 0·66 | 0·57 |
| (n) | (148) | (199) | (149) | (496) | (148) | (199) | (149) | (496) |

n = number of sample farms.

TABLE 5.4

JAVA'S OPERATED FARM SIZE DISTRIBUTION, 1963

| *Location* | *Farming households (per cent)* | *Class 1 less than 0·1 hectare (per cent)* | *Class 2 0·1–0·49 hectare (per cent)* | *Class 3 0·5–0·99 hectare (per cent)* | *Class 4 1·0–1·99 hectares (per cent)* | *Class 5 2·0–5·0 hectares (per cent)* | *Class 6 more than 5·0 hectares (per cent)* |
|---|---|---|---|---|---|---|---|
| West Java | 79·9 | 25·0 | 42·0 | 18·2 | 10·4 | 3·7 | 0·3 |
| Central Java | 90·2 | 23·5 | 40·1 | 21·2 | 11·3 | 3·3 | 0·2 |
| East Java | 73·1 | 15·6 | 40·5 | 24·2 | 14·2 | 4·8 | 0·3 |

*Source:* 1963 Agricultural Census data, Djakarta.

*Land tenure*: the AES sample of 'other strata' farmers in the three provinces were mostly (70-82 per cent) owner operators, a further 10-15 per cent were farming on a combination of their own land and additional leased land (rented or share cropped) while 8-17 per cent were farming on wholly leased land (rented or share cropped). On the basis of the 1963 Agricultural Census data (excluding the 'less than 0·1 hectare' farmers) the same three classes contained 48 per cent, 28 per

TABLE 5.5

PERSONAL TRAITS OF THE SAMPLE FARMERS, JAVA

| *Strata by Location* | *Average size of household* | *Farmers' average age years* | *Farmers' average schooling (general + religious) years* |
|---|---|---|---|
| West Java: | | | |
| large | 7·3 | 39·7 | 4·0 |
| other | 5·4 | 39·8 | 3·2 |
| Central Java: | | | |
| large | 6·8 | 47·2 | 5·5 |
| other | 6·3 | 43·8 | 3·9 |
| East Java: | | | |
| large | 8·5 | 49·6 | 6·2 |
| other | 5·8 | 46·2 | 4·0 |

cent and 4 per cent respectively of all the farmers on Java. Thus by comparison the AES sample included many more (50 per cent more) owner-operators, less (only half) owner-lessees and two to four times the number of wholly renters or share cropper farmers than would be typical for Java as a whole. AES 'large' and 'other' farmers showed the same proportions for the first two classes but for the third the 'large' farmers included only 4 per cent.

*Size of household, age, schooling*: again, compared with average household size for Java as a whole (four persons in West and East Java, five in Central Java), AES sample households were larger, especially the 'large' stratum. It is striking that the average age of AES sample farmers in 1969 was over forty years. Their schooling (range 3·2 to 6·2 years) must therefore have been in the 1930s and their farming practice probably dated from the 1940s, and in some instances extended through some twenty four years.

The AES sample farmers can thus be seen as being in the top 30-40 per cent in terms of farm size, they farm in better irrigated paddy areas, and are in the older, more established group in their villages. However, one must remember that we purposely selected areas with relatively good irrigation facilities in the major rice producing areas while the Agricultural Census of 1963, was actually on a sample basis and had all the farmers in each province in Java as its population.

### *Farmers Adopting the New Varieties*

In Java it is known that the large-scale Bimas paddy-package-credit programs at the village level put pressure on farmers to plant new varieties, to use fertilisers and pesticides and to accept credit. Even with pressures, however, many farmers clearly did not sow all their rice plots with the new varieties. In this connection it is best to compare adoption trends over time. One method is to classify sample farmers into three groups: those who in any one season planted only IR (a good sign of it being the chosen variety), those with only national improved (NI) varieties and those with only local ones. A second method is to group sample farmers into those planting at least IR (together with other varieties), those planting at least national improved varieties and those planting at least local varieties. The first and second categories of farmers at least are trying out IR and NI varieties respectively, while the last category is still clinging to local varieties. Using this second method, farmers may be counted in more than one category.

With these two methods of comparison we may look first at adoption data over time, then examine them in relation to certain characteristics of the farmers and of their operations: size of rice operation, land tenure status, outside farm work and marketing of rice, also whether participation in Bimas programs was a factor in their continued planting of IR varieties.

Trends of variety plantings by AES sample farmers in six seasons from wet season 1968/9 to dry season 1971 (Tables 5.5 and 5.6) provide indications of farmers' variety adoption patterns and of their cropping patterns, though it should be kept in mind that even with available irrigation water, not all farmers grew a second rice crop in the dry season. For example, sample farmers in Central Java showed a greater tendency to include rice as part of their dry season cropping pattern than those in the two other regions. In both the wet and dry seasons they also showed the lowest rates of IR adoption with the two indicators planting 'at least IR' and 'only IR in any season', with an uneven trend of adoption over time. The highest rates of adoption were shown by East Java sample farmers who, for at least IR showed a growth of 41-58 per cent in the wet seasons and a decline of 44-32 per cent in the dry seasons. For only IR planted there was positive growth in adoption in both seasons over the three years. Proportions of West Java farmers who had adopted at least IR and only IR were lower than for East Java farmers but they increased for both seasons over the three years.

Preference for local varieties and national improved ones (in that order) amongst sample farmers, in wet as well as in dry seasons, was highest amongst the Central Java farmers, followed by the West Java sample. This was in notable contrast to East Java farmers on whom IR varieties had had the most convincing impact (Table 5.7).

*Size of paddy farm operations*: the average size of paddy farm amongst farmers who planted only an IR variety was noticeably larger than the average amongst those who planted only a local variety

TABLE 5.6

ADOPTION PATTERNS FOR IR VARIETIES, JAVA, 1968–71

| *AES Sample Farmers* | *Wet Season* 1968/9 | 1969/70 | 1970/1 | *Dry Season* 1969 | 1970 | 1971 |
|---|---|---|---|---|---|---|
| | (percentage of farmers) | | | | | |
| *West Java:* 180 farmers | | | | | | |
| Planted at least IR | 8 | 17 | 23 | 15 | 15 | 27 |
| Planted only IR | 2 | 3 | 8 | 4 | 6 | 12 |
| Did not plant rice | 0 | 11 | 3 | 27 | 20 | 25 |
| *Central Java:* 240 farmers | | | | | | |
| Planted at least IR | 6 | 3 | 12 | 18 | 2 | 15 |
| Planted only IR | 0·4 | 0 | 5 | 3 | 1 | 7 |
| Did not plant rice | 0 | 2 | 6 | 0·4 | 5 | 6 |
| *East Java:* 180 farmers | | | | | | |
| Planted at least IR | 41 | 57 | 58 | 44 | 35 | 32 |
| Planted only IR | 21 | 23 | 41 | 12 | 27 | 29 |
| Did not plant rice | 0 | 2 | 3 | 26 | 31 | 30 |

TABLE 5.7

ADOPTION PATTERNS FOR NATIONAL IMPROVED AND LOCAL VARIETIES, JAVA, 1968–71

| *AES Sample Farmers* | *Wet Season* 1968/9 | 1969/70 | 1970/1 | *Dry Season* 1969 | 1970 | 1971 |
|---|---|---|---|---|---|---|
| | (percentage of farmers) | | | | | |
| *West Java:* 180 farmers | | | | | | |
| Planted at least NI | 20 | 21 | 26 | 14 | 21 | 8 |
| Planted only NI | 14 | 12 | 17 | 12 | 17 | 4 |
| Planted at least local | 79 | 73 | 67 | 57 | 51 | 56 |
| Planted only local | 72 | 52 | 50 | 47 | 39 | 43 |
| *Central Java:* 240 farmers | | | | | | |
| Planted at least NI | 32 | 33 | 29 | 34 | 23 | 23 |
| Planted only NI | 27 | 29 | 20 | 26 | 21 | 15 |
| Planted at least local | 73 | 68 | 65 | 65 | 73 | 69 |
| Planted only local | 62 | 64 | 57 | 53 | 72 | 60 |
| *East Java:* 180 farmers | | | | | | |
| Planted at least NI | 42 | 44 | 20 | 31 | 10 | 5 |
| Planted only NI | 23 | 20 | 12 | 7 | 6 | 4 |
| Planted at least local | 39 | 19 | 39 | 34 | 33 | 36 |
| Planted only local | 34 | 19 | 29 | 14 | 27 | 34 |

TABLE 5.8

PADDY FARM SIZE AND PLANTING OF RICE VARIETIES, JAVA, 1968–70

| | *Average paddy farm size (hectares)* | | | | | | | | |
|---|---|---|---|---|---|---|---|---|---|
| | *West Java sample farmers* | | | *Central Java sample farmers* | | | *East Java sample farmers* | | |
| | *Local* | *NI*[a] | *IR*[b] | *Local* | *NI* | *IR* | *Local* | *NI* | *IR* |
| *Wet season 1968/9* | | | | | | | | | |
| Only one variety | 1·01 | 0·89 | 1·50 | 0·56 | 0·96 | — | 0·87 | 1·56 | 1·46 |
| (n)[b] | (130) | (27) | (3) | (149) | (64) | (0) | (62) | (41) | (37) |
| At least one variety | 1·12 | 1·06 | 2·03 | 0·62 | 1·02 | 0·76 | 1·03 | 1·58 | 1·51 |
| (n) | (145) | (36) | (14) | (174) | (77) | (14) | (72) | (74) | (73) |
| *Dry season 1969* | | | | | | | | | |
| Only one variety | 0·81 | 0·40 | 1·39 | 0·56 | 0·86 | 0·57 | 0·52 | 0·37 | 0·90 |
| (n) | (88) | (20) | (7) | (127) | (62) | (9) | (25) | (14) | (27) |
| At least one variety | 0·95 | 0·79 | 1·69 | 0·65 | 0·94 | 1·06 | 0·51 | 0·54 | 0·72 |
| (n) | (106) | (23) | (27) | (157) | (80) | (43) | (60) | (56) | (80) |
| *Wet season 1969/70* | | | | | | | | | |
| Only one variety | 0·78 | 0·76 | 0·78 | 0·58 | 0·80 | — | 0·62 | 0·81 | 1·68 |
| (n) | (93) | (21) | (6) | (153) | (70) | (0) | (34) | (38) | (40) |
| At least one variety | 0·96 | 0·99 | 1·34 | 0·62 | 0·83 | 2·19 | 1·01 | 1·40 | 1·67 |
| (n) | (132) | (37) | (28) | (162) | (78) | (7) | (62) | (80) | (101) |

[a] NI = national improved, IR = IRRI rice varieties.
[b] n = number of observations.

in East and West Java in all three seasons, with only one exception (Table 5.8). This was not evident for Central Java, though sample numbers were sufficient for comparison in only one season for those with only an IR variety. The average size differential was also noticeable for farmers who planted at least one local variety in East, West and Central Java. This latter comparison is the more meaningful since sample numbers of those adopting an IR variety were larger. Thus in general, IRRI high yielding varieties have been adopted on relatively larger sized farms on which previously used local varieties are still retained.

A preference for IRRI higher yielding over national improved rice varieties amongst larger rice farm operators in each of the seasons was most apparent in the West Java sample, was evident in the East Java sample (except in the wet season 1968/9) but was not for the Central Java sample, where there were too few IRRI adopters to allow a comparison to be made.

A preference for national improved over local rice varieties in each of the three seasons amongst larger rice farm operators was apparent for the Central Java sample; it was less clearly so in the East Java

TABLE 5.9

LAND TENURE STATUS AND PLANTING OF RICE VARIETIES, JAVA, 1968–70

| | Percentage of Farmers[d] | | | | | | | | |
|---|---|---|---|---|---|---|---|---|---|
| | *West Java sample* | | | *Central Java sample* | | | *East Java sample* | | |
| | *Local* | *NI* | *IR* | *Local* | *NI* | *IR* | *Local* | *NI* | *IR* |
| *Wet season 1968/9* | | | | | | | | | |
| 1[a] Owner-operator | 87 | 85 | 100 | 80 | 89 | — | 87 | 87 | 89 |
| lessee[c] | 32 | 26 | 0 | 39 | 25 | — | 43 | 30 | 27 |
| (n) | (130) | (27) | (3) | (149) | (64) | (0) | (62) | (41) | (37) |
| 2[b] Owner-operator | 88 | 89 | 93 | 83 | 91 | 100 | 89 | 93 | 95 |
| lessee | 32 | 28 | 28 | 40 | 29 | 71 | 40 | 26 | 20 |
| (n) | (145) | (36) | (14) | (174) | (77) | (14) | (72) | (74) | (73) |
| *Dry season 1969* | | | | | | | | | |
| 1 Owner-operator | 88 | 85 | 71 | 84 | 87 | 89 | 72 | 86 | 93 |
| lessee | 28 | 25 | 29 | 41 | 26 | 33 | 44 | 21 | 26 |
| (n) | (88) | (20) | (7) | (127) | (62) | (9) | (25) | (14) | (27) |
| 2 Owner-operator | 90 | 91 | 93 | 83 | 88 | 81 | 83 | 86 | 86 |
| lessee | 31 | 35 | 33 | 42 | 23 | 40 | 33 | 30 | 34 |
| (n) | (106) | (23) | (27) | (157) | (80) | (43) | (60) | (56) | (80) |
| *Wet season 1969/70* | | | | | | | | | |
| 1 Owner-operator | 89 | 90 | 83 | 81 | 91 | — | 94 | 89 | 98 |
| lessee | 20 | 14 | 33 | 44 | 11 | — | 32 | 19 | 26 |
| (n) | (93) | (21) | (6) | (153) | (70) | (0) | (34) | (38) | (40) |
| 2 Owner-operator | 90 | 92 | 89 | 81 | 90 | 86 | 87 | 91 | 91 |
| lessee | 23 | 27 | 25 | 42 | 13 | 14 | 40 | 29 | 32 |
| (n) | (132) | (37) | (28) | (162) | (78) | (7) | (62) | (80) | (101) |

[a] Refers to farmers planting only one variety.
[b] Refers to farmers planting at least one variety.
[c] Renting and/or share cropping.
[d] Some farmers are both owner-operators and lessees and are thus counted twice, once in each category. Total of the two categories can thus exceed 100 per cent.

sample (e.g. not in the dry season 1969) and was not shown at all for the West Java sample.

Of the three paired comparisons of preference by larger farmers among the three categories of rice varieties in the three seasons the last mentioned (national improved over local varieties) is the least convincing. Such preferences are supported by data on rice yields which show that national improved rice varieties generally did not outyield local varieties by much and at times were inferior.

*Land tenure*: calculations were made for three seasons of the proportions of owner-operators and lessees amongst paddy farmers planting various rice varieties[7] (Table 5.9). These showed that the incidence of leasing was possibly somewhat more common for plantings of local varieties than for NI or IR varieties and most noticeably so in the Central Java sample. There appeared to be a general stability in the proportions both for owner operators and lessees over the period for the three variety categories. The only exceptions were declines in the proportions of lessees over the three seasons in the Central Java sample for NI and IR varieties. Leaving aside these particular features, the average proportions of lessees[8] and owner-operators were around 30 and 90 per cent respectively for all variety categories. No special characteristics or trends could be discerned which might be ascribed to the adoption of IR varieties. Data did not, for example, suggest that the adoption of these varieties was influenced by the form of land tenure.

*Commercial rice farming*: an indication of the extent to which rice farming was commercialised was obtained from AES survey data by calculating the proportions of sample farmers that purchased and/or sold paddy. This was done for three seasons and by categories of rice varieties which included IR varieties (Table 5.10). The existing level of commercialisation, given by the marketed proportions amongst sample farmers who planted at least local rice varieties, was clearly high for East Java farmers where as many as 87 per cent were recorded as selling rice in one season. Central and West Java samples showed lower but still substantial proportions of sellers. Calculations for the IR adopters, using the 'at least one variety' measure, showed for West Java an increased proportion of sellers in the second wet season (1969/70) observations, suggesting a possibly positive effect of IR adoption upon the degree of commercialisation. In the East Java sample, the proportion of paddy sellers in the 1969/70 wet season was no greater than that for local variety growers.

If the proportion of buying to selling rice by these paddy farm operators is taken as a measure of the level of commercialisation and the 1969 dry season and 1969/70 wet season figures are compared, there is little difference found for East Java between seasons either for the local or IR variety categories, with ratios of 36 and 39 per cent respectively.[9] The West Java sample showed higher buying proportions in the dry season both for smaller (local variety) and larger (at

[7] The single measure 'planting at least one variety' is referred to in this discussion.

[8] In West Java, leasing commonly took the form of share cropping. In the other two region samples, it was more typically land renting.

[9] The measure of 'at least planting local' and 'at least planting IRRI varieties' is again used here.

TABLE 5.10

FARMER PURCHASE AND SALE OF RICE AND PLANTING OF RICE VARIETIES, JAVA, 1968–70

| | Percentage of Farmers | | | | | | | | |
|---|---|---|---|---|---|---|---|---|---|
| | *West Java sample* | | | *Central Java sample* | | | *East Java sample* | | |
| | *Local* | *NI* | *IR* | *Local* | *NI* | *IR* | *Local* | *NI* | *IR* |
| *Wet season 1968/9* | | | | | | | | | |
| 1[a] sell | 55 | 33 | 33 | 60 | 48 | — | 74 | 68 | 84 |
| buy | n.a. | n.a. | n.a. | n.a. | n.a. | n.a. | n.a. | n.a. | n.a. |
| (n) | (130) | (27) | (3) | (149) | (64) | (0) | (62) | (41) | (37) |
| 2[b] sell | 54 | 31 | 64 | 65 | 57 | 79 | 78 | 77 | 49 |
| buy | n.a. | n.a. | n.a. | n.a. | n.a. | n.a. | n.a. | n.a. | n.a. |
| (n) | (145) | (36) | (14) | (174) | (77) | (14) | (72) | (74) | (73) |
| *Dry season 1969* | | | | | | | | | |
| 1 sell | 60 | 40 | 29 | 57 | 34 | 44 | 72 | 64 | 85 |
| buy | 34 | 35 | 14 | 21 | 11 | 0 | 4 | 21 | 41 |
| (n) | (88) | (20) | (7) | (127) | (62) | (9) | (25) | (14) | (27) |
| 2 sell | 58 | 35 | 48 | 61 | 44 | 74 | 87 | 75 | 83 |
| buy | 38 | 39 | 33 | 18 | 10 | 0 | 28 | 20 | 34 |
| (n) | (106) | (23) | (27) | (157) | (80) | (43) | (60) | (56) | (80) |
| *Wet season 1969/70* | | | | | | | | | |
| 1 sell | 55 | 57 | 67 | 61 | 37 | — | 76 | 61 | 80 |
| buy | 28 | 43 | 0 | 14 | 3 | — | 56 | 29 | 40 |
| (n) | (93) | (21) | (6) | (153) | (70) | (0) | (34) | (38) | (40) |
| 2 sell | 59 | 54 | 79 | 63 | 37 | 57 | 76 | 70 | 77 |
| buy | 27 | 41 | 14 | 16 | 8 | 0 | 32 | 26 | 28 |
| (n) | (132) | (37) | (28) | (162) | (78) | (7) | (62) | (80) | (101) |

[a] Refers to farmers planting only one variety.
[b] Refers to farmers planting at least one variety.

TABLE 5.11

OFF-FARM INCOME EARNING AND PLANTING OF RICE VARIETIES, JAVA, 1968/9

| | Percentage of farmers with outside farm income | | | | | | | | |
|---|---|---|---|---|---|---|---|---|---|
| | *West Java sample* | | | *Central Java sample* | | | *East Java sample* | | |
| | *Local* | *NI* | *IR* | *Local* | *NI* | *IR* | *Local* | *NI* | *IR* |
| Farmers planting only one rice variety | 19 | 37 | 33 | 43 | 30 | — | 53 | 37 | 41 |
| (n)[a] | (130) | (27) | (3) | (149) | (64) | (0) | (62) | (41) | (37) |
| Farmers planting at least one variety | 23 | 31 | 43 | 42 | 31 | 43 | 50 | 39 | 40 |
| (n) | (145) | (36) | (14) | (174) | (77) | (14) | (72) | (74) | (73) |

[a] number of observations in each category.

least IR) farmers (65-8 per cent). In the 1969/70 wet season the proportion for smaller farmers was a lower 45 per cent for smaller farmers and a low 17 per cent for larger farmers, showing the occurrence of more 'real rice surplus' from the wet season harvest. In the Central Java sample, the ratios were low (29 and 25 per cent) for smaller (at least local varieties) in the two seasons while the 1969/70 wet season the larger (at least planting IRRI) farmers were not shown in the sample to have purchased any rice at all.

*Off-farm income*: survey data on the incidence of income from sources outside of their own farms for one season (Table 5.11) showed, for the West Java sample, that farmers planting at least IR were more active than those growing at least local and national improved varieties.[10] The reverse situation was found in East Java, 50 and 53 per cent of the local variety producers had outside income compared with only 40 and 41 per cent of the IR producers. Previously, we found (Table 5.8) that these farmers with IR varieties in West and East Java had on the average larger rice field operations than the local growers. The fact that more of the farmers in East and Central Java had outside income may be related to the greater population pressure in these areas.

*Participation in Bimas credit package programs*: by distinguishing between sample farmers who were participating in a Bimas package (credit in the form of production inputs) program in any season, and those who were outside such programs, the spread of IRRI higher yielding varieties in the two groups can be compared. The question is on the one hand whether better seed has been supplied in enough quantities through the Bimas program to enable its participants to plant the new varieties, and on the other, whether outside such programs, by other means, farmers have also been able to adopt the new varieties. See Table 5.12.

East Java sample data provides most evidence of a positive relation between IR adoption and the Bimas programs. The sample farmers who were in a Bimas program and planted local and IRRI varieties in the wet season 1968/9 were 15 and 49 per cent respectively compared with 65 per cent and 6 per cent amongst farmers outside such programs. In the second wet season, however, the proportion of IRRI varieties' adopters outside the Bimas programs had risen to almost that of farmers within the programs. This was at the expense of the proportion using local varieties (29 per cent). The 1969 dry season showed intermediate proportions which were 22 per cent local and

10 There were too few sample farmers who grew only IR to include this measure in the comparisons for West Java and Central Java.

TABLE 5.12

SAMPLE FARMERS' PARTICIPATION IN BIMAS CREDIT PROGRAMS AND THEIR PLANTING OF RICE VARIETIES, JAVA, 1968–70

| | *Percentage of farmers planting at least one variety* | | | |
|---|---|---|---|---|
| | *Local* | *NI* | *IR* | *Number of observations* |
| *Wet season 1968/9* | | | | |
| *West Java:* | | | | |
| In Bimas[a] | 68 | 25 | 6 | 111 |
| Outside Bimas | 82 | 9 | 8 | 85 |
| *Central Java:* | | | | |
| In Bimas | 62 | 31 | 7 | 175 |
| Outside Bimas | 71 | 28 | 1 | 90 |
| *East Java:* | | | | |
| In Bimas | 15 | 36 | 49 | 140 |
| Outside Bimas | 65 | 29 | 6 | 79 |
| *Dry season 1969* | | | | |
| *West Java:* | | | | |
| In Bimas | 66 | 16 | 18 | 69 |
| Outside Bimas | 69 | 15 | 16 | 87 |
| *Central Java:* | | | | |
| In Bimas | 58 | 27 | 15 | 163 |
| Outside Bimas | 53 | 31 | 16 | 117 |
| *East Java:* | | | | |
| In Bimas | 22 | 30 | 48 | 102 |
| Outside Bimas | 39 | 28 | 33 | 94 |
| *Wet season 1969/70* | | | | |
| *West Java:* | | | | |
| In Bimas | 65 | 20 | 15 | 115 |
| Outside Bimas | 69 | 19 | 12 | 82 |
| *Central Java:* | | | | |
| In Bimas | 63 | 33 | 4 | 157 |
| Outside Bimas | 68 | 32 | 0 | 90 |
| *East Java:* | | | | |
| In Bimas | 23 | 34 | 43 | 151 |
| Outside Bimas | 29 | 32 | 39 | 92 |

[a] Bimas program: for these three seasons, it was primarily the Bimas Gotong Rojong which in most cases were foreign companies involved in supplying farm production inputs to the rice farmers. To a lesser degree it was the National Bimas.

48 per cent IRRI varieties in Bimas programs and 39 per cent local and 33 per cent IRRI varieties outside such programs.

The West Java sample shows a high level of preference for retaining local varieties, both by farmers in and outside the Bimas programs. More than 65 per cent of the sample farmers in the Bimas programs

TABLE 5.13

AVERAGE PRE-HARVEST LABOUR USE IN PADDY FARMING, BY RICE VARIETIES, JAVA, WET SEASON, 1969/70

| | *West Java sample* | | | *Central Java sample* | | | *East Java sample* | | |
|---|---|---|---|---|---|---|---|---|---|
| | *Local* | *NI* | *IR* | *Local* | *NI* | *IR* | *Local* | *NI* | *IR* |
| Pre-harvest labour use (man/days hectare) | 217·7 | 206·3 | 339·6 | 234·2 | 238·8 | 197·3 | 253·3 | 258·4 | 246·9 |
| Hired labour (per cent) | 79 | 66 | 71 | 72 | 87 | 87 | 84 | 82 | 87 |
| Farmers using hired labour (per cent) | 97 | 89 | 96 | 100 | 100 | 100 | 100 | 99 | 100 |
| Average size of farm operation (hectares) | 0·83 | 0·76 | 0·59 | 0·62 | 0·77 | 0·80 | 0·67 | 0·94 | 1·11 |
| Number of sample farm observations | (121) | (35) | (25) | (161) | (77) | (7) | (61) | (75) | (99) |

planted local varieties in the three seasons. By 1969/70 IRRI varieties still were not adopted by more than 15 per cent of the farmers. Apparently, Bimas programs produced no significantly better performance in the adoption of the high yielding varieties.

The slowest overall spread of the new varieties was shown in the Central Java sample with fewer than 10 per cent of the sample farmers growing IR varieties in the two wet seasons and only 18 per cent of sample farmers in 1969 dry season (Table 5.6). Of these farmers, the ones in the Bimas program, only 7 per cent in wet season 1968/9, 15 per cent in the dry season 1969, and 4 per cent in the wet season 1969/70 planted IR varieties. Bimas programs apparently had little influence on their adoption of new varieties over this period. Data for national improved rice varieties showed levels of preference inside and outside the programs which were quite similar, again suggesting a lack of influence.

## LABOUR USE IN PADDY FARMING

Data from one wet season (1969/70) on pre-harvest labour use in paddy farming by the sample farmers (Table 5.13) show the high intensity of labour use prior to the adoption of newer IRRI rice varieties. In planting local varieties the West Java sample recorded an average of 217 man days, while in the Central and East Java samples the averages were 234 and 253 man days per hectare (6 hours/day basis) respectively. The high levels of hired labour use (even with average farm size operations ranging from 0·59 to 1·11 hectares) was another characteristic. The practice of relying mostly on hired labour and utilising a relatively small proportion of family labour was followed by the majority of farm operators. The main explanation here is that Java's small farmers in most cases earn back from their fellow farmers in the village the wages paid to hired labour, when they in turn work as hired labour for fellow farmers.

Vink's wet paddy farm data on pre-war labour use in East Java (the same rice areas as in our sample from eight surveys conducted in the 1920s and 1930s) are relevant here. In summary they showed an average pre-harvest labour input per hectare of 151 man days (on a six hours/day basis) in a sample of some 89 farmers on wet paddy farms of over 0·58 hectares, with a range of 105-240 man days. Of that labour, an average of 41 per cent was family labour, and 59 per cent was hired: the range for hired labour was 41-72 per cent (Vink 1941: ch. 4 and Annexes I-IV).

A comparison with Vink's data suggests that (probably not only in East Java) in the course of four to five decades, wet paddy farming in Java has become even more labour intensive, with a higher component of hired wage labour. How much of this latter practice is in fact a means whereby small farmers work in turn for each other, for money wages and how much it reflects the realities of employment opportunities for landless villagers who work for paddy owner-operators, remains to be analysed.[11]

In the Agro-Economic Survey sample, no data on harvest labour inputs have been recorded because of the difficulty in estimating this input. In most cases paddy operators will not decide how many to accept for harvest work. People are already standing in his field waiting for his signal to start. Wages are given in kind (one fifth to one twentieth, varying in different regions) and the paddy operator knows only his net share of the harvest, not the gross yield. On the input of harvest labour, Vink arrived at an average additional 98 man days/hectare or 65 per cent over pre-harvest labour use which was for harvesting paddy with a small knife, cut ear by ear. Four to five decades later, as shown in the survey, this harvesting technology was still used by most farmers in Java, so that Vink's figures may still be relevant.

[11] AES data collected in wet season 1971/2 on this topic are also still to be analysed.

TABLE 5.14

INPUTS OF HIRED AND FAMILY LABOUR, USED BY THE SAMPLE PADDY FARMERS, JAVA, WET SEASON, 1969/70

| *Labour inputs* | *Rice varieties* Local | *NI* | *IR* |
|---|---|---|---|
| *Average family labour* | | | |
| (man days/hectare) | 36·8 | 53·65 | 43·00 |
| Standard deviation | 33·42 | 54·65 | 36·62 |
| Number of sample farmers[a] | 106 | 54 | 57 |
| *Average hired labour* | | | |
| (man days/hectare) | 195·7 | 228·2 | 250·3 |
| Standard deviation | 84·92 | 80·69 | 121·31 |
| Number of sample farmers | 110 | 54 | 58 |

[a] Some sample farmers did not use family labour, which explains the difference in the number of sample farmers for family and hired labour by varieties.

Has the adoption of the new IRRI varieties led to higher levels of labour use compared to labour inputs with local varieties? Family and hired labour use were compared by varieties for sample farmers in three districts, representing rice areas in three river basins (two in East Java: Ngawi and Sidoardjo, one in Central Java: Banjumas) in Table 5.14. Family labour use for local and IRRI rice varieties showed no significant difference at the 5 per cent level. However for hired labour there was a significant difference at the 1 per cent level. For the IRRI rice varieties, sample farmers in the three districts in that wet season used 28 per cent more hired labour (250·3 man days/hectare) which with family labour inputs gave total pre-harvest labour inputs of 293·3 man days/hectare for IRRI varieties.[12] For local varieties the figures were 195·7 man days/hectare (hired) and 232·5 man days/hectare (pre-harvest total).

## USE OF NEW FARM INPUTS

Use of the new farm inputs, especially fertilisers and pesticides, was examined with data from two wet seasons, 1969/70 and 1970/1 (Table 5.15). Nitrogen fertiliser use (urea) reached about the same level (110-20 kilograms/hectare) with local varieties in the three provinces,

[12] The t-test for two independent samples was used to determine whether family labour and hired labour means for the two varieties differed significantly in the sample. For family labour means, t = 1·10 with 161 degrees of freedom and was not significant at 5 per cent level. For hired labour means, t = 3·42 with 166 degrees of freedom which was significantly different at 1 per cent level.

Table 5.15

FARMERS' AVERAGE USE OF FERTILISERS AND PESTICIDES, JAVA, WET SEASON, 1969/70 COMPARED WITH WET SEASON, 1970/1

| | *Rice variety* | | | | | |
|---|---|---|---|---|---|---|
| | *Local* | | *NI* | | *IR* | |
| | *1969/70* | *1970/1* | *1969/70* | *1970/1* | *1969/70* | *1970/1* |
| *West Java sample:* | | | | | | |
| Farmers using N fertiliser (per cent)[a] | 78 | 68 | 82 | 42 | 74 | 80 |
| N fertiliser use (kilograms/hectare) | 119 | 109 | 100 | 71 | 122 | 158 |
| Farmers using both N and P fertiliser (per cent)[b] | n.a. | 68 | n.a. | 42 | n.a. | 80 |
| N and P fertiliser use (kilograms/hectare)[c] | n.a. | 159 | n.a. | 93 | n.a. | 252 |
| Farmers using pesticides (per cent) | 20 | 43 | 8 | 36 | 15 | 75 |
| Number of sample farmers | 125 | 121 | 38 | 36 | 27 | 40 |
| *Central Java sample:* | | | | | | |
| Farmers using N fertiliser (per cent) | 88 | 86 | 83 | 80 | 100 | 92 |
| N fertiliser use (kilograms/hectare) | 88 | 111 | 96 | 117 | 173 | 148 |
| Farmers using both N and P fertiliser (per cent) | n.a. | 86 | n.a. | 80 | n.a. | 92 |
| N and P fertiliser use (kilograms/hectare) | n.a. | 113 | n.a. | 118 | n.a. | 150 |
| Farmers using pesticides (per cent) | 5 | 29 | 6 | 49 | 0 | 73 |
| Number of sample farmers | 161 | 160 | 78 | 69 | 7 | 26 |
| *East Java sample:* | | | | | | |
| Farmers using N fertiliser (per cent) | 87 | 87 | 76 | 64 | 98 | 98 |
| N fertiliser use (kilograms/hectare) | 142 | 118 | 123 | 159 | 182 | 204 |
| Farmers using both N and P fertiliser (per cent) | 45 | 87 | 22 | 64 | 36 | 98 |
| N and P fertiliser use (kilograms/hectare) | 48 | 160 | 84 | 203 | 97 | 247 |
| Farmers using pesticides (per cent) | 45 | 38 | 24 | 24 | 41 | 54 |
| Number of sample farmers | 60 | 70 | 76 | 34 | 101 | 104 |

[a] N fertiliser was urea.
[b] P fertiliser was Tri-sulphate.
[c] Fertiliser quantity (kilograms/hectare) are actual, not nutrient, weights.

but showed a wide range with national improved varieties from around 70 kilograms/hectare (West), to 120 kilograms/hectare (Central) and 160 kilograms/hectare (East Java). The highest levels of nitrogen fertiliser use were with IRRI varieties, varying from 150 kilograms/hectare in West and Central Java to 200 kilograms/hectare in East Java.

Total N and P fertiliser use with local varieties in the 1970/1 wet season reached levels of 160 kilograms/hectare, in the West and East Java samples but only 110 kilograms/hectare in the Central Java sample where P fertiliser use is very low. With national improved varieties total N and P fertiliser use again shows a wide variation, from 90 kilograms/hectare in West Java to 120 kilograms/hectare in Central Java (again where P fertiliser use is not noticeable) and to the highest level of 200 kilograms/hectare in East Java. Use with IRRI varieties showed the highest levels for total N and P fertiliser: 150 kilograms/hectare in Central Java (very low phosphate use) to 250 kilograms level in West and East Java samples, where phosphate use has reached levels as high as 100 kilograms/hectare (Table 5.15). Overall, East Java sample farmers showed the widest spread for IRRI varieties and the highest levels of fertiliser use. Of the three categories of rice varieties, IRRI varieties received the highest doses in all three samples.

An important aspect of rice farming on Java is the large number of the farmers who use fertiliser on local rice varieties, a development of only the last twelve years. In the three provinces the percentages using fertiliser were between 68 and 87 for local varieties and only slightly higher for the IRRI varieties. If this is combined with the high level of hired labour use for local varieties, it is clear that farmers on Java have been using intensive methods of production even without the high yielding varieties.

The use of pesticides by farmers has spread rather quickly in the second wet season, especially with IRRI varieties (75 per cent) in the West and Central Java samples. East Java sample farmers already showed a higher rate of adoption of pesticides in the first wet season for both the new and local rice varieties (41.5 per cent). Proportionally fewer East Java sample farmers (54 per cent) seem to use pesticides on IRRI varieties by comparison with farmers in the other areas. It is not clear how much farmer 'gambling' there was with possibly more favourable local climatic conditions, in needed preventive measures.

## RICE YIELDS BY VARIETIES

What has been the yield effect of the large number of the sample farmers applying relatively high levels of fertilisers and insecticides on

TABLE 5.16

PERCENTAGE INCREASE IN YIELDS FOR FIVE SEASONS, JAVA, 1968–71

| | Percentage increases[a] | | |
|---|---|---|---|
| | *West Java sample* | *Central Java sample* | *East Java sample* |
| *Wet season 1968/9* | | | |
| IR over NI | 184 | 46 | 54 |
| IR over local | 202 | 39 | 40 |
| NI over local | 10 | −4 | −9 |
| *Dry season 1969* | | | |
| IR over NI | 51 | 70 | 16 |
| IR over local | 68 | 72 | 36 |
| NI over local | 11 | 1 | 17 |
| *Wet season 1969/70* | | | |
| IR over NI | 36 | 20 | 38 |
| IR over local | 53 | 30 | 18 |
| NI over local | 17 | 8 | −14 |
| *Dry season 1970* | | | |
| IR over NI | 0·1 | 51 | 11 |
| IR over local | 12 | 38 | 26 |
| NI over local | 11 | −8 | 13 |
| *Wet season 1970/1* | | | |
| IR over NI | 192 | 5 | 17 |
| IR over local | 111 | 35 | 38 |
| NI over local | −28 | 28 | 18 |

[a] These percentages are based on the yields in Table 5.17.

local varieties as well as high yielding varieties of rice? Although comparisons are difficult to make by provinces,[13] the percentage yield increases of IRRI varieties over the local varieties varied greatly as shown in Table 5.16 for five seasons. The most reliable comparisons can be made for East Java because of the more balanced number of observations for each variety. Table 5.17 shows the average yields in paddy and the number of observations for each estimate. The percentages increases of IR over local varieties in the five seasons were 40, 36, 18, 26, and 38 respectively. These increases are not spectacular owing to the levels of input use for local varieties.

[13] The best approach would be to compare yields for individual farmers, but this is impossible because of the low percentage growing both varieties. Even by villages it is difficult because in any one village a large majority grows one or the other and not both local and IR. If comparisons are made by province, differences in soil, climate, etc., are hidden.

To appreciate the level of these yields by varieties, all the farmer estimates in each province for four seasons and our crop cutting yield estimates for two seasons were combined. Of course this eliminates differences by season but does help to make the comparisons clearer. In East Java the rice yields in paddy (metric tons/hectare) were 3·43 tons for local varieties, 3·37 tons for national improved varieties, and 4·50 tons for IRRI varieties. Confirming these differences by varieties, the crop cutting estimates were higher but the increases were similar (Table 5.18).

The survey also involved interviews with rice farmers for the five seasons on Sumatra, Bali and Sulawesi. If all these data on yields are combined, the averages are 3·49 tons/hectare, 3·35 tons/hectare, and 4·75 tons/hectare for the varieties respectively, based on farmer interviews. The yields are 4·51, 4·31 and 5·91 tons/hectare for the three types of variety if based on the crop cutting estimates (Table 5.18). It can be seen that the percentage change in yield of IR over local varieties, 36·3 based on the interviews and 31 based on the crop cuttings, were close.

## BENEFITS OF PADDY FARM OPERATION

For one wet season (1970/1), data from 181 paddy farm operators in nine villages (three villages in each of the West, Central and East Java samples) were analysed to obtain a measure of the comparative net benefits obtained from growing the three types of rice varieties. These calculations showed that the West Java IRRI rice adopters, on a per hectare basis, had the highest margin of net benefits over those planting local or national improved varieties ($US119.42 or 52 per cent higher). East Java paddy operators had the next highest and still substantial margin. There, IRRI rice adopters earned a net benefit of $US77.29 per hectare or 47 per cent higher than those planting local or national improved varieties. In the West and East Java samples, cash expenditures per hectare for IRRI and the other rice varieties were no different. The least convincing case for IRRI rice adoption was shown in the Central Java sample. While IRRI rice adopters received a high net benefit of $US96.30, they were only 11 per cent better off than those planting local or national improved varieties. Central Java farmers incurred extra cash expenditures for the IRRI varieties which were higher than in the other two areas. See Table 5.19.

Such comparisons of performance by IRRI rice adopters and farmers planting other varieties can be made most safely among farmers in the same village where the level of wage labour and prices of farm inputs are the same and where rice price differentials (received

TABLE 5.17

RICE YIELDS BY VARIETIES FOR FIVE SEASONS, JAVA, 1968–71

| | *Rice yields in paddy (m. tons/hectare)*[a] | | |
|---|---|---|---|
| *Sample by Location and Season* | *Local* | *NI* | *IR* |
| *Wet season 1968/9* | | | |
| *West Java:* | | | |
| Yield | 3·01 | 3·30 | 9·09 |
| Number of observations | 105 | 13 | 7 |
| *Central Java:* | | | |
| Yield | 3·25 | 3·11 | 4·54 |
| Number of observations | 175 | 77 | 13 |
| *East Java:* | | | |
| Yield | 3·14 | 2·85 | 4·41 |
| Number of observations | 73 | 70 | 70 |
| *Dry season 1969* | | | |
| *West Java:* | | | |
| Yield | 2·69 | 2·98 | 4·52 |
| Number of observations | 87 | 15 | 16 |
| *Central Java:* | | | |
| Yield | 2·76 | 2·79 | 4·75 |
| Number of observations | 163 | 79 | 43 |
| *East Java:* | | | |
| Yield | 2·99 | 3·51 | 4·07 |
| Number of observations | 54 | 56 | 76 |
| *Wet season 1969/70* | | | |
| *West Java:* | | | |
| Yield | 3·03 | 3·54 | 4·83 |
| Number of observations | 99 | 18 | 9 |
| *Central Java:* | | | |
| Yield | 3·66 | 3·96 | 4·78 |
| Number of observations | 161 | 79 | 7 |
| *East Java:* | | | |
| Yield | 4·25 | 3·64 | 5·05 |
| Number of observations | 61 | 80 | 100 |
| *Dry season 1970* | | | |
| *West Java:* | | | |
| Yield | 2·82 | 3·13 | 3·17 |
| Number of observations | 90 | 9 | 6 |
| *Central Java:* | | | |
| Yield | 4·14 | 3·80 | 5·72 |
| Number of observations | 175 | 53 | 5 |
| *East Java:* | | | |
| Yield | 3·35 | 3·78 | 4·21 |
| Number of observations | 58 | 15 | 59 |

*CARRIED OVER*

| *Sample by Location and Season* | *Rice yields in paddy (m. tons/hectare)* [a] *Local* | | *NI* | | *IR* | |
|---|---|---|---|---|---|---|
| *Wet season 1970/1* | | | | | | |
| *West Java:* | | | | | | |
| Yield | 3·27 | | 2·36 | | 6·88 | |
| Number of observations | | 122 | | 38 | | 42 |
| *Central Java:* | | | | | | |
| Yield | 3·02 | | 3·87 | | 4·06 | |
| Number of observations | | 156 | | 70 | | 27 |
| *East Java:* | | | | | | |
| Yield | 3·66 | | 4·31 | | 5·04 | |
| Number of observations | | 69 | | 34 | | 105 |

[a] The rice yields are in paddy which means the stalk has not yet been removed. One ton of paddy is equivalent to 0·52 tons of milled rice. These yields are based on the farmer's estimate of his yield.

by farmers) for the two rice varieties are of one scale.[14] Comparisons on a provincial scale can be made less confidently since these conditions do not hold.

## SUMMARY AND CONCLUSIONS

On an all-Java basis sample farmers in our AES study were the more established farmers in their villages; they were in the older age group (forty years and more) and were in the top 30 to 40 per cent of the farm size classes (more than 0·5 hectares). They operated paddy farms in the best irrigated areas in Java, with more villages chosen from the northern coastal plains of West and Central Java than from the southern coastal plains (Central and East Java), with three river basin areas represented (Central and East Java) and one mountain area (West Java). These localities also had a better infrastructure which facilitated wider marketing relations.

For the smaller paddy farm operators in the sample (0·47 to 0·66 hectares) paddy occupied some 66 per cent, and for larger operators some 78 per cent, of total farm land. In terms of tenure a high proportion of owner-operators was found in the sample (70 to 82 per cent). Of the rest, about half were only share croppers and renters who owned no land. Share croppers were concentrated in the West Java sample and renters in the Central and East Java samples.

[14] In the three villages of the West Java sample a t-test (not shown here), village by village, was significant. The IRRI rice growers obtained higher net benefits in each locality. In the three villages cash expenditures were $US33.54, $53.80 and $87.90, mainly reflecting the respective levels of wage labour costs.

TABLE 5.18

SAMPLE FARMERS' AVERAGE YIELDS FOR FOUR SEASONS, JAVA, 1969–71 (BASED ON FARMERS INTERVIEWS AND FOR TWO SEASONS BASED ON CROP CUTTINGS)

| Location | *Estimated rice yields*[a] *by variety* | | | | | |
|---|---|---|---|---|---|---|
| | *Local* | | *NI* | | *IR* | |
| | *Farmers* | *Crop cutting* | *Farmers* | *Crop cutting* | *Farmers* | *Crop cutting* |
| *West Java*[b] | | | | | | |
| Yield[c] | 2·90 | 3·12 | 3·26 | 3·40 | 5·22 | 6·21 |
| Number of observations | 381 | 102 | 55 | 28 | 38 | 1 |
| *Central Java* | | | | | | |
| Yield | 3·46 | 4·14 | 3·38 | 4·35 | 4·78 | 5·04 |
| Number of observations | 674 | 210 | 288 | 111 | 68 | 6 |
| *East Java* | | | | | | |
| Yield | 3·43 | 4·39 | 3·37 | 4·31 | 4·50 | 5·24 |
| Number of observations | 246 | 99 | 221 | 113 | 305 | 160 |
| *Java*[d] | | | | | | |
| Yield | 3·29 | 3·95 | 3·36 | 4·22 | 4·61 | 5·24 |
| Number of observations | 1301 | 411 | 564 | 252 | 411 | 167 |
| *All Provinces*[e] | | | | | | |
| Yield | 3·49 | 4·51 | 3·35 | 4·31 | 4·76 | 5·91 |
| Number of observations | 2185 | 628 | 773 | 314 | 1087 | 518 |

[a] These averages are composed of yield data gathered in four seasons for the farmer estimates and in two seasons for the crop cutting estimates.

[b] The two *desas* in Serang Kabupaten were not included because of their frequent disease, insect, and irrigation problems.

[c] All yield estimates are in metric tons paddy per hectare (1 ton paddy = 0·52 tons milled rice).

[d] The Java averages are based on sample farmer data from the three provinces on Java.

[e] The all province averages are based on the sample farmer data from North Sumatra, West Sumatra, Lampung, Java, Bali, and South Sulawesi (Celebes).

In this study East Java sample farmers showed the widest acceptance (50 per cent) of the new IRRI rice varieties during the 1968-71 period, in both the wet and dry seasons, followed by West Java sample farmers (20 per cent). Both samples followed a cropping pattern where, in the dry season, a quarter or more of farmers grew crops other than rice in their *sawah* (wet fields). In both samples too the larger paddy farm operators had progressed most in terms of adopting the new varieties. In general the adoption of higher yielding IRRI varieties was by farmers who also still planted local and national improved varieties. This was least apparent in the East Java sample (39 per cent) but was still at high levels (66 per cent) amongst West

TABLE 5.19

AVERAGE NET BENEFITS OF PADDY FARM OPERATION, BY RICE VARIETIES, JAVA, WET SEASON, 1970/1

| | Paddy operators in | | | | | |
|---|---|---|---|---|---|---|
| | *West Java* | | *Central Java* | | *East Java* | |
| | *Only IR*[c] | *Only Local/NI*[d] | *Only IR* | *Only Local/NI* | *Only IR* | *Only Local/NI* |
| Average size of paddy farm | | | | | | |
| (hectares) | 0·66 | 0·54 | 0·56 | 0·86 | 1·22 | 0·56 |
| (number of observations) | (13) | (43) | (7) | (56) | (9) | (53) |
| Cash expenditures[a] | 58·43 | 51·39 | 61·86 | 53·34 | 53·51 | 53·93 |
| Net benefits | | | | | | |
| ($US/hectare)[b] | 119·42 | 89·75 | 96·30 | 86·66 | 77·29 | 52·46 |
| Benefit/cash expenditure | 2·07 | 1·74 | 1·55 | 1·59 | 1·44 | 0·97 |

[a] Cash expenditures included wages of hired non-family labour and purchases of other inputs.

[b] Net benefits were the farmers' net share in money value of total input value (farm gate price) after deducting cash expenditures and the value of harvest costs in kind. An exchange rate of $US1.00 = Rupiah 415 was used (post-August 1971 rate). This return can be thought of as a net return to the farm family because family labour costs have not been deducted.

[c] Refers to farmers planting only IRRI varieties.

[d] Refers to farmers planting only local or national improved varieties. National improved has been grouped with local because of our belief that the seed quality at the farm level has deteriorated.

and Central Java sample farmers. The lowest rate of spread of IRRI rice varieties was in the Central Java sample (10 per cent).

The most direct explanation of such trends of adoption may possibly be found in the higher net benefits (per hectare) from growing the new IRRI rice varieties compared to other varieties. For local varieties the net benefit was lowest in the East Java sample ($US52/hectare); higher in the Central and West Java samples (respectively $US86 and $89/hectare). By comparison, IRRI rice growing has reached the appreciably higher levels of net benefits in the East Java ($US77/hectare) and West Java samples ($US119/hectare), with roughly the same level of cash expenditures (respectively $US58 and $51/hectare) for wage labour and the new farm inputs, fertiliser and pesticides. Total fertiliser use was highest for the East and West Java samples (some 250 kilograms/hectare for IRRI varieties) and was much lower in the Central Java sample (respectively 160 and 110 kilograms/hectare for IRRI and local varieties). Use of pesticides spread more quickly in the 1970/1 wet season than in previous seasons.

Bimas programs have not had much direct impact (i.e. in the actual supply of seeds) in the spread of the new IRRI varieties. This was apparent both in the low proportion of farmers using the new varieties who had joined the program (especially West and Central Java samples) and in the same quick spreading of the new varieties in the East Java sample both inside and outside the Bimas programs.

Rice farming on Java as shown by the sample farmers was highly commercialised, both in terms of selling rice which was the highest in the East Java sample (80 per cent) and 60 per cent in the other samples, and in buying back rice. The proportion of farmers buying to selling rice was a high 65 per cent in the dry season in West Java sample (lower in the wet season, 45 per cent) but was on the same level for the two seasons in the East Java sample (36-9 per cent). The lowest proportion was for Central Java farmers (25 per cent) who had a cropping pattern of double paddy with very few exceptions of non-paddy production in the dry season. They were either the most paddy subsistence oriented or, worse still, bought other food of lesser quality.

The trait of commercial farming may be linked with the fact that Java's farmers are with some regional variations mostly part-time. Larger farmers in the West Java sample (most adopting the new varieties) were more active in seeking outside farm income while in the East Java sample that group was less active. On the other hand smaller farmers in the Central Java sample were more active in outside farm activities than larger farmers in the same localities. This pattern is probably influenced by the individual farmland base, the local off-farm opportunities and levels of income achieved in each locality.[15]

The high labour intensity of rice culture in Java was also evident and was even higher than that shown by pre-war survey data. Another trait of commercial farming was the high rate of hired wage labour (already noted in surveys some forty years earlier in Java). This accounted for 72-84 per cent of pre-harvest labour and was a practice with 90 per cent or more of the sample farmers. Such farms can scarcely be called family farms.

In some sample areas, farmers growing the new IRRI varieties were found to have used more labour than those growing other varieties. While the level of family labour inputs was the same, 28 per cent, more labour was hired, which provides some evidence that the new varieties have created more work opportunities for some people. The characteristic of family farms using so much hired labour may again be linked with off-farm cash income earning opportunities,

[15] Wet season 1970/1 data were collected on this topic and are currently under analysis.

for the latter is one important source of money income which can be used, along with farm cash income, to pay wage labour. Wage labour payment formed a high proportion of total cash expenditures (64-85 per cent) with a smaller part allocated for purchase of farm inputs such as fertiliser and pesticides. The new rice varieties did have higher yields per hectare. Comparing the sample farmers for all of Indonesia, the increase in yield of IRRI varieties over local varieties for four seasons was 36 per cent.

So far as land tenure and adoption were concerned the results were mixed. Amongst the minority of the sample farmers who were lessees, those who share cropped (West Java) showed a similar pattern of adoption to that of owner operators while those who rented land in order to grow the new varieties showed two differing trends. The East Java sample suggested that renting land for the high yielding varieties was becoming attractive, while the Central Java sample showed the reverse.

In summary, although sample farmers have responded as well as local situations and incentives permitted in adopting the new higher yielding varieties, there is no picture yet of any green revolution for Java's paddy farm operators.

6

# The Impact of New Technology on Rice Production: A Study of Change in Three Philippine Municipalities, 1966-9

*Randolph Barker and Violeta Cordova*

JUST before the high yielding semi-dwarf rice varieties were introduced, Liao (1968) made a study of rice farming in three municipalities of Laguna, Philippines. The study was based upon a survey of farm production practices, costs, and returns on 155 farms for the crop year 1966/7. This survey provided an excellent benchmark. Subsequent surveys have been made annually after the wet season harvest to examine the pattern of adoption of the high yielding varieties and the changes that have occurred in farming practices, rice production, costs, and returns.

This article is based on surveys made from 1966 to 1969. It first describes the survey area, with emphasis on the variation in farm resources and farm practices that existed in 1966. It then traces the pattern of varietal diffusion through the four-year period and shows the resulting changes in yields and farm income for the wet season crop. It also examines in detail the changes that have occurred in individual inputs—fertiliser, chemicals, labour, and credit—and it analyses how the various claimants have shared the returns.

In his survey, Liao chose seven *barrios* (villages) from among those that Von Oppenfeld et al. (1957) had surveyed in 1954/5 (Figure 6.1). Liao drew a random sample of farms from each of three municipalities —Biñan, Cabuyao, and Calamba. He selected these three principally on the basis of their differences in water resources. Calamba has a gravity irrigation system and water is available throughout the year. Most Cabuyao farms are irrigated by low lift pumps. In Biñan, water from a gravity system is available only for three crops in two years.

Eighty eight of the 155 farms Liao studied were in the 1954/5 sample of Von Oppenfeld et al. Information obtained from Biñan shows clearly that the irrigation system has deteriorated since the mid-1950s and that the deterioration has influenced crop production

Table 6.3

RETURNS AND COSTS PER HECTARE FOR RICE FARMS IN THREE MUNICIPALITIES OF LAGUNA, WET SEASONS, 1966 AND 1969

| *Year* | *Farm size (hectares)* | *Land in HYV (percent)* | *Grain yield (tons/hectare)* | *Standard deviation of yield (tons/hectare)* | *Paddy price (pesos/kilogram)* | *Gross returns (pesos/hectare)* | *Variable costs*[a] *(pesos/hectare)* | *Net return*[b] *(pesos/hectare)* | *Net return based on 1966 prices*[c] *(pesos/hectare)* |
|---|---|---|---|---|---|---|---|---|---|
| | | | | Biñan (45 farms) | | | | | |
| 1966 | 3·3 | 0 | 1·93 | 0·78 | 0·40 | 772 | 247 | 525 | 525 |
| 1969 | 3·4 | 80 | 2·54 | 1·34 | 0·41 | 1042 | 477 | 565 | 509 |
| Increase (per cent) | | | 32 | | | 35 | 93 | 8 | 0 |
| | | | | Cabuyao (56 farms) | | | | | |
| 1966 | 2·5 | 2 | 2·23 | 0·72 | 0·40 | 892 | 267 | 625 | 625 |
| 1969 | 2·4 | 68 | 4·25 | 1·41 | 0·45 | 1914 | 557 | 1357 | 1223 |
| Increase (per cent) | | | 91 | | | 115 | 109 | 117 | 96 |
| | | | | Calamba (51 farms) | | | | | |
| 1966 | 1·8 | 1 | 2·98 | 0·67 | 0·41 | 1221 | 266 | 955 | 955 |
| 1969 | 1·8 | 61 | 4·02 | 1·92 | 0·45 | 1809 | 526 | 1283 | 1156 |
| Increase (per cent) | | | 35 | | | 48 | 98 | 34 | 21 |

[a] Includes costs for seed, fertiliser, chemicals, and hired labour.

[b] Gross return less variable costs.

[c] Net return deflated by the consumer price index of all items including rice (January 1967 = 100; January 1970 = 111).

Differences in variable costs of production between municipalities were not large although in all municipalities variable costs rose by about 90 per cent during the four-year period (Table 6.3). The increase in costs occurred mainly because farmers applied more fertiliser and chemicals, contracted for tractor preparation of land more often, hired more labour to harvest the increased production, and had to pay higher farm wages. The last seems to be more a result of inflation than of an increased demand for agricultural labour.

Gains in yield per hectare and gross returns have differed significantly between municipalities. Biñan has had the smallest increase in both absolute and relative terms, while Cabuyao has had the largest. Gross returns in Cabuyao and Calamba were higher in 1969 not only because of a higher yield but also because of higher prices for glutinous rice.

Returns net of variable costs have followed the same pattern as gross returns. In 1966 net returns per hectare in Cabuyao were only slightly higher than those in Biñan. However, in the 1969 wet season, returns on Cabuyao farms matched those of Calamba farms. If net returns are deflated to take account of the rise in prices of items other than rice, the real income of the farmers surveyed in Biñan did not rise between 1966 and 1969 in spite of the widespread adoption of high yielding varieties (Table 6.3). But Biñan farmers were not necessarily wrong in adopting the high yielding varieties. The modest gains in yield were offset by rising costs. Had they continued to plant traditional varieties, their real income might have declined.

Most farmers in the survey areas increased their returns from rice production between 1966 and 1969 (Table 6.3). But the varieties they chose to plant and their success in increasing their returns appear to be related to the resources of the particular location. Because the price of glutinous rice fluctuates widely (partly as a result of the introduction of high yielding varieties), the best variety to choose for high returns has not always been obvious.

## USE OF FERTILISERS AND CHEMICALS

Higher yields resulting from the adoption of the high yielding varieties are normally associated with the increased application of fertilisers, particularly nitrogen. The recommended package of inputs for the new varieties usually includes herbicides and insecticides.

*Fertilisers*: over 70 per cent of the farmers in the study area used fertiliser in 1966. Farmers typically applied one to three bags (45 kilograms/bag) of ammonium sulphate per hectare. In the past two to three years, however, urea has become readily available and is being

TABLE 6.4

FERTILISER USERS, NITROGEN DOSAGE AND RICE YIELDS BY LOCATION AND BY ADOPTION OF HIGH YIELDING VARIETIES, LAGUNA, WET SEASONS 1966 AND 1969

| *Category* | *Fertiliser users (per cent)* | | *Nitrogen applied (kilograms/hectare)* | | *Grain yield (tons/hectare)* | |
|---|---|---|---|---|---|---|
| | *1966* | *1969* | *1966* | *1969* | *1966* | *1969* |
| *Location* | | | | | | |
| Biñan | 80 | 98 | 8 | 47 | 1·9 | 2·5 |
| Cabuyao | 75 | 91 | 16 | 51 | 2·2 | 4·3 |
| Calamba | 84 | 94 | 22 | 44 | 2·9 | 4·0 |
| *Adoption of HYV 1969* | | | | | | |
| Full adopters | 73 | 94 | 14 | 49 | 2·3 | 3·8 |
| Partial adopters | 84 | 96 | 17 | 46 | 2·5 | 3·5 |
| Non-adopters | 95 | 90 | 19 | 45 | 2·5 | 2·8 |
| *Full adopters by location 1969* | | | | | | |
| Biñan | 72 | 100 | 8 | 52 | 2·0 | 2·9 |
| Cabuyao | 65 | 86 | 15 | 44 | 2·1 | 4·6 |
| Calamba | 83 | 95 | 21 | 51 | 2·9 | 4·3 |

supplied to farmers by both domestic producers and importers.[1] More than 90 per cent of the fertiliser users in the sample area applied urea in the 1969 wet season. The typical rate of application was two to four bags per hectare. Because of the switch to urea, the price paid for nitrogen fell from about ₱1.70 to ₱1.20/kilogram. Farmers normally apply fertiliser two or three times during a season, but the time of application varies widely.

The amount of fertiliser applied per hectare has doubled or tripled on most farms (Table 6.4). In 1966, the Calamba farmers who were growing glutinous rice were applying the most fertiliser, and Biñan farmers the least. In 1969, however, the amounts of fertiliser used, either by location or by level of adoption were not significantly different. Farmers applied an average of 45 to 50 kilograms/hectare of nitrogen. They spent about ₱50 to ₱55/hectare for fertiliser and only rarely more than ₱100/hectare. Table 6.4 also shows that on farms adopting the high yielding varieties (principally IR8 and C4-63) the lower yields experienced in Biñan clearly did not result from lower nitrogen application.

[1] Complete-mix fertiliser is not widely used because Laguna lowland rice soils usually do not show a yield response to the application of phosphorus and potassium.

TABLE 6.5

PERCENTAGES OF USERS AND AVERAGE EXPENSES PER HECTARE FOR HERBICIDES AND INSECTICIDES, BY LOCATION AND ADOPTION OF HIGH YIELDING VARIETIES, LAGUNA, WET SEASONS 1966 AND 1969

| Category | *Herbicides* | | | | *Insecticides* | | | |
|---|---|---|---|---|---|---|---|---|
| | *Users (per cent)* | | *Expenses (Peso/hectare)* | | *Users (per cent)* | | *Expenses (Peso/hectare)* | |
| | *1966* | *1969* | *1966* | *1969* | *1966* | *1969* | *1966* | *1969* |
| *Location* | | | | | | | | |
| Biñan | 71 | 100 | 2 | 6 | 38 | 96 | 2 | 9 |
| Cabuyao | 75 | 96 | 6 | 9 | 52 | 84 | 3 | 10 |
| Calamba | 86 | 84 | 7 | 5 | 71 | 76 | 4 | 6 |
| *Adoption of HYV 1969* | | | | | | | | |
| Full adoptors | 74 | 93 | 4 | 7 | 53 | 90 | 3 | 9 |
| Partial adoptors | 89 | 98 | 6 | 6 | 51 | 84 | 2 | 7 |
| Non-adoptors | 80 | 85 | 5 | 9 | 70 | 65 | 3 | 8 |
| *Full adoptors by location 1969* | | | | | | | | |
| Biñan | 70 | 100 | 2 | 6 | 41 | 97 | 2 | 10 |
| Cabuyao | 70 | 97 | 6 | 9 | 52 | 93 | 3 | 9 |
| Calamba | 83 | 78 | 5 | 4 | 69 | 78 | 4 | 8 |

*Herbicides*: in 1966, nearly as many farmers used herbicides as fertiliser. Liquid 2,4-D has been the most popular herbicide throughout the area. Although the number of herbicide users has risen (Table 6.5), the level of use has not increased as dramatically as that of nitrogen. Farmers typically applied under ₱10 of herbicide per hectare. Occasionally they spent as much as ₱20/hectare. Nearly all farmers apply herbicides in liquid form after weeds emerge. In 1969, they spent less than ₱10/hectare.

The widespread adoption of herbicides undoubtedly reflects in part the relatively high cost of labour in this area. Even in 1966 only 27 per cent of the farms used hand weeding.

*Insecticides*: the major insect pest in the three municipalities is the yellow stem borer. In 1966, about half of the farmers used insecticides. The number of insecticide users has increased sharply since then. As a result the average insecticide expenditure per hectare in the municipalities has more than doubled (Table 6.5). But the increased expenditure per hectare by those who were already applying insecticide in 1966 has been far less dramatic. Farmers apply liquid insecticides such as folidol and thiodan to protect seed beds and to kill severe infestations of stem borers. Few farmers use granular insecticides. Typically, farmers spend less than ₱10/hectare for insecticides although a few spend as much as ₱50-60.

As with fertilisers and herbicides, in 1966 Biñan farmers applied the smallest quantities of insecticides; they have shown the largest increase since then. Conversely, Calamba farmers had the highest level of use in 1966, but showed the smallest increase.

## LABOUR UTILISATION

The only significant changes in labour utilisation were a decline in labour used for land preparation and a rise in labour used for harvesting (Table 6.6). Farmers employ less labour for land preparation because more and more hand tractors are being used for harrowing. The increase in harvesting and threshing labour is related to the increase in yield per hectare. While yields increased by over 70 per cent, labour for harvesting and threshing increased by about 50 per cent.

From findings in other surveys, one might have expected an increase in the use of labour for weeding. In these Laguna municipalities, however, more weed control was used, but labour input did not increase because labour-saving chemicals and mechanical weed control were substituted for some hand weeding.

Although the total labour input for rice production increased only slightly since 1966, hired labour increased significantly. Farmers are

TABLE 6.6

LABOUR USED AND PERCENTAGE HIRED FOR SPECIFIC PRACTICES ON FORTY TWO FARMS PLANTED IN HIGH YIELDING VARIETIES 1969, LAGUNA, WET SEASONS 1966 AND 1969

| *Farm Operation* | *1966* Man days (per hectare) | *1966* Percentage hired (per hectare) | *1969* Man days (per hectare) | *1969* Percentage hired (per hectare) |
|---|---|---|---|---|
| Land preparation[a] | 22 | 5 | 16 | 38 |
| Transplanting | 10 | 35[b] | 10 | 100 |
| Weeding | 14 | | 14 | 79 |
| Other pre-harvest | 10 | | 11 | 54 |
| Harvesting/threshing | 23 | 100 | 34 | 100 |
| Total | 79 | 46 | 85 | 79 |
| Yield in rough rice tons/hectare | 2·5 | | 4·3 | |

[a] Tractor users: 1966, 31 per cent; 1969, 62 per cent.

[b] Transplanting, weeding and other pre-harvest operations are combined in the 1966 records.

TABLE 6.7

MONEY ADVANCED BY LANDLORDS FOR SPECIFIC ITEMS, LAGUNA, 1965 AND 1969

| *Item* | *1965* | | *1969* | |
|---|---|---|---|---|
| | *Number reporting* | *Pesos/hectare* | *Number reporting* | *Pesos/hectare* |
| Fertiliser | 18 | 28 | 24 | 65 |
| Chemicals | 10 | 5 | 23 | 16 |
| Land preparation | 0 | 0 | 14 | 108 |
| Transplanting | 13 | 34 | 19 | 52 |
| Weeding | 4 | 15 | 10 | 55 |
| Others | 1 | 39 | 9 | 74[a] |

[a] Mostly fuel for irrigation pumps.

Based on a survey of thirty tenants. Fifty per cent of the money advanced represents a loan to the tenant under normal sharing relationships except land preparation cost which is not shared.

hiring more labourers at harvest and they have substituted hired labour for some family labour in land preparation and weeding.

### EXPANSION OF CREDIT

Thirty tenants from the Liao's original sample and their landlords were interviewed in 1969 to learn more about decision making and the use of credit. The tenants and the landowners were questioned separately. They agreed almost completely on who was responsible for decisions about the kind and amount of resource use. In twenty four out of the thirty cases, the tenant made the decision regarding variety, fertiliser, and chemicals. Normally, the landowner visits the farm only three or four times a crop season. Sometimes he goes only at harvest to participate in the sharing. Occasionally, the landowner actively participates in the decision making and visits the farm at least once a week. One half of the landowners surveyed had holdings of 10 hectares or less. The other half had holdings of 10-50 hectares. Those interviewed were asked to recall the amount of credit obtained for specific items in 1965, before the introduction of high yielding varieties, and in 1969. Most loans were made for fertiliser, chemicals, and labour for transplanting (Table 6.7). In 1969, both the number of loans and the average size of loan increased substantially. Although lending practices varied widely from farm to farm, the money extended by the landowner (tenant and landlord share) increased about two to three times from around ₱50/hectare to ₱175/hectare on the average.

All but three of the tenants interviewed borrowed money from the landowner. Only eight tenants borrowed from a source other than the landowner. Usually the landowner supplies inputs such as fertiliser at the time they are needed, and makes a cash payment to the hired labour. The tenant makes the repayment at harvest time. The tenant sets aside enough rice to cover the cost of the inputs supplied before the one-to-one sharing takes place. Under this system, it is difficult or impossible to determine the rate of interest charged. Many landowners indicate that they charge no interest. Some based the amount of repayment upon the price of rice at the time the input was purchased.

Most often the landowner himself did not find it necessary to borrow the money to lend to the tenants. Eight landowners borrowed money from the banks for loans on cash operating expenses; another eight borrowed to invest in pumps or other capital items. Many of the landowners are engaged in some form of business or profession outside of agriculture. Even funds borrowed from banks were frequently used for business activities not connected with the farm.

Although the landowner was not actively engaged in the decision making on most of the farms surveyed, he could exercise some control over decisions since he normally provided the credit for the purchase of inputs. Nine of the thirty tenants interviewed said that there was an upper limit on the amount of credit they could borrow. Frequently, this limit was specified in terms of a specific item. For example, for fertiliser purchases, seven farmers were limited to an amount equal to about ₱60/hectares.

TABLE 6.8

OUTPUT DISTRIBUTION, BY FACTORS, ON FORTY TWO FARMS SHIFTING FROM LOCAL VARIETIES IN 1966 TO HIGH YIELDING VARIETIES IN 1969, LAGUNA, WET SEASONS

| Output to: | *Local 1966* | | *HYV 1969* | |
|---|---|---|---|---|
| | *Pesos* | *per cent* | *Pesos* | *per cent* |
| Land[a] | 374 | 38 | 609 | 36 |
| Labour[b] | 259 | 26 | 439 | 26 |
| Capital | 175 | 17 | 295 | 18 |
| Residual | 185 | 19 | 333 | 20 |
| Total | 993 | 100 | 1676 | 100 |

[a] Land share based on rental rate.
[b] Family labour valued at Pesos3.50/day.

The sample was too small to obtain much impression about investment in capital equipment since 1965. The increase in the number of hand weeders from only seven before 1965 to twenty nine in 1969, is worth noting however. In terms of peso value, the major investments were made by landowners in pumps and by tenants in hand tractors and *carabao.*

These findings underscore the important role played by the landowner as the primary supplier of operating capital. Landowners in Laguna apparently found no difficulty in extending a substantial amount of additional credit for what was obviously a profitable investment for them. Usually, they themselves did not have to borrow to meet the increased demand for credit.

## SHARING OF RETURNS

The returns on investment can be partitioned between the various claimants in two ways. The conventional method is to determine the share of the return going to each factor—land, labour, and capital. A second, less conventional, but in these circumstances perhaps more meaningful, approach is to divide the returns among three groups of people—landowner, tenant, and hired labourer. This procedure was applied to forty two of the 155 farms in the original survey. All forty two farms were full adopters of high yielding varieties in 1969.

Returns to inputs for 1966 and 1969 can be compared in Table 6.8. The share of the gross return assigned to land is determined by the rental rate and not the land value. Land values do not necessarily reflect the value of land for agricultural use because of the proximity of the area to Manila. On the other hand, the one-to-one share rental

TABLE 6.9

OUTPUT DISTRIBUTION, BY RECIPIENTS, ON FORTY TWO FARMS SHIFTING FROM LOCAL VARIETIES IN 1966 TO HIGH YIELDING IN 1969, LAGUNA, WET SEASONS

| *Output share* | *Local 1966* | | *HYV 1969* | | *Increase 1966–9* |
|---|---|---|---|---|---|
| | *Pesos* | *per cent* | *Pesos* | *per cent* | *per cent* |
| Operating capital | 69 | 7 | 147 | 9 | 113 |
| Hired labour | 183 | 18 | 341 | 20 | 86 |
| Tenant | 348 | 35 | 540 | 32 | 65 |
| Landowner | 393 | 40 | 648 | 39 | 55 |
| Total | 993 | 100 | 1676 | 100 | |

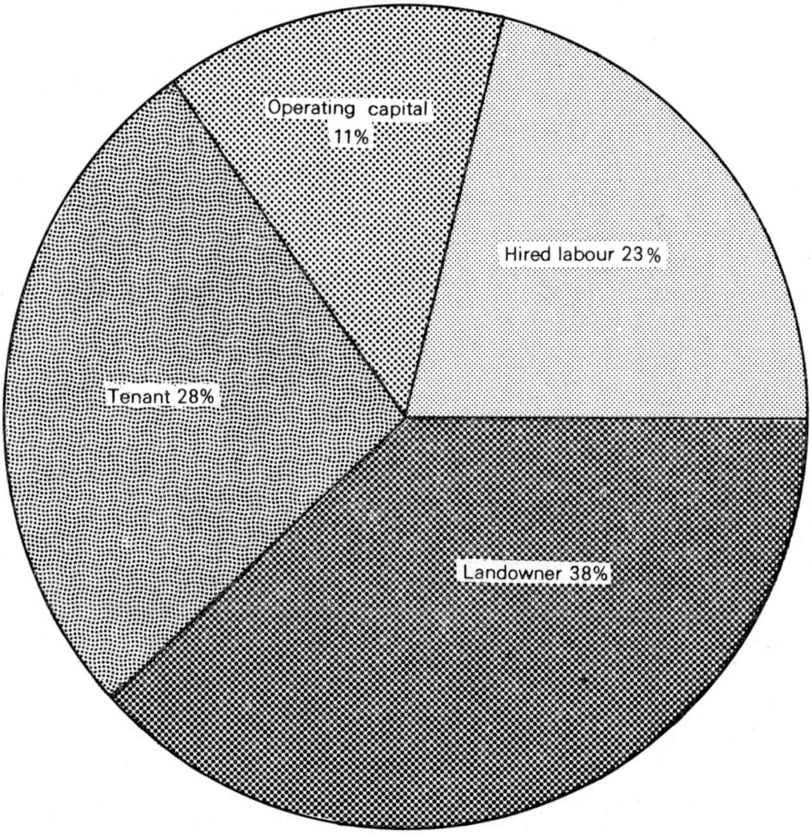


Figure 6.3 Distribution of increase in profits on forty two Laguna farms, wet seasons, 1966-9.

may not be an accurate reflection of land's contribution to total output either. There is a fairly large unexplained residual, and labour's share seems to have been understated.

The gross return is divided between landlord, tenant, hired labour, and operating capital in Table 6.9. The landowner receives payment for contributing land and capital in the form of credit. The tenant is paid for his labour and his family's labour, and for supplying capital equipment such as the *carabao*, plough, harrow and small tools. Hired workers are paid for contributing their labour. Operating capital consists of such items as fertiliser, chemicals, and fuel.

Between 1966 and 1969 capital and labour showed the largest percentage increase (Table 6.9). In spite of this, the share of the gross returns to capital and hired labour changed only slightly. It is not surprising to find that the landowner received the largest share of the increase, followed by the tenant, hired labour, and operating capital in that order. The division of the increase in gross returns from 1966 to 1969 is illustrated in Figure 6.3.

## CONCLUSIONS

The incomes of farmers have generally risen as a result of the changes that have taken place in rice production in the study area. There have, however, been important differences by location in both the choice of inputs by farmers and the yield and profit response.

Biñan farmers, who have the poorest soil-water resources, made the largest change, 1966-9. The area was almost completely converted to high yielding varieties. Use of fertiliser and chemicals, which was lower than that in the other municipalities in 1966 increased, until by 1969, the levels were the same. Biñan farmers, however, have had the smallest increases in yield and income.

Cabuyao, an area served principally by pump irrigation, tended to fall between Biñan and Calamba in terms of yield and input levels and income in 1966. However, yields have responded favourably to higher inputs in this area, and by 1969 income level matched that of Calamba.

Calamba, traditionally regarded as the best farming area of the three, was well in the forefront in use of modern technology in 1966. Farm practices in this municipality over the three-year period changed less than in the other two. The area planted to glutinous varieties has declined substantially, but it still remains large. As a result of favourable prices for malagkit, which is well suited for this area, there perhaps has been reason for change.

The resource situation of Biñan is more typical of Philippine rice growing areas than that of Calamba or Cabuyao. This raises a number of important questions which deserve further investigation:

1 Is the comparatively low response of yields to nitrogen input in Biñan related to the water resources, or is it due to other factors such as soil?

2 Why do the input levels of the farmers in Biñan match those of Cabuyao and Calamba when the yield responses are so different?

3 Are the input levels in Biñan too high, or are they too low in Cabuyao and Calamba? Possibly this situation will adjust itself over time.

4 Related to this question, do the input levels of farmers in Cabuyao and Calamba represent the economic optimum, or can yields and incomes be pushed higher?

5 Could the farmers in Calamba, who have water available throughout the year, use the water during the other months to grow two and a half crops of rice a year or to grow rice in combination with other crops?

These are the types of question that should be answered if the information presented in this study is to be of value to policy makers as well as to those concerned with problems at the farm level.

All farm groups in this area have shared initially in the benefits of the new rice technology. Landowners have clearly been the largest beneficiaries, but they have played an instrumental role in making credit available to farmers. There has been a sharp rise in the demand for hired labour, particularly at harvest. As a result hired labour has increased its share of the benefits. Farmers on the whole are anxious for new knowledge and information that will help them increase their returns further.

7

# Background to the Introduction of High Yielding Varieties of Rice in Thailand[1]

*Delane E. Welsch and Sopin Tongpan*

THIS chapter describes Thailand's experience to date with new technology in rice, namely new high yielding varieties with associated inputs, which form one part of the green revolution in Asia. First the historical background in exports, production, and domestic marketing and pricing is presented. Next the significant developments in rice research in Thailand are described, followed by a brief outline of the adoption of new varieties. In the final section, some implications are drawn for future developments, based on the experience gained until now.

The considerable attention paid thus far to the green revolution and high yielding rice varieties has dealt only with two categories of countries. One comprises those developing countries which were food deficient, and are trying to reduce food imports or even gain self-sufficiency. Their problems with the new technology include production and marketing as direct effects and employment and income distribution as indirect effects. The second category consists mostly of developed countries which are historically large exporters of food grain, together with Japan which has recently switched from a rice importer to a rice surplus producer and the EEC which is nearly self-sufficient. These countries face problems of shrinking commercial export markets and/or surplus disposal problems. Thailand represents a third and largely neglected category, namely historically food surplus, rice exporting developing countries, which in a sense now face, because of the green revolution, all of the problems faced by the first two categories. This chapter is intended to bring some of these special problems to light in the Thai context.

## EXPORTS

Thailand has been an important exporter since 1855 when the Bowring Treaty with Great Britain opened up the country to international

1 This chapter draws heavily on Welsch 1971a and b and Welsch and Tongpan 1971.

TABLE 7.1

VOLUME OF RICE EXPORTS FROM THAILAND, 1857–1944

| *Period* | *Average Quantity ('000 m. tons/year)* |
|---|---|
| 1857–70 | 59 |
| 1860–4 | 110 |
| 1865–9 | 98 |
| 1870–4 | 112 |
| 1875–9 | 212 |
| 1880–4 | 215 |
| 1885–9 | 319 |
| 1890–4 | 435 |
| 1895–9 | 480 |
| 1900–4 | 668 |
| 1905–9 | 886 |
| 1910–14 | 913 |
| 1915–19 | 947 |
| 1920–4 | 1061 |
| 1925–9 | 1043 |
| 1930–4 | 1543 |
| 1935–9 | 1522 |
| 1940–4 | 795 |

*Source:* Ingram 1971: 38.

trade on a significant scale (Corden and Richter 1967: 128). By the early 1930s, Thailand was exporting one half of its annual production. The volume of exports from 1857 to 1944 is shown in Table 7.1. World War II greatly disrupted rice production and trade in Southeast Asia, but by 1949 Thai rice exports had recovered to 1·2 million metric tons of milled rice, or 27 per cent of annual production. Exports reached a peak in 1965, when 1·9 million tons, or 30 per cent of total production (Corden and Richter 1967: 129) were exported. Exports since then have declined to about one million tons per year, or 10-12 per cent of production (Table 7.2).[2]

In 1965, rice was exported to fifty nine different countries, but only nine countries took five per cent or more of the total (Table 7.3). The distribution of exports among various qualities is shown in Table 7.4 for the period 1966-9.[3] Rice white 100 per cent is the highest quality

[2] Exports in 1971 at the time of writing were reaching a high level. By 1 October, they had reached 1·07 million metric tons with some prospect of achieving a level of 1·5 millions tons by the end of the year (*Financial Post,* 21 October 1971: 9).

[3] There are thirty eight distinct and clearly defined grades of milled rice in Thailand, and export prices are reported weekly for each. The grading system was instituted and is maintained and policed by the private sector.

TABLE 7.2

VOLUME AND VALUE OF RICE EXPORTS FROM THAILAND, 1946–71

| | *Volume (million m. tons)* | *Value (million baht)* | *Exports as per cent of Production*[a] |
|---|---|---|---|
| 1946 | 0·46 | 267 | 19 |
| 1947 | 0·39 | 385 | 13 |
| 1948 | 0·81 | 1255 | 22 |
| 1949 | 1·22 | 1869 | 27 |
| 1950 | 1·42 | 1672 | 32 |
| 1951 | 1·47 | 1824 | 33 |
| 1952 | 1·55 | 2629 | 32 |
| 1953 | 1·36 | 3747 | 31 |
| 1954 | 1·00 | 3087 | 18 |
| 1955 | 1·24 | 3133 | 33 |
| 1956 | 1·27 | 2861 | 26 |
| 1957 | 1·57 | 3622 | 29 |
| 1958 | 1·13 | 2968 | 31 |
| 1959 | 1·09 | 2576 | 23 |
| 1960 | 1·20 | 2570 | 27 |
| 1961 | 1·58 | 3598 | 30 |
| 1962 | 1·27 | 3240 | 24 |
| 1963 | 1·42 | 3424 | 23 |
| 1964 | 1·90 | 4389 | 29 |
| 1965 | 1·90 | 4334 | 30 |
| 1966 | 1·51 | 4001 | 25 |
| 1967 | 1·48 | 4653 | 19 |
| 1968 | 1·07 | 3775 | 17 |
| 1969 | 1·02 | 2945 | 14 |
| 1970 | 1·05 | 2520 | 12 |
| 1971 | 1·55 | n.a. | 18 |

[a] Calculated from this table and Table 7.5, with exports lagged one year behind production, i.e., exports of 1970 divided by production of 1969, with production in paddy converted to milled basis by multiplying by 0·66.

*Sources:* 1946–9: Ingram 1971: 38; 1950–67: Thailand, Ministry of Agriculture 1967: 127; 1968–70: Bangkok Bank Monthly Review 1971: 177; exports as per cent of production from Corden and Richter 1967; 1971: unpublished data from Board of Trade, Bangkok.

of rice, for which Thailand has a reputation in world markets. It is usually traded only by the private sector and Hong Kong, Malaysia, and Singapore are the major buyers. The chief competitor is the People's Republic of China, and Thai exporters claim that quotas set by the importers which split the market between Thailand and China are politically determined. The largest export grade is parboiled rice, with India the major buyer followed by Ceylon, Mid-Eastern and

TABLE 7.3

VOLUME OF THAI RICE EXPORTS, PERCENTAGE OF TOTAL BY COUNTRY OF DESTINATION, 1957–69

| *Year:* | *1957* | *1958* | *1959* | *1960* | *1961* | *1962* | *1963* | *1964* | *1965* | *1966* | *1967* | *1968* | *1969* |
|---|---|---|---|---|---|---|---|---|---|---|---|---|---|
| *Country:* | | | | | | | | | | | | | |
| *Southeast Asia* | | | | | | | | | | | | | |
| Hong Kong | 12·3 | 15·1 | 14·9 | 14·9 | 12·4 | 17·2 | 13·5 | 10·8 | 10·8 | 14·2 | 14·5 | 12·3 | 16·9 |
| Malaysia | 14·3 | 17·0 | 24·4 | 19·9 | 15·7 | 15·6 | 17·7 | 15·4 | 14·3 | 10·2 | 13·7 | 17·8 | 14·7 |
| Singapore | 17·2 | 19·6 | 20·0 | 17·4 | 12·3 | 13·5 | 12·4 | 12·1 | 9·1 | 9·5 | 8·1 | 12·3 | 13·1 |
| Indonesia | 11·4 | 11·6 | 6·8 | 11·4 | 23·8 | 21·0 | 23·9 | 23·8 | 5·7 | 11·1 | 12·0 | 4·2 | 7·9 |
| Philippines | 6·6 | 4·1 | 0·04 | — | 8·9 | —0·03 | 5·1 | 5·7 | 6·8 | 3·2 | 6·7 | — | — |
| Sub-total | 61·8 | 67·4 | 66·14 | 63·6 | 73·1 | 67·3 | 72·6 | 67·8 | 46·7 | 48·2 | 55·0 | 46·6 | 52·6 |
| *East Asia* | | | | | | | | | | | | | |
| Japan | 9·3 | 6·4 | 8·5 | 7·5 | 3·2 | 5·0 | 7·5 | 6·7 | 8·1 | 6·1 | 9·3 | 9·3 | 6·2 |
| Taiwan | — | — | 0·4 | 1·6 | 4·8 | — | 0·1 | 1·2 | — | 0·1 | 0·4 | 0·01 | 0·7 |
| Sub-total | 9·3 | 6·4 | 8·9 | 9·1 | 8·0 | 5·0 | 7·6 | 7·9 | 8·1 | 6·2 | 9·7 | 9·3 | 6·9 |
| *West Asia* | | | | | | | | | | | | | |
| India | 0·2 | 0·02 | 0·01 | 0·01 | — | 0·01 | — | 1·8 | 11·3 | 12·1 | 12·4 | 19·5 | 11·2 |
| Ceylon | 2·9 | 0·02 | — | 0·7 | 3·9 | 3·5 | 2·5 | 1·6 | 9·8 | 7·4 | 6·5 | 5·4 | 3·0 |
| Sub-total | 3·1 | 0·04 | 0·01 | 0·7 | 3·9 | 3·5 | 2·5 | 3·4 | 21·1 | 19·5 | 18·9 | 24·9 | 14·2 |
| Middle East | 5·4 | 5·9 | 10·9 | 9·3 | 4·6 | 7·4 | 6·6 | 5·5 | 5·2 | 4·7 | 5·5 | 7·8 | 7·6 |
| Europe | 4·9 | 7·8 | 5·0 | 2·8 | 2·9 | 4·4 | 1·5 | 2·8 | 2·0 | 2·2 | 0·5 | 0·4 | 1·0 |
| Other | 15·5 | 12·4 | 9·0 | 14·4 | 7·4 | 12·3 | 9·1 | 12·5 | 16·8 | 19·1 | 10·4 | 11·0 | 17·6 |
| Total | 100 | 100 | 100 | 100 | 100 | 100 | 100 | 100 | 100 | 100 | 100 | 100 | 100 |

*Source:* Thailand, Department of Customs annual issues.

TABLE 7.4

VOLUME OF THAI RICE EXPORTS BY GRADE, PERCENTAGE OF TOTAL, 1966–9

| Grade[a] | *1966* | *1967* | *1968* (*percentage of total export volume*) | *1969* | *1970* |
|---|---|---|---|---|---|
| Rice white 100 per cent | 14·2 | 13·4 | 17·3 | 21·2 | 23·8 |
| Rice white 5 and 10 per cent | 6·4 | 3·2 | 1·8 | 2·0 | 1·9 |
| Rice white 15, 25, 35 per cent | 14·0 | 10·2 | 11·9 | 7·5 | 22·3 |
| Rice broken | 19·7 | 22·9 | 19·4 | 18·3 | 16·8 |
| Rice glutinous | 5·2 | 8·1 | 8·9 | 12·7 | 8·2 |
| Rice parboiled | 26·4 | 23·1 | 22·0 | 31·7 | 25·3 |
| All others | 14·1 | 19·2 | 18·7 | 6·6 | 1·7 |
| Total | 100·0 | 100·0 | 100·0 | 100·0 | 100·0 |

[a] Standard International Trade Classification (SITC) code 0420200 series.

1 Code 0420210.

2 Code 0420211, 12.

3 Code 0420213, 15, 17.

4 Code 0420230, 31.

5 Code 0420242, 47, 49.

6 Code 0420251.

7 Code 0420229, 01, 09, 14, 16, 18, 19, 20, 32, 33, 34, 35, 36, 38, 39, 02, 03, 04, 05, 06, 07, 08, 50, 52, 53, 43, 44, 45, 46, 48.

After 1969, based on the Brussels Tariff Nomenclature, Chapter 10.

*Source:* Thailand, Department of Customs annual issues.

African countries. This type is frequently traded on a government to government basis. Rice broken is another major grade. Hong Kong, Malaysia, and Singapore are the principal buyers, usually through private channels. Rice white 5 per cent and 10 per cent are the most important grades in domestic consumption, and are the grades on which the local consumer price index is based. Rice white 15 per cent, 25 per cent and 35 per cent are normally traded only by government to government, with Indonesia as the major purchaser. Rice glutinous faces a very narrow market, based on special tastes. Japan has been the major buyer, with Laos as second, although there is probably considerable volume of unrecorded exports of this type to Laos. In any given year, the proportions which are exported as 100 per cent, broken, and parboiled depend partly on the weather and partly on external demand. Adverse weather conditions during harvest can lead to more cracks than normal in the dried paddy. Parboiling of such paddy before milling will usually result in more head rice. So the relative strengths of the broken and parboiled markets partially determine the relative proportions going to each.

## PRODUCTION

Rice is grown in every one of the seventy one *changwats* (provinces) in Thailand, with the most concentrated area in the Central Plains, particularly on the Chao Phya delta. About 75 to 80 per cent of the Thai people live in rural areas and are involved in some aspect of agricultural production. Probably 80 to 90 per cent of farms produce some rice. Rice is the staple food with per capita consumption steady at 155 kilograms (Thailand Government, 1966: 108-10).

Rice production data for the whole kingdom during the period 1907-67, by year and ten-year averages, are shown in Table 7.5. Although there may be valid questions raised about the data for earlier years, this table illustrates the problem of stagnation in yield per unit area in Thai rice production.[4] Some work has been done to separate the yield depressing effect of extending production to areas marginal for rice production from the aggregate data in order to measure yield improvement in areas well suited to rice. Hsieh and Ruttan (1967: 310) estimate that the 2·9 per cent per year growth in total rice production from 1907 to 1964 consisted of a 2·5 per cent per year increase in area harvested and a 0·4 per cent per year increase in yield. Silcock (1970: 180) estimated the components of the increase in output of rice from 1951-3 to 1962-4 as a 13 per cent increase in area and a 15 per cent increase in yield.

Data for each of the four main geographical regions (Table 7.6) illustrate the wide regional variation in yield, with those in the north being more than double those in the northeast. If data were available by agro-climatic or agronomic zones, they would show even greater differences in yield among zones. Yields were highest in the north, at 2·9 metric tons/hectare, with yields by *changwat* in the north ranging 2·4-3·7 metric tons/hectare. Yields in the northeast and central regions were the same (1·6 tons/hectare), with ranges of 1·1-2·9 and 1·2-3·6 tons/hectare respectively. Ranges in yields in the south were 1·4-2·8 tons. It is becoming increasingly clear that environmental factors, particularly water control, are among the most important factors affecting increase in yield per unit area. Adoption of the new high yielding varieties of rice and wheat that are the basis of the green revolution have occurred almost entirely on land with good water control.

Conceptually one can classify rice areas as irrigated, rainfed, and upland. On the first two, rice is grown in a submerged condition, with the paddies bunded on all sides to maintain at least a certain level of water. Irrigated usually means that a supplementary source

[4] Part of the controversy over validity of data deals with increases in output in the Central Plains. See Hsieh and Ruttan 1961; Ruttan, Soothipan and Venegas 1966; Trescott 1968; and Ruttan and Soothipan 1970.

TABLE 7.5

RICE AREA HARVESTED, DAMAGED AREA, PRODUCTION AND YIELD, THAILAND, 1907-67

| *Year* | *Harvested Area* ('000 *hectares*) | *Damaged Area* (*percentage of planted area*) | *Production* ('000 *m. tons*) | *Yield* (*m. tons/hectare harvested*) |
|---|---|---|---|---|
| 1907 | 1385 | 7·2 | 2582 | 1·86 |
| 1908 | 1253 | 14·4 | 2368 | 1·89 |
| 1909 | 1753 | 1·3 | 3044 | 1·74 |
| 1910 | 1452 | 3·2 | 2953 | 2·03 |
| 1911 | 1512 | 3·8 | 2881 | 1·91 |
| 1912 | 1972 | 0·1 | 3670 | 1·86 |
| 1913 | 2008 | 7·1 | 2853 | 1·42 |
| 1914 | 2018 | 1·0 | 3102 | 1·54 |
| 1915 | 2057 | 0·7 | 3267 | 1·59 |
| 1916 | 2079 | 4·2 | 3786 | 1·82 |
| Average 1907–16 | 1749 | 4·0 | 3051 | 1·74 |
| 1917 | 1757 | 21·0 | 2989 | 1·70 |
| 1918 | 2009 | 6·9 | 3384 | 1·68 |
| 1919 | 1404 | 43·4 | 2270 | 1·62 |
| 1920 | 2239 | 8·4 | 4266 | 1·91 |
| 1921 | 2356 | 9·2 | 4235 | 1·80 |
| 1922 | 2402 | 4·9 | 4340 | 1·81 |
| 1923 | 2365 | 12·0 | 4399 | 1·86 |
| 1924 | 2558 | 7·9 | 4942 | 1·93 |
| 1925 | 2387 | 12·8 | 4193 | 1·76 |
| 1926 | 2785 | 3·8 | 5226 | 1·88 |
| Average 1917–26 | 2226 | 12·8 | 4024 | 1·79 |
| 1927 | 2554 | 12·8 | 4564 | 1·79 |
| 1928 | 2386 | 16·3 | 3882 | 1·63 |
| 1929 | 2445 | 19·5 | 3875 | 1·59 |
| 1930 | 2909 | 8·5 | 4826 | 1·66 |
| 1931 | 2581 | 16·5 | 4068 | 1·58 |
| 1932 | 3011 | 6·3 | 5116 | 1·70 |
| 1933 | 3014 | 7·1 | 5008 | 1·66 |
| 1934 | 2933 | 12·1 | 4598 | 1·59 |
| 1935 | 2971 | 12·0 | 4727 | 1·59 |
| 1936 | 2226 | 31·7 | 3380 | 1·52 |
| Average 1927–36 | 2703 | 10·7 | 4404 | 1·63 |
| 1937 | 2943 | 12·7 | 4556 | 1·54 |
| 1938 | 3129 | 10·8 | 4524 | 1·44 |
| 1939 | 3072 | 11·3 | 4560 | 1·48 |
| 1940 | 3235 | 15·0 | 4923 | 1·52 |
| 1941 | 3628 | 8·6 | 5120 | 1·41 |
| 1942 | 2876 | 34·3 | 3854 | 1·34 |
| 1943 | 3813 | 8·1 | 5536 | 1·45 |

*CARRIED OVER*

TABLE 7.5 — CONTINUED —

| | *Harvested Area* ('000 *hectares*) | *Damaged Area* (*percentage of planted area*) | *Production* ('000 *m. tons*) | *Yield* (*m. tons/hectare harvested*) |
|---|---|---|---|---|
| 1944 | 3813 | 6·1 | 4928 | 1·29 |
| 1945 | 2847 | 24·3 | 3572 | 1·26 |
| 1946 | 3509 | 11·9 | 4442 | 1·27 |
| Average 1937–46 | 3286 | 14·5 | 4602 | 1·40 |
| 1947 | 4304 | 10·8 | 5506 | 1·28 |
| 1948 | 4930 | 5·4 | 6835 | 1·39 |
| 1949 | 4963 | 5·8 | 6684 | 1·34 |
| 1950 | 5295 | 4·4 | 6782 | 1·28 |
| 1951 | 5736 | 3·7 | 7325 | 1·28 |
| 1952 | 5130 | 4·4 | 6602 | 1·29 |
| 1953 | 5931 | 3·9 | 8239 | 1·39 |
| 1954 | 4524 | 18·6 | 5709 | 1·26 |
| 1955 | 5376 | 6·8 | 7334 | 1·36 |
| 1956 | 5762 | 4·3 | 8297 | 1·44 |
| Average 1947–56 | 5195 | 6·7 | 6931 | 1·33 |
| 1957 | 4287 | 15·5 | 5570 | 1·30 |
| 1958 | 5169 | 10·2 | 7053 | 1·34 |
| 1959 | 5263 | 13·2 | 6770 | 1·29 |
| 1960 | 5643 | 4·7 | 7835 | 1·39 |
| 1961 | 5656 | 8·5 | 8177 | 1·47 |
| 1962 | 5191 | 7·0 | 9279 | 1·50 |
| 1963 | 6354 | 3·7 | 10,029 | 1·58 |
| 1964 | 5971 | 8·7 | 9559 | 1·60 |
| 1965 | 5960 | 9·1 | 9199 | 1·54 |
| 1966 | 6878 | 5·9 | 11,846 | 1·73 |
| Average 1957–66 | 5737 | 8·4 | 8532 | 1·49 |
| 1967 | 5601 | 12·6 | 9595 | 1·71 |
| 1968 | 6259 | 12·4 | 10,771 | 1·72 |
| 1969 | 7246 | 5·1 | 13,346 | 1·84 |
| 1970 | 7131 | 8·6 | 13,401 | 1·88 |

*Sources:* 1907–62: Isrankura 1967 and Thailand, Ministry of Agriculture, Rice Department 1962; later years: from Thailand, Ministry of Agriculture, Division of Agricultural Economics 1967; 1968: unpublished data, Division of Agricultural Economics, Ministry of Agriculture, Bangkok; 1969–70: unpublished data, Department of Agricultural Extension, Ministry of Agriculture, Bangkok.

of water is available, while rainfed usually means that the only source of water to the individual paddy is rainfall, with perhaps a little run-off from the immediately surrounding area. Upland rice is produced as a dryland crop, i.e., it is not grown in a submerged condition. Data on production and area are not readily available for comparing yields on irrigated and non-irrigated or rainfed land in Thailand. The

TABLE 7.6

THAI RICE AREA PLANTED AND YIELD PER PLANTED HECTARE BY REGION, 1950/1 TO 1966/7

| Year | Area Planted ('000 hectares) North | Northeast | Central | South | Yield (m. tons/hectare planted) North | Northeast | Central | South |
|---|---|---|---|---|---|---|---|---|
| 1950 | 368 | 2031 | 2679 | 462 | 1·4 | 0·9 | 1·4 | 1·3 |
| 1951 | 374 | 2364 | 2758 | 464 | 1·5 | 1·0 | 1·4 | 1·3 |
| 1952 | 367 | 1882 | 2675 | 444 | 1·5 | 1·0 | 1·4 | 1·4 |
| 1953 | 379 | 2538 | 2778 | 476 | 1·8 | 1·1 | 1·5 | 1·4 |
| 1954 | 376 | 1951 | 2743 | 487 | 1·7 | 0·8 | 1·1 | 1·3 |
| 1955 | 388 | 2323 | 2632 | 428 | 1·9 | 0·9 | 1·4 | 1·6 |
| 1956 | 378 | 2483 | 2706 | 457 | 2·1 | 1·1 | 1·6 | 1·4 |
| 1957 | 387 | 1659 | 2600 | 430 | 2·1 | 0·9 | 1·0 | 1·2 |
| 1958 | 393 | 2159 | 2730 | 476 | 1·8 | 0·9 | 1·4 | 1·3 |
| 1959 | 410 | 2467 | 2732 | 456 | 1·9 | 0·8 | 1·2 | 1·4 |
| 1960 | 419 | 2329 | 2723 | 450 | 1·9 | 1·0 | 1·5 | 1·4 |
| 1961 | 412 | 2465 | 2822 | 479 | 2·1 | 0·9 | 1·5 | 1·5 |
| 1962 | 411 | 2851 | 2892 | 504 | 2·1 | 1·1 | 1·6 | 1·5 |
| 1963 | 424 | 2715 | 2961 | 501 | 2·6 | 1·1 | 1·8 | 1·6 |
| 1964 | 434 | 2475 | 3126 | 504 | 2·3 | 1·1 | 1·7 | 1·2 |
| 1965 | 437 | 2453 | 3129 | 534 | 2·4 | 0·9 | 1·6 | 1·4 |
| 1966 | 438 | 3116 | 3268 | 553 | 2·6 | 1·2 | 1·9 | 1·5 |
| 1967 | 439 | 2255 | 3165 | 552 | 2·9 | 1·0 | 1·7 | 1·3 |
| 1968 | 440 | 2860 | 3278 | 567 | 2·9 | 1·2 | 1·6 | 1·8 |
| 1969 | 458 | 3245 | 3353 | 581 | 2·7 | 1·5 | 1·9 | 1·9 |
| 1970 | 470 | 3390 | 3383 | 559 | 2·9 | 1·6 | 1·6 | 1·9 |

*Sources:* 1950–67: Thailand, Ministry of Agriculture, Division of Agricultural Economics 1967: 48; 1968: unpublished data, Division of Agricultural Economics, Ministry of Agriculture, Bangkok; 1969–70: unpublished data, Department of Agricultural Extension, Ministry of Agriculture, Bangkok.

production and area data on irrigated land are collected by a different method from 'all rice' data, and therefore calculation of non-irrigated yields by a residual method is questionable. The top section of Table 7.7 shows the percentage distribution among the four geographic regions of the area that the Royal Irrigation Department considers as irrigated. The low proportion irrigated in the northeast and the south is a likely partial explanation for the low relative yields of these two regions as shown in Table 7.6.

'Irrigated' is, moreover, not really a very specific term. Dwarf varieties, with a height of around one metre, require a rather carefully controlled water level. Not only is a supplemental source of water required so that water can be brought in when needed, but an adequate drainage system is needed so that excess water can be taken off

TABLE 7.7

AREAS PLANTED AND IRRIGATED, METHOD OF PLANTING AND TYPE OF RICE, BY GEOGRAPHIC REGION, THAILAND, 1959–70

| | *North* | *Northeast* | *Central* | *South* |
|---|---|---|---|---|
| Area planted (1966) as per cent of total kingdom | 6·0 | 42·0 | 44·0 | 8·0 |
| Area irrigated as per cent of (1966) total kingdom | 10·0 | 10·0 | 77·0 | 3·0 |
| Area irrigated as per cent of area planted | 40·4 | 7·9 | 45·0 | 9·5 |
| Method of planting (1959) | | | | |
| Transplanted | 99·6 | 98·4 | 54·8 | 71·3 |
| Broadcast | 0·4 | 1·6 | 45·2 | 28·7 |
| Type of rice (1959) | | | | |
| Non-glutinous | 8·5 | 28·2 | 96·7 | 95·7 |
| Glutinous | 91·5 | 71·8 | 3·3 | 4·3 |
| Type of rice (1970) | | | | |
| Non-glutinous | 15·2 | 36·3 | 96·0 | 96·1 |
| Glutinous | 84·8 | 63·7 | 4·0 | 3·9 |

*Sources:* area planted, Thailand, Ministry of Agriculture, Division of Agricultural Economics 1967: 48; area irrigated, Thailand, Ministry of Agriculture, Division of Agricultural Economics 1967: 174–5 (includes state irrigation projects, people's irrigation projects, and tank irrigation projects, irrigation area only); method of planting, Kulthongkham and Ong 1964:7; types of rice, for 1959 Kulthongkham and Ong 1964: 6, for 1970 unpublished data from Department of Agricultural Extension, Ministry of Agriculture, Bangkok.

the paddy, i.e., the water level must be controllable. Vast areas of the Chao Phya delta of Thailand, however, have no provisions for drainage. Water may reach depths of 1-3 metres, and depths of 5 metres have been observed. In these areas, floating rice, which is usually defined as a type of rice possessing the genetic ability to elongate rapidly under rising water conditions, is planted (Yantasast, Prechachat and Jackson 1970a). Most floating rice is broadcast sown in late May and early June following the first few showers that mark the approach of the monsoon season. The rice seeds germinate and grow as the monsoon showers increase in intensity and frequency. Water levels usually do not increase rapidly until late August or early September at which time the level may change as much as 3-5 centimetres per day. The proportion of broadcast rice in each of the four geographic regions in Table 7.7 are a measure of the floating rice area but not an exact one since some farmers also routinely broadcast some of their non-floating rice, usually because of labour restrictions at planting time (NEDECO 1969) for transplanting requires perhaps as much as twenty to thirty times more labour than broadcasting. Farmers with fairly good water control may start by transplanting, but when time runs short for finishing planting they broadcast the rest.

The area in the central region planted to the true floating varieties has been estimated at 800,000 hectares, with another million hectares planted to tall non-floating varieties that are not sensitive to water depths of 1 metre and occasional flooding (Yantasast, Prechachat and Jackson 1970a).

The last two sections of Table 7.7 show the relative proportions of area planted to non-glutinous and glutinous rice in 1959 and 1970, respectively. The proportions in the central and south regions remained constant, with glutinous being only a small fraction of the total. This reflects the nature of glutinous rice as a specialty in consumption in those two regions. However glutinous rice is the staple food in the north and part of the northeast. The increased proportion of non-glutinous, especially in the north, can be interpreted as increased production for sale and export out of the region, rather than a change in taste.

## MARKETING

The internal distribution system for rice in Thailand is remarkable. It handles part of the production of most of the 3-3·5 million farms that produce rice. It distributes it among the 35 million consumers, who are very discriminating in their tastes and preferences for rice, at the rate of about 155 kilograms per person per year, and channels the surplus of over one million tons per year into world markets. It does this with marketing margins that are low by international standards, with farmers receiving something over 70 per cent of the retail price paid by domestic consumers (Thailand Government, 1970: 246).

The rice marketing system as it now operates in Thailand is essentially private enterprise, although there have been some important government involvements in the past, particularly in pricing. The physical task of carrying out the marketing functions is undertaken by private firms with the exception of one small government rice milling concern, a small amount of storage operated in conjunction with a minor price support program, and one retail outlet intended for low income consumers in Bangkok.

The first steam powered rice mill was built in Bangkok in 1858, and by 1877, mills were being located in the rice growing areas (Ingram 1971: 70). By 1910, there were about sixty mills in Bangkok, and this number remained nearly constant until a decline commenced after World War II. In 1956 it was estimated that there were 6067 mills in the whole kingdom, with 3518 of these with a capacity of less than 5 tons per day, 2179 of 5-30 tons, 273 of 31-65 tons, fifty six of

66-100 tons, and forty one of over 100 tons per day capacity (Kulthongkaham and Ong 1964: 64). The mills are generally scattered throughout the country and fairly well located on the basis of production. Most of the larger mills are of the abrasive stone disc-sheller type, with steam engines burning rice husks for fuel as the common source of power.

A common milling payment is for the milling to be done for the bran. The farmer brings the paddy and receives all of the milled rice, while the miller keeps the bran, and sometimes the smallest brokens, as a milling fee. Thus, millers frequently also have a pig feeding enterprise along with their milling business (Welsch and Tongpan 1971). Most mills also have equipment to grade the rice. Japanese type rubber roller mills have been tested but are not currently economic because the cost of rollers is greater than added value of whole rice recovered from the rubber roller type mills. Another factor is that the present steam powered mills have few parts to wear out; the millers usually make their own abrasive stones, and one frequently finds operating mills that are 30-40 years old or older.

Rice is usually stored in paddy form, as milled rice does not keep as well, and must be reconditioned if it is not consumed within a few months after milling. Small quantities are usually stored on farms and larger quantities in the mills. Most of the milled rice that moves to Bangkok for consumption or export is transported by river barge at low cost. Grading equipment is used to blend grades to specification for export. Retail shops commonly offer a wide range of grades to consumers.

Internal pricing of rice from the end of World War II to 21 April, 1971, operated through a constrained free market system. Internal prices at all points in the market channel were set by normal market forces, and generally moved up and down in concert with world rice prices. However, they differed from world prices by the amount of a rice export tax, called the rice premium. The history of the rice premium and its economic effects has been well documented and analysed by Silcock and Ingram, so only a brief summary will be given here (see Silcock 1970: Appendix I and Ingram 1971: Chapters 3, 4 and 11).

Because of circumstances regarding Thailand's status during World War II, the government was required post-war to provide the United Nations with 1·5 million tons of milled rice at a price considerably below existing world levels. A Rice Bureau was set up and made the only legal exporter of rice, and so as to avoid heavy treasury costs, domestic prices of rice were kept very low. By the end of 1949, the United Nations allocation of rice ended, but the system was maintained because a rise in domestic prices to world levels at that time would have reduced real incomes of urban consumers to politically intolerable levels (Silcock 1970: 217). The high world prices during the Korean conflict enabled the government to obtain considerable money by buying at the low domestic price and exporting at world price. However, a rapid expansion in world production resulted in a buyer's market for rice from 1954 onwards and the Rice Bureau had difficulty in exporting. Private traders were allowed to export, but were required to pay a premium for the privilege. The rate of that premium was roughly set at the difference between world market price and Thai domestic price. At the beginning of 1955, the government returned all of the rice trade to the private sector. The rice premium was however retained as the mechanism for keeping the domestic price at the desired level below the world price. It is therefore essentially an export tax (Corden 1967).

The level of the rice premium was set by the government as a fixed amount per ton of milled rice loaded aboard ship, and it varied by grades. The exporter procured his rice in the domestic market and made his own sales negotiations abroad. When the rice was loaded aboard ship and certified by customs inspectors, the exporter then paid the premium to the government.

The level of the premium was supposedly flexible to take account of changes in world price levels. In practice, it rarely changed more than once or twice per year, and in fact, remained constant from May 1963 to January 1967. In addition changes in the level were very small during the period 1956-66. In 1967 the procedure was changed from a specific rate to an *ad valorem* rate.

As an example, white rice 100 per cent considered to be the top quality Thai rice, sold for $US250 per metric ton, f.o.b. Bangkok at the peak in world rice prices in October 1967. The rice premium on it at that time was $US82 per ton. During the week ending 5 April, 1971, f.o.b. price was $US120 per ton, and the premium was $US38 per ton for this grade. Numerous statistical studies have shown that domestic wholesale prices usually are very nearly equal to f.o.b. prices less the rice premium and exporting costs (Chuchart and Tongpan 1965).[5]

The Thai Government also exports rice on a government to government basis. Terms are negotiated between governments after which the Thai Government calls for bids from the private sector to provide the agreed upon quantity and quality aboard ship. The difference between the cost of procurement to the government and the sale price

[5] There is an additional export tax on rice at 5·7 per cent *ad valorem*, which is collected on all agricultural exports.

is also called rice premium and is paid to the Treasury. Since 1963, the proportion of rice exported under such agreements fluctuated from 25 to 50 per cent of total exports.

Over the years, the rice premium has been discussed, researched, and debated more than any other economic topic in Thailand. The proponents claim seven benefits to Thailand. They are:

1 It provides a major source of government revenue.

2 It is the only effective method of taxing agriculture.

3 It promotes agricultural diversification by making the returns from other crops more attractive relative to rice.

4 It maintains domestic food price stability.

5 It aids low income groups by keeping prices of rice low.

6 It promotes industrialisation by keeping cost of living and thereby labour costs low.

7 It provides bargaining power in exporting.

Ingram (1971: 243-61) provides a detailed analysis and discussion of each. It should be noted that the slump in world rice prices and consequent drop in domestic paddy prices caused the government to remove the rice premium on all grades except 100 per cent and 5 per cent white, parboiled and cargo on 21 April, 1971. Although the premium on parboiled was later removed, it still remains on white rice 100 per cent, which represented one quarter of 1970 rice exports. This would appear to be a disincentive for quality rice, particularly to millers. Exporters, however, argue that it will not be, and that their price competition comes with lower grades, not 100 per cent.

## RICE RESEARCH IN THAILAND

The first rice research station was established in Thailand in 1916 at Rangsit, about 30 kilometres from Bangkok. Head selections and variety yield trials were started, with emphasis on grain quality and yield. This work, plus normal selections by farmers, produced a large number of traditional varieties, each best suited to the specific conditions of a local area. The stress on quality is illustrated by Thailand's winning the first honour and ten other prizes for high quality long grain rice at the World Grain and Seed Exhibition in Regina, Canada in 1933 (Dasananda 1968).

Rice breeding work began on an intensive scale in 1950. Initial work involved the identification of superior types from the existing material collected from farmer fields. Concentration was on pure line selection, based on the identification of high yielding, high quality, long-grain types adapted to local conditions. A big stimulus was given to rice research in 1954 with the establishment of a separate Rice Department within the Ministry of Agriculture. Hybridisation and mutation breeding were initiated in that year. Research in soil fertility, plant protection, and mechanisation were also started or intensified during this period. Work on blast disease (*Piricularia oryzae*), bakanae disease (*Gibberella fujikuroi*), yellow-orange leaf virus (tungro), bacterial leaf blight (caused by *Xanthomonas oryzae*), and gall midge (*Pachydiplosis oryzae*) was also started. During this period, varietal selection was based, in addition to the quality criterion, on the ability of a variety to produce higher yield without the addition of fertiliser.

The 1960s turned out to be a significant period of change for rice in Asia and Thailand. The International Rice Research Institute (IRRI) was established in 1960, and Thailand entered into fully fledged co-operation with it. Thai rice breeding objectives and methods were re-evaluated, and new lines of work were started, taking advantage of the training, facilities, and germ plasm which became available through IRRI. Hybridisation efforts were revived, and a number of crosses were made, three of which turned out to be very important.

1 In 1964 a cross was made at IRRI between the Thai-recommended glutinous variety, gam pai-15 and taichung native-1 (Jackson, Panichapat and Awakul 1969).

2 Also in 1964 a cross was made at IRRI between the Thai-recommended floating variety, leb mue nahng-111 and a dwarf experimental line originating from peta/2 x taichung native-1 (IR95). This cross was designated IR442 (Yantasast, Prechachat and Jackson 1970a).

3 In April, 1966, a cross was made in Thailand between the Thai-recommended non-glutinous variety, leuang tawng and IR8 (Jackson, Panichapat and Awakul, 1969).

A new glutinous variety, RD2, was developed from the first cross listed above and was approved by the Variety Release Committee[6] in December, 1969. At the same time, the Committee approved two new non-glutinous varieties, RD1 and RD3, that were developed from the third cross mentioned above. 'Their release marks a new era in rice breeding (in Thailand) because they are the first hybrid varieties possessing the short, stiff straw and plant type, the main characteristics of IR8 variety' (Jackson, Panichapat and Awakul 1969: 83).

Although experimental lines resulting from the second cross mentioned above are still under evaluation, results thus far indicate that

> the ability to withstand deep water and occasional flooding can be incorporated into future stiff-strawed short height varieties much in the same manner that disease and insect resistance is being developed. The successful incorporation of this character could

[6] This committee is composed of Thai government officials and persons in the private rice trade, primarily millers and exporters. Although the committee uses several criteria, exhaustive quality tests appear to be the most important measure of approval.

open up large areas prone to flooding which presently prevent the production of dwarf varieties (Yantasast, Prechachat and Jackson 1970b).

The reader will recall that, after several years of testing, the famous rice variety, IR8, was released by IRRI in November 1966, followed by IR5 in December 1967. Thailand participated in the early testing of IR8, and extensively tested it in both the dry and wet seasons of 1967 (Sopanaratna 1968). Generally IR8 performed well under Thai conditions, exhibiting high yield, fertiliser responsiveness and good plant type. However its grain quality was so poor compared to Thai standards that merchants were reported to have discounted its price by 30 to 40 per cent (Jackson, Panichapat and Awakul 1969). IR8 has been available to farmers who asked for it but it was not taken to the Variety Release Committee, primarily because of the quality factor.

The situation in Thailand, a rice surplus and major exporting country, was quite different from that in some of the food deficit countries whose decisions to encourage IR8 was a contributory factor in the green revolution. As mentioned previously, Thai consumers are very discriminating in their taste for rice. With per capita consumption already at high levels, it appears that increased incomes result in increased expenditures for the same quantity but of a higher quality of rice. Owing to normal weather and other factors, Thailand already had ample quantities of low grades of rice, while in world trade, Thailand has a reputation for high quality long grain white rice, which usually commands a substantial price advantage over lower grades. A second factor was that, through the extensive cooperation with IRRI, the potential of IR8 and other materials as parent stock in the hybridisation program was recognised early, and progeny from the crosses using these and high grain quality Thai varieties were already in the testing stage.

One further high yielding variety should be mentioned. The variety C4-63, developed by the University of the Philippines, was tested at about the same time as the IR8. Although C4-63 was resistant to *tungro*, RD1 and RD3 were already in the testing stage and they produced somewhat better yields and possessed superior grain quality. Therefore C4-63 was not submitted to the Variety Release Committee, but it has become moderately popular with farmers in limited areas with good water control.

## DISSEMINATION OF NEW VARIETIES

The main research units of the Rice Department are the Breeding, Technical and Engineering Divisions. Extension was an integral part of the Rice Department until 1968, when it was made into a separate department. The main units of the Breeding Division are the Breeding, Statistics, Seed Multiplication, and Regional Yield Trial Sections (Jackson n.d.). The Breeding Division is also responsible for maintenance and supervision of twenty one rice experiment stations, which are located in all major rice growing areas of the Kingdom.

The Rice Department identifies four classes of seed: breeders, foundation, stock and multiplication (Pookamana and Jackson 1970). Each class of seed is limited to one generation. The Seed Multiplication Section is responsible for the production and maintenance of the breeders and foundation seed of all recommended varieties, promising experimental hybrids and old varieties which continue to be valuable for special areas of research. The network of rice experiment stations is used for this production. Extension workers assist Farmer Associations at the village level to promote multiplication of stock and multiplication seed. This system of classification and the implementation program was first organised in 1954 and is still in operation.

The Regional Trial Section conducts yield trials in farmers' fields using the most advanced experimental lines. This section is also responsible for assisting the extension workers by preparing seed for yield trials and inspecting their yield trials at least once during the growing season.

Pathologists, entomologists and agronomists in the Technical Division have extensive research programs underway in those subject matter areas. They also assist in the testing and evaluation of new experimental hybrids both in the laboratory and in the field. Thus weaknesses are rapidly identified and the information incorporated into the breeding program.

The Rice Department has other programs which directly and indirectly contribute to the spread of new varieties (Dasananda 1968). Pest control units which were transferred to the new Department of Extension when it was formed in 1968 are maintained in major growing areas to monitor and assist farmers with insect and disease problems. Soil fertility experiments are conducted at the outlying rice experiment stations, and fertiliser trials are conducted in farmers' fields in at least 200 locations each year. Large scale demonstration plots were initiated in 1964 where the Rice Department undertakes to assist selected villages for a three year period. Technical advice, fertiliser on credit, and in some cases chemicals, are provided for farmers who wish to participate. After three years, the farmers, with the assistance of the local extension worker, are expected to continue on their own. A yield contest among farmers was initiated in 1963, with the winner receiving a special award from the King during the annual ploughing ceremony. Since 1964, the winners have all achieved yields of 7 tons/hectare or more. A rice grain quality contest among farmers is also conducted annually.

The Engineering Division is also active in research and development of mechanisation. Machines which they have developed include a special outboard motor for boats in shallow water, a water pump for shallow water and low lift, a rice soil puddling machine and a small tractor. Centrifugal rice hullers, small threshing machines and a harvester are under development. Experiments on aerial seeding of clouds to control rainfall are also underway.

### ADOPTION OF NEW HIGH YIELDING VARIETIES

With approval of the three new varieties in late 1969, the 1970 wet season (June to November) was the first opportunity for most farmers to adopt them. Only 25 tons of Foundation seed were available to the Extension Department in February, 1970, and another 30 tons in June (Thailand, Rice Department 1970). Regular rice statistical data collection procedures do not permit identification of variety, so that the only estimates of adoption available are those made by experienced observers. The best guess is that 100,000 hectares were planted to the new varieties, primarily RD1, in the 1970 wet season and 250,000 hectares in the 1971 wet season. These plantings were primarily in the western and northern parts of the Chao Phya delta, and in localised areas in the Chiang Mai valley in the far north of Thailand. Those farmers who are able to grow RD1 successfully report approximately double the yields over conventional varieties. RD2 has not been superior to the old type varieties in the north and has not been adopted in that region. High prices for fertiliser, depressed prices for glutinous paddy since its introduction, and lack of a dependable water supply have limited adoption of RD2 in the northeastern region.

Evaluation of the adoption thus far presents a mixed picture. The reader should keep in mind several crucial points. First, the new varieties were introduced at a time of the lowest rice prices and the highest fertiliser prices in Thailand since 1950 (Welsch 1971a; Welsch and Tongpan 1971). As a major exporter, Thailand experienced directly the decline in world rice prices, and with an export tax (the rice premium) that held farm prices for paddy at about one half of world levels, farmers received little for their rice. At the lowest point, in early April, 1971, farm prices for non-glutinous paddy fell as low as $US35/ton, and for glutinous paddy, below $US25/ton.[7] Paddy became a feed grain for some hog producers. On the other hand the cash price for ammonium sulphate in rural areas was $US0.10/kilogram or about $US0.50/kilogram of elemental nitrogen, which must be nearly the

[7] Farm price for corn at the same time was $US50/ton.

highest level in the rice growing world. Fertiliser rice price ratios are very important in the adoption of the new varieties, and thus the Thai experience cannot be compared directly with food deficit countries in which rice prices are kept above world levels and fertiliser is subsidised. Second, the new varieties may occupy nearly 5 per cent of the rice area in Thailand, but perhaps only 10 per cent of the rice area has the degree of water control necessary for the short, stiff-strawed plant types.

## CONCLUSIONS

It seems that the Thai experience with new high yielding varieties should be evaluated in the context of 'the evolutionary nature of the new rice technology' (Barker 1971). As such, it would receive a high score. A solidly based national rice research program is well established with multi-disciplinary teams hard at work on a single commodity with a clear problem focus (Welsch and Sprague 1970). An open flow of information and training with regional and international rice research is maintained. High yielding varieties are available that have the IR8 yielding potential but with the preferred Thai grain quality. A scientific breakthrough with dwarf varieties that can tolerate deep water or occasional deep flooding is imminent, and could have Asia-wide impact. However, there are several policy measures that need to be taken if Thai farmers are to gain the full potential of the rice technology that is now available and even newer technology that will be forthcoming from the research program.

The first has to do with input price policies. The present high fertiliser prices are a direct barrier to the gaining of the full benefits of the new varieties on areas to which they are suited. Second, rice cannot be expected to carry all of the development burden in Thailand. If yields increase as rapidly and to the level that they are capable of doing on land suited only to rice, then market realities will require some contraction of the area now in rice but which could produce something else (Welsch 1971b). Other farm enterprises will be needed to improve farm income in the areas coming out of rice, and an expanded research effort will be needed to develop such enterprises. Third, the level and the distribution of income, both regional and personal, are of increasing concern in Thailand. In some agro-climate zones, rice may be the development tool to solve these income problems. But high yielding rice varieties have their limitations, and policy makers should not be surprised if the new varieties have no impact on the poorest agro-climate zones. The problems in such areas are severe, and it will take a number of co-ordinated policies to solve them. Fourth, research is a long-run proposition. Although much has been accomplished in developing high yielding rice varieties specific to certain Thai situations, the task is not finished. RD1 and RD3 are not the final answer for Thai rice growers, but merely represent the prototype upon which improvements can be made. Continued strong support for rice research is required.

# 8

# The High Yield Program in Burma

*H. V. Richter*

THERE has so far been only a very limited adoption of the new productivity-raising agricultural technology in Burma. Thus far, the country has probably lost more than it has gained from the higher yielding techniques which have recently become available in tropical crop production. It has lost part of the natural comparative advantage that it formerly enjoyed in rice growing. At the same time, it has not itself been able to take full advantage of the new technology.

The reasons why Burma has not taken such advantage are partly that the new techniques are not at their present stage well suited to the country's basic resources of land and water control. This will be discussed in the earlier part of this chapter, which will also draw on available data to show what progress has been made in the high yield program. The latter part will discuss the economic and socio-political factors which have been as important as the resource factors in slowing down development.

The analysis will focus mainly on rice, because of its dominating role in the economy. Paddy occupied 55 per cent of the gross sown area in 1970/1, somewhat less than in neighbouring Thailand. But rice's role as an export earner is far more important to Burma than to Thailand. In spite of falls in both the volume and price of shipments, rice provided over half of export earnings in 1970, compared to 17 per cent in Thailand.

Much of the present cultivated area is suitable only for the single cropping of traditional varieties of rice. Deep flood conditions prevent the growing of the new short-strawed varieties in at least a third of the total cropped area and of a much higher proportion of areas commercially oriented, and will continue to do so until improved tall and floating varieties are bred which are suited to the environment.

One tenth of the total sown area, and 15 per cent of the gross sown rice area, was irrigated in 1970/1. But comparatively little irrigated land is used commercially, the bulk being planted to rice and other foods for subsistence. Moreover, most irrigation is of the gravity flow type. Water levels cannot be carefully controlled and fields drained and reflooded, as is required to take full advantage of the new technology.

TABLE 8.2

POPULATION AND RICE PRODUCTION, BURMA, 1965/6

| | *Rice, Gross Sown Area* ('000 *hectares*) | *Production* ('000 *m. tons*) | *Yield* (*kilograms/hectare*) | *Population*[a] *1966* ('000) | *Production per head* (*kilograms/paddy*) |
|---|---|---|---|---|---|
| Upper Burma | 770 | 950 | 1234 | 8295 | 114 |
| Lower Burma | 3575 | 6306 | 1764 | 12,577 | 501 |
| States | 669 | 799 | 1194 | 4374 | 183 |
| Total | 5014 | 8055 | 1606 | 25,246 | 319 |

[a] Estimate based on extrapolation of the 1941 census, the last countrywide census completed.

*Sources:* rice area and output by district, *Statistical Yearbook 1967*, Tables 67 and 68; population, *Statistical Pocketbook* 1967, Table 11.

TABLE 8.3

IRRIGATED AREAS, BY CROPS[a], BURMA, 1970/1

| | *Rice* | *Pulses* | *Sugar Cane* | *Other Edible* ('000 *hectares*) | *Cotton* | *Other Industrial* | *Total* |
|---|---|---|---|---|---|---|---|
| Upper Burma | 464 | 26 | 1 | 75 | 46 | 6 | 618 |
| Lower Burma | 52 | — | 1 | 10 | — | 27[b] | 89 |
| States | 219 | 1 | 5 | 11 | — | — | 236 |
| Total | 734 | 27 | 7 | 97 | 46 | 32 | 943 |

[a] Includes double cropping.

[b] Mainly jute, pump irrigation.

*Source:* Burma, *Report to the People 1971/2*, Appendix 10, in Burmese.

varieties. Moreover the lack of water control renders unprofitable the use of fertilisers by individual farmers and inhibits the growing of fertiliser-responsive high yielding rice strains. There are, however, some areas, especially in the northern portion, or mid-zone, where flood depth is shallow enough to permit growing of short statured rices.

In many districts groundnuts, beans, sugarcane and jute are grown and high value crops such as tobacco and vegetables are planted on fertile riverine land (*kaing*). Paddy, however, occupies about 84 per cent of the gross sown area. Moreover the sharp seasonality of rainfall and the heavy nature of the soils severely restricts the possibility of double cropping. In 1970/1, 215,000 hectares were sown to two crops, this being 5·4 per cent of the net sown area of four million hectares. Just over 2 per cent of the cultivated area was irrigated. It is estimated that, with pump irrigation and tractor ploughing, the area sown to

TABLE 8.4

LOWER BURMA SOWN AREAS[a], 1970/1

| | *Pegu* | *Irrawaddy* | *Tenasserim and Arakan* | *Total* |
|---|---|---|---|---|
| Sown, net ('000 hectares) | 1628 | 1506 | 878 | 4012 |
| Sown, gross ('000 hectares) | 1739 | 1596 | 892 | 4227 |
| Irrigated, gross ('000 hectares) | 36 | 42 | 11 | 89 |
| Irrigated, rice ('000 hectares) | 28 | 14 | 10 | 52 |
| No. of farms ('000)[b] | 638 | 597 | 569 | 1804 |
| Net sown per farm (hectares) | 2·5 | 2·5 | 1·5 | 2·2 |

[a] Excludes rubber and other tree crops.

[b] 1969/70.

*Source:* Burma, *Report to the People 1970/1*, Appendix 4, and *1971/2* (in Burmese) Appendices 6, 9, 10 and 11.

jute after the main paddy crop could be expanded from its present 40,000 hectares to 400,000, bringing the share of land double cropped in Lower Burma to 14 per cent.

*Upper Burma: Mixed Cropping*

Different physical constraints are found in Upper Burma, which has inadequate rainfall and irrigation systems of rather limited utility. Rainfall is spasmodic and sparse, ranging 31-40 inches a year in the different districts. Partly to conserve moisture, more than one fourth of the occupied area is left fallow each year. Rice accounts for only a quarter of the gross sown area. The region has a rice deficit and supplements local production from Lower Burma. The remaining three quarters of the sown area is planted to groundnuts, sessamum, pulses, millet, wheat, vegetables, cotton and tobacco. The region is the main source of the country's vegetable oils and domestically grown cotton. Up to a quarter of output of pulses and maize are exported. All other crops are either consumed locally or marketed in Lower Burma.

Soils are generally rather light and poor, and there is considerable erosion on hills where forest cover has been cut. Over half the rice area is irrigated (Table 8.1). Separate data are not available for irrigated and rainfed rice, but low yields of the latter pull average yields in almost all districts below those of the Lower Burma rainfed plains (Burma, *Statistical Yearbook*, 1965, Table 74). *Kaing* land (overflow land along streams) is sometimes used for rice, but is more often planted to high value crops such as vegetables and tobacco. Dry land agriculture is hazardous and in a bad season farmers may have to replant three or four times. Inter-planting of crops such as maize is also practised as drought insurance. Farmers take advantage of late rains

TABLE 8.5

UPPER BURMA SOWN AREAS, 1970/1

| | *Sagaing* | *Mandalay* | *Magwe* | *Total* |
|---|---|---|---|---|
| Sown, net ('000 hectares) | 1138 | 1025 | 779 | 2942 |
| Sown double cropped ('000 hectares) | 265 | 363 | 267 | 895 |
| Irrigated net ('000 hectares) | 210 | 227 | 87 | 523 |
| Irrigated double cropped ('000 hectares) | 23 | 52 | 20 | 95 |
| Irrigated rice, gross ('000 hectares) | 199 | 189 | 76 | 464 |
| No. of farms ('000)[a] | 522 | 503 | 581 | 1606 |
| Net sown per farm (hectares) | 2·2 | 2·0 | 1·3 | 1·8 |

[a] 1969/70.
*Sources:* as Table 8.4.

to plant a second, drought resistant crop, such as beans or groundnuts, which will produce low but worthwhile yields without further rain provided the seed germinates. The area double cropped ranges up to 30 per cent of the net sown area, and although some of this is irrigated, the bulk of it is not (Table 8.5). Most second crops are grown on dry or *kaing* land.

Irrigation systems, some dating back nearly a thousand years, have been designed mainly for single cropping. According to official statistics only 18 per cent of the net irrigated area was double cropped in 1970/1. This may well have ecological advantages (Whitcombe 1972) as may the fact that the use of artificial fertilisers is so far limited. What is certain is that the technological revolution in agriculture has not yet reached the region. Agronomic practices are backward. Little is done in green manuring or crop rotation to ensure soil fertility and failure to weed further reduces yields.

The irrigated area fluctuates annually by up to a tenth according to the height of the streams. In 1970/1 the gross irrigated area was 618,000 hectares, 16 per cent of the gross sown area of 3·8 million hectares. The region is divided into three administrative Divisions: Mandalay, which had the highest proportion, 20 per cent, of the gross sown area irrigated in 1970/1, Sagaing, 16·5 per cent and Magwe 10 per cent. 75 per cent of irrigated land was devoted to rice for local consumption and sale in this season, and 11 per cent to industrial crops. The planning authorities are exerting pressure on the peasants to put more irrigated land under long staple cotton needed for the newly constructed Upper Burma textile complex. The irrigated area sown to cotton was raised from 34,000 hectares in 1969/70 to 46,000 in the following season, but output is still barely sufficient to supply existing State factories and only half the level which will be needed when two nearly completed mills come into production.

TABLE 8.6

THE STATES AND CHIN SPECIAL DIVISION SOWN AREAS[a], BURMA, 1970/1

| | *Kachin* | *Shan* | *Chin and Kayah* | *Kawthoolei* | *Total* |
|---|---|---|---|---|---|
| Sown, gross ('000 hectares) | 144 | 493 | 111 | 227 | 976 |
| Irrigated, gross ('000 hectares) | 54 | 161 | 11 | 11 | 237 |
| Irrigated, rice ('000 hectares) | 54 | 144 | 10 | 10 | 218 |
| No. of farms ('000)[b] | 103 | 563 | 81 | 134 | 881 |
| Net sown per farm (hectares) | 1·3 | 0·8 | 1·4 | 1·7 | 1·1 |

[a] Excludes area under fruit, tea, mulberry, other tree crops and opium.
[b] 1969/70.
*Sources:* as Table 8.4.

### *The States: Subsistence Crops*

Rice cultivation in the States and the Chin Special Division is wholly for the farmers' own use or local sale, as is the bulk of other field crop production. Two thirds of the cultivated area is planted to rice, and the rest to maize, wheat, millet, pulses, groundnut, sugar cane, tea and tobacco. Farms are smaller than elsewhere in Burma, although there are enormous areas of cultivable land available. No detailed land surveys have been made but cultivable wastes (excluding forests) are thought to be nearly five times the present occupied area. One third of this latter is fallowed annually (Burma, *Report to the People 1970/1*, p. 16).

Poor transport in these mountainous regions makes unprofitable the commercial cultivation of most field crops, with the notable exception of opium. Cattle are raised and tea and some fruit and vegetables are grown for markets in Burma Proper. Roughly a third of the net sown area (almost all planted to rice) is irrigated in three of the States. Less than 4 per cent of the net sown area of 941,000 hectares was double cropped in 1970/1, and an equally small share of irrigated land. The central government is trying to persuade the Kachins on the northern frontier to grow cotton as a second crop after paddy on the 5600 hectares irrigated in the dry season by the Washaung Dam near Myitkyina, completed in 1967/8. Last season 400 hectares were planted.

## RESEARCH AND EXTENSION

Government agricultural research, extension and development work is divided between two agencies: the Department of Agriculture and the Agricultural and Rural Development Corporation (ARDC). The Department was founded in 1880 and initially dealt only with rice.

After 1907 it extended its interest to crops grown in Upper Burma and by the 1930s had set up eight experimental stations in the districts. The Corporation is a State board, founded in 1951, which is intended to be run on commercial lines. For a long time the Corporation was a more lively and important institution than the Department, but the balance has recently become more even, mainly because the Department has been made responsible for the State plan for agriculture, but also because some of its long-term research is now yielding useful results.

Responsibilities of the two institutions has been altered from time to time, especially in relation to the States. Broadly the Department conducts most fundamental scientific research and deals with rice and most other food crops such as maize, wheat and beans. The Corporation handles all industrial crops such as jute, cotton, rubber and tobacco and foodcrops such as groundnuts[2] and sugarcane which are required by government factories. It is also responsible for land reclamation, tractor service and some village works. Both agencies distribute quality seed and fertilisers.

Agriculture has been relatively neglected compared to industry in government spending since independence, both on capital and current account. It received, together with livestock, fisheries and irrigation, a maximum of 15·2 per cent of the State capital budget in 1962/3. In the two fiscal years ending in 1971-2 these sectors were allotted an average of K83 million ($17 million)[3] of the capital budget annually, which was 8 per cent of the total, compared to 38 per cent allocated to industry. Current expenditure estimates averaged K94 million ($20 million) for the Department of Agriculture in these two years, 6 per cent of the Departmental total.[4] Current funds for the Corporation were larger and separately budgeted. However a large share of these funds was for employment by the Corporation in activities other than research and extension.

Considering the comparatively slender means at their disposal, both agencies have sizable programs and staffs. The Department employs nearly 5000 people, of whom 500 are conducting research. It controls a central research station at Gyogon and thirteen affiliated experimental stations throughout the country, which specialise on different crops according to the locality. State farms, linked with these stations, covers 6000 hectares. The Corporation employs 30,000 workers, including

2 In 1971 responsibility for groundnut quality seed production was switched back to the Department of Agriculture.

3 The exchange rate used was that in force up to December 1971 of Kyat 4·76 = $US1.00.

4 Includes the Ministry of Land Nationalisation, in effect a land-use bureau.

seasonal workers. Many of these, however, are engaged in the mechanisation program or in public works projects. The two State institutions together employ nearly 4000 field assistants, about 1500 in the Upper Burma dry zone, 1300 in the Delta Divisions and the remainder evenly spread throughout the coastal Divisions and the States.

*Government Seed Distribution*

The government quality seed program is not extensive enough to provide farmers with regular supplies of improved seed to plant a substantial proportion of any one crop, with the exception of groundnut. Its aim is rather to introduce good quality seed over a series of years and then to leave the breeding of the bulk of fresh supplies to the farmers themselves. This is especially so as regards rice. During the 1960s paddy seed distributed would have covered 0·6-4·1 per cent of acreage planted each year, if used for transplanting, less if broadcast.

The principal rice research station is at Hwambi near Rangoon, which is conducting field trials on imported and domestic varieties, both hybrid and pure line. Extension is a three tiered process, the station sending out seed to State-owned multiplication farms, which supply 95 per cent certified pure seed to neighbouring foundation stock farmers. These are private farmers who get premium prices for producing selected seed, which must be 90 per cent pure and registered. Since irrigation facilities have been improved on some of the State farms, double cropping is now possible, so that the time taken to get a new variety into mass production has halved from eight to four years.

In 1970/1 nearly 12,000 tons of selected paddy seed was distributed. The largest part of this was improved local varieties, with yields in the range of 1600-2200 kilograms/hectare rather than high yield strains. One important objective of the Department has always been to encourage the growing of pure lines of selected seed in the various districts to produce uniform deliveries for the mills and so improve milling quality and market prospects. It has had some success in this since 1959, when the caretaker government of General Ne Win reintroduced the quality differentials for each of the main varieties of paddy, which had been an important element in marketing in the colonial period. At the same time work is going on to combine the best qualities of local varieties with the high yielding characteristics of imported strains. It is hoped to breed heavier cropping tall strawed varieties. Hwambi sends research workers occasionally to institutes in the Philippines, Japan and Israel, but employs no foreign scientists.

## HIGH YIELD RICE STRAINS

The high yield rice program has had a checkered history. It started in 1967 when 200 tons of IR8 were imported from the Philippines

TABLE 8.7

HIGH YIELD RICE PRODUCTION, BURMA, 1967–71

| | *Yagyaw I* | *Yagyaw II* | *Ngwetoe* | *Total* |
|---|---|---|---|---|
| *Sown area* (hectares) | | | | |
| 1967/8 | 3431 | — | — | 3431 |
| 1968/9 | 166,900 | — | — | 166,900 |
| 1969/70 | 129,241 | 4777 | 9900 | 143,918 |
| 1970/1 | — | 172,900 | 16,989 | 189,889 |
| *Production* (m. tons) | | | | |
| 1967/8 | 10,186 | — | — | 10,186 |
| 1968/9 | 510,331 | — | — | 510,331 |
| 1969/70 | 364,250 | 15,755 | 22,859 | 402,864 |
| 1970/1 | — | 472,591 | 38,356 | 510,947 |
| *Yields* (kilograms/hectare) | | | | |
| 1967/8 | 2969 | — | — | 2969 |
| 1968/9 | 3058 | — | — | 3058 |
| 1969/70 | 2818 | 3298 | 2309 | 2799 |
| 1970/1 | — | 2733 | 2258 | 2691 |
| *Prices* ($/m. ton) | | | | |
| 1967/8 | 37·5 | — | — | |
| 1968/9 | 36·0 | — | — | |
| 1969/70 | 36·0 | 37·5 | 37·5 | |
| 1970/1 | — | 36·0 | 36·0 | |

*Sources:* Burma, *Report to the People 1970/1*, Appendix 26 and *1971/2*, Appendix 28.

(Barker, 1969). This variety was named yagyaw I in Burma. IR5 and C53-39 were obtained the following year and named yagyaw II and ngwetoe respectively. Table 8.7 shows that progress was rapid at first, the area sown to yagyaw I rising from just over 3000 hectares in 1967/8 to 167,000 in 1968/9. Government authorities then planned for an increase in the area sown to the various high yield strains to 400,000 hectares the following season and imported exceptionally large quantities of fertilisers to provide for this production.[5] In the event only 144,000 hectares were sown to high yield rices in 1969/70 and yagyaw I production was later phased out. In 1970/1, 190,000 hectares were planted to yagyaw II and ngwetoe, this being 3·8 per cent of the total area sown to rice. Small areas of C-4-63 were planted in the Mandalay Division (*Working People's Daily*, 2 June, 1971).

[5] 236,000 tons of fertilisers were imported in 1967/8, at a cost of $18·4 million, compared with an average of 33,000 tons over the previous five years. Only 71,000 tons were used on all crops in that season according to official data.

TABLE 8.8

DISTRIBUTION OF HIGH YIELD RICES, BURMA, 1969–71

| | *Yagyaw I 1969/70* Hectares sown | *Yagyaw I 1969/70* Yield (kilograms/hectare) | *Yagyaw II 1970/1* Hectares sown | *Yagyaw II 1970/1* Yield (kilograms/hectare) |
|---|---|---|---|---|
| Sagaing | 26·844 | 2767 | 33·578 | 2524 |
| Mandalay | 25·737 | 2384 | 20·777 | 1604 |
| Magwe | 13·354 | 2627 | 9·758 | 2323 |
| Upper Burma | 65·936 | 2589 | 64·112 | 2195 |
| Pegu | 39·398 | 3138 | 63·994 | 2903 |
| Irrawaddy | 20·332 | 3057 | 38·281 | 2481 |
| Tenasserim and Arakan | 1·366 | 2651 | 3·328 | 3477 |
| Lower Burma | 61·095 | 3100 | 105·604 | 2768 |
| States | 2·209 | 1873 | 1·681 | n.a. |
| Total | 129·241 | 2818 | 172·900[a] | 2733 |

[a] Includes 1504 unspecified.
*Source:* as in Table 8.7.

### *District Distribution*

Over 60 per cent of yagyaw rice production comes from Lower Burma, the bulk from the mid-zone regions of Pegu. But a higher share of district rice lands have been put under the new strains in Upper Burma. Table 8.8 shows distribution by district in 1969/70 and 1970/1. About 8 per cent of the rice area of Upper Burma was sown to yagyaw in the two seasons analysed. In Lower Burma the share was 2-3 per cent and in the States only 0·3-0·2 per cent. The reasons for the slow spread of high yield rices in Lower Burma are partly geographic, because of uncontrolled flooding, as already explained. Similarly, in Upper Burma water control systems are not flexible enough to create ideal conditions for them.

Nevertheless, this is not sufficient to explain the reluctance of the farmers to take up the new varieties. The non-geographic explanation for this reluctance seems to be threefold. First, yagyaw yields per hectare, although 80 per cent higher than the all-Burma average, were only 25-40 per cent above those of the best indigenous varieties. Second, the price paid by State procurement agents (Table 8.7) was at the bottom of the scale, equivalent to $36-37·5 per ton, compared to $35 for the lowest grade of ordinary paddy and $46 for the highest.

Taking extra fertiliser costs into account, it was clearly unprofitable to devote good paddy lands to the new varieties. Third, the eating quality of the high yield rices is not acceptable to the very discriminating Burmese farmers.

Gross returns per hectare of yagyaw were K524 ($110) in 1968/9 according to official data (*Report to the People, 1969/70,* 43). An average of 293 kilograms/hectare of artificial fertiliser was used. The minimum price of urea was K541 ($114) a metric ton at this date.[6] This would cost K159, reducing returns per hectare to K365 ($77), not taking into account other extra costs, such as additional labour for the larger harvest. In contrast, the gross value of the best quality rice on paddy land yielding 2200 kilograms/hectare was K481 ($101) and other costs would have been much lower, since fertiliser is used only on nurseries.

The usage of chemical fertilisers on rice has declined steadily since 1968/9, when 49,000 tons were used on yagyaw alone. In 1969/70 total consumption was 61,200 tons, of which rice accounted for 36,000 and other crops for 24,600. In 1970/1 total usage was 50,500 tons, of which rice was 24,600 and other crops 25,900 (Burma, *Report to the People, 1971/2,* Appendix 18).

Taste preference resulting in marketing difficulties explains the failure of the authorities to promote the high yield strains vigorously after the initial impetus and the comparatively low procurement prices offered. Burmese officials explain that yagyaw has only one sizable market, and that is amongst poor labourers, who find both its low price and slowness in digestion an advantage.[7] For this reason India is the best export market but is becoming increasingly self-sufficient.[8] It is therefore hoped that as much of yagyaw as possible will be absorbed domestically, both to release supplies of better quality rices for export and to reduce the State's losses on subsidised domestic sales.[9]

### *Foreign Exchange Repercussions*

The net impact of high yield grains on Burma's export earnings has probably thus far been negative, although it is difficult to measure the country's loss with any precision. On the one hand the volume of the

[6] Urea prices were reduced by 20 per cent in December 1971 to K433 ($91) a metric ton.

[7] This latter characteristic is shared by parboiled rices.

[8] Recent figures on the destination of rice exports are not available. In 1969/70 India and Pakistan took under 9 per cent of the value of total exports. In 1940/1 India accounted for 52 per cent.

[9] Medium quality rice, ngasein 35 per cent, is sold retail by government People's Shops at K0.305/kilogram, equivalent to $64/ton.

country's rice supplies was raised. Production of high yield paddy averaged 474,000 tons in the three seasons ending in 1970/1, seasons in which climatic conditions were normal. Let us assume that this incorporates a 50 per cent increase on yields in the districts where it was grown, or about 160,000 tons on an original 314,000. Converted to rice, 160,000 tons of paddy would be around 100,000 tons at present Burmese milling ratios. Exports of whole rice in the three seasons 1967/8 to 1969/70 averaged 400,000 tons, or 300,000 tons net of the extra production attributable to the high yield rices.

On the other hand unit prices of rice exports fell. They averaged $140 a ton in the two seasons 1967/8, and 1968/9, and $86 a ton in 1970/1. Part of this fall may be attributable to the impact of the new grain technologies on world wheat, maize and rice production and on world market prices, although it is difficult to isolate other elements such as climatic conditions in the importing countries and non-commercial sales under aid programs. If, for the sake of illustration, one attributes the whole of the decline in world rice prices to high yield grains, then it can be calculated that Burmese sales of 400,000 tons at $86 a ton rather than 300,000 at $140, would result in a loss of 18 per cent in the country's gross rice export earnings. The net foreign exchange loss would be larger after taking into account the foreign exchange cost of fertiliser imports, which cost an average of $6·3 million between 1967 and 1971, and of written off fertiliser factory investment.

## OTHER FIELD CROPS

Conditions for raising yields per hectare of crops other than rice are even less propitious than for rice. Forty five per cent of the sown area is devoted to growing a variety of such crops, most of them food crops. But over two thirds of it lies in Upper Burma, where drought is a constant hazard, and only 5 per cent is irrigated. Although much can be, and is, being done to raise the productivity of irrigated land, this is too small a share of the whole to affect overall yields greatly.

Only in the case of two industrial crops, jute and cotton, is a significant share of cultivated area irrigated. These two crops, together with sugar cane, are the main ones to receive substantial dosages of artificial fertiliser. The result of this, and of natural hazards, is that yields are generally very low (Table 8.10). There are no data distinguishing irrigated from dry land yields.

Up to 16 per cent of the sown areas of upland crops fail seasonally to the extent that they are written off for land revenue purposes. But even if one takes this into account and discusses yields on harvested area, yields are still low. A comparison of yields per hectare on matured areas with those of Thailand show Burma's results ranging from just over a fifth to two thirds of those of its neighbour's.

TABLE 8.9

AREA[a] OF FIELD CROPS: OTHER THAN RICE, BURMA, 1966–71

| | Total ('000 *hectares*) | 1966/7 Irrigated ('000 *hectares*) | Per cent Irrigated | 1970/1 Irrigated ('000 *hectares*) |
|---|---|---|---|---|
| Upper Burma | 2630 | 147 | 5·6 | 154 |
| Lower Burma | 679 | 19 | 2·8 | 37 |
| States | 350 | 10 | 2·9 | 19 |
| Total | 3660 | 176 | 4·8 | 209 |

[a] Gross sown area.
*Sources:* as for Tables 8.1 and 8.3.

TABLE 8.10

UPLAND CROPS: IRRIGATION, FERTILISER AND YIELDS, BURMA, 1970/1

| | Sown Area ('000 *hectares*) | Irrigated Area ('000 *hectares*) | Per cent irrigated | Fertiliser usage[a] tons | Fertiliser (kilograms/ hectare) | Yields (kilograms/ hectare) |
|---|---|---|---|---|---|---|
| Sessamum | 1016 | [b] | [b] | [b] | [b] | 130 |
| Groundnuts | 702 | [b] | [b] | 2397 | 3·4 | 754 |
| Pulses | 638 | 26·7 | 4·2 | 720 | 1·1 | 428 |
| Maize | 151 | 0·8 | 0·5 | 366 | 2·4 | 315 |
| Sugar cane | 96 | 6·6 | 6·9 | 5776 | 60·2 | 14,960 |
| Wheat | 70 | 1·3 | 1·9 | 794 | 11·3 | 573 |
| Other edibles[c] | 1107 | 95·7 | 8·6 | 4717 | 4·3 | — |
| Cotton | 189 | 46·3 | 24·5 | 8563 | 45·3 | 226 |
| Jute | 47 | 26·7 | 56·8 | 1953 | 41·6 | 604 |
| Other non-edibles[d] | 49 | 5·3 | 10·8 | 15 | — | — |
| Total | 4065 | 209·4 | 5·2 | 25,301 | 6·2 | |

[a] Mainly urea and superphosphate.
[b] Included in 'other edibles'.
[c] Mainly vegetables. Includes millet and other cereals 219,000 hectares in 1966/7.
[d] Mainly tobacco.
*Sources:* Calculated from Burma, *Report to the People 1971/2*, Appendices 9, 18 and 25.

The difference in productivity is especially striking in the case of maize, which has over the past decade become one of Thailand's principal export earners. The area under maize in Thailand has grown from 56,000 hectares in 1955/6 to 744,000 in 1967/8. But in spite of this expansion, yields per hectare have been raised by nearly 50 per cent. Thai production has shifted from the soft local variety to the hard Guatemalan quality in demand abroad. It is concentrated in the region north of the Central Plain, where the flood plain becomes

TABLE 8.11

YIELDS[a] OF CERTAIN CROPS, BURMA, 1967/8

| | *Maize* | *Pulses* | *Groundnut* | *Sessamum* | *Cotton* | *Jute* | *Sugar* |
|---|---|---|---|---|---|---|---|
| Thailand | | | | | | | |
| (kilograms/hectare)[a] | 1813 | 892[b] | 1263 | 681 | 838 | 1069 | 31·875 |
| Burma | | | | | | | |
| (kilograms/hectare)[a] | 397 | 435 | 737 | 167 | 287 | 737 | 14·975 |

[a] Harvested areas.

[b] Mung beans, soy beans and castor beans.

*Sources:* Calculated from Burma, *Report to the People, 1970/1,* Appendices 23 and 24; Thailand, *Agricultural Statistics of Thailand,* 1967, Table 16.

TABLE 8.12

VALUE OF QUALITY SEED DISTRIBUTED, BURMA, 1961–71

| | *1961/2* | *1964/5* | *1968/9* | *1969/70* | *1970/1* |
|---|---|---|---|---|---|
| Groundnuts ($'000) | 954 | 4128 | 2069 | 513 | 37 |
| Rice ($'000) | 74 | 403 | 533 | 244 | 602 |
| Sugar ($'000) | 92 | 297 | 209 | 455 | 354 |
| Other crops ($'000) | 259 | 387 | 339 | 391 | 512 |
| Total ($'000) | 1378 | 5214 | 3149 | 1604 | 1504 |

*Source:* Burma, *Report to the People, 1970/1,* Appendix 16.

shallow. (This is a region not dissimilar by nature to the mid-zone of Burma in the northern portions of the Irrawaddy and Pegu Divisions.) It is now protected against flood by the Chainat dam, although not irrigated, and served by an excellent transport system (Silcock 1970). Availability of good quality seed, efficient transport, and favourable maize prices encouraged Thai farmers to diversify away from rice.

### *Distribution of Quality Seed*

By contrast, the Burmese government has concentrated until recently on import replacement crops in its agricultural extension services. Over the past decade it has allotted a very large share of funds to groundnuts, in a crash program to reduce the country's dependence on vegetable oil imports. In this it has been fairly successful, and although yields per hectare are still low, they are more than a quarter above those of the late 1930s, in spite of a doubling of the area under the crop. Groundnut seed production is now being left mainly to the farmers themselves. This accounts for the cutback in government funds spent on all seed distribution to less than a third of the mid-1960s level (Table 8.12). The emphasis of the program has now swung to other crops, rice, maize, wheat, sugar, jute, cotton and pulses.

### AGRARIAN AND PRICE POLICIES

Government policy since independence has affected farm productivity in two main ways: through agrarian reform and through price policies. Agrarian reform programs have had their positive as well as negative aspects. Income distribution in the rural sector is more equal than before. Many former landless now have their own plots and holdings are much more equal in size than formerly. Nearly half of the cultivated area has been subject to rent abolition or expropriation since 1948. The area cultivated has, however, risen by much less than the rural population over the past two decades, so that average farm sizes have halved since the 1930s and the marketable surplus has declined sharply (Richter, 1969). Under a 1953 land nationalisation law farmers may not rent or buy land and so raise production extensively. Procurement prices have been generally set too low in relation to agricultural inputs to create incentives to raising output intensively.

Policies are, however, changing. It seems to be generally held by those concerned with planning that agrarian reform has gone far enough, and that while maintaining agrarian policies, the main effort should be shifted from egalitarianism to efficiency within a socialist framework. The shift in emphasis is partly being forced on the government because rice price policies are failing to achieve their objectives. The principal policy aims were to offer modest but guaranteed paddy prices to growers; to maintain low retail prices to domestic consumers at minimum cost to public funds; and to obtain State revenue for investment through exporting large volumes of rice at prices well over procurement levels. Procurement prices have been raised gradually since 1959, so that the margins between them and present export prices is quite narrow. The government's domestic sales, which are subsidised, have exceeded exports in volume since 1966. The rice trade is no longer a net contributor of government revenue for investment, although still bringing in half of export income.

#### *Government Planning*

Policy changes are occurring in three ways. First, government planning is becoming more widespread, especially in the irrigated areas. In some districts it has become mandatory, in that farmers are required to grow planned crops in the main season or face loss of land holdings. Second, price relationships have been altered by raising procurement prices of some export and industrial crops[10] and by reducing the price

[10] In March 1971 government purchase prices for the different varieties of beans were raised 8-20 per cent, and of various types of cotton 7-50 per cent. Paddy procurement prices have not been altered.

of urea, the main fertiliser used. Third, co-operatives are to be reorganised and greatly stimulated under a law of May 1970, under which nearly all heads of farm households will become members of co-operative societies. The marketing of rice and other crops and of farm inputs is to be gradually passed over by the government to the societies, which will also purchase tractors and pumps from the State for hire. The aim is to stimulate 'self help' projects amongst the farmers, which will gradually raise rural prosperity.

The government hopes that a combination of State leadership and farm incentives will increase output. According to the 1971/2 to 1974/5 plan recently adopted, agricultural production is to rise in four years by 16 per cent in constant prices. This is an average of 3·78 per cent a year, a fairly modest growth rate in view of the annual rise of 2 per cent in population. It is to be achieved both by area extension and yield increases. The sown area is to be enlarged by 285,000 hectares, or 3·6 per cent, mainly by bringing fallow land into production. An additional 78,000 hectares is scheduled to be irrigated, which would bring the share of upland crop land irrigated up from 5 to 7 per cent. According to the plan, rice yields per hectare are to be raised 7·5 per cent in four years, and groundnut yields by 4·2 per cent. These two crops contributed 40 and 17 per cent respectively to the gross value of agricultural production in 1969/70. Yields of certain upland crops,[11] which together contributed 7·5 per cent to gross value in that year are to be raised by 20 per cent on the average. The rest of the increase will presumably be achieved either by additional double cropping, by switching to higher value crops or by raising yields on the remaining dry land crops.

11 Wheat, maize, pulses, cotton, kenaf, sugar, Virginia tobacco, chillies and tapioca.

9

# Mechanisation, Industrialisation, and Technical Change in Rural Pakistan[1]

*Hiromitsu Kaneda*

IN reviewing the recent developments in rural Punjab of Pakistan designated by the wide-ranging title of this paper, it is useful to provide at the outset the underlying analytical framework to be used. This is composed of four interlocking factors of particular significance for the question of agricultural development in the content of the overall development of Pakistan's economy. These factors, not listed necessarily in any priority order, are the demographic conditions, relating particularly to the labour absorption problem in the agricultural sector, as suggested by the so-called 'arithmetic of population growth'; the types of innovation in agricultural production and their relation to the sources of growth in labour productivity; the 'backward linkage' of the agricultural sector, as it interacts and influences the development of indigenous industry; and the choice between the so-called 'bi-modal' and 'uni-modal' patterns of agricultural development, or, in relation to Pakistan, the relative emphasis to be given to the 'growth rate' and the 'misery rate'.

The implications of the 'arithmetic of population growth' for the expected growth of the potential labour force in the agricultural and in the non-agricultural sectors are quite as relevant to the situation in Pakistan, as to many other less developed countries of Asia. The harsh reality of a rapidly rising population at about 3 per cent per year, and the fact that 43 per cent of the population in 1961 (last census year) were children under 14 years of age, dictate that the potential growth in labour supply is an essential component of any discussion relating to the strategy of development. Furthermore, the relative share of the agricultural labour force in Pakistan stood at 59 per cent in 1961, although this did represent a decline from the 1951 figure of 65 per cent. Given the high growth rate of the total labour force, and the very large share of the agricultural labour force in the

1 Unless otherwise stated, Pakistan here refers to what was formerly West Pakistan.

total, it would take decades for the absolute size of the agricultural labour force to begin to decline, even if non-agricultural employment expanded as rapidly as it did in the 1950s. Such expansion in the non-agricultural sectors, however, is unlikely to have remained as high in recent years and will probably decline further in the coming years. Since relatively capital-intensive investment is inevitable in building up the infrastructure and in expanding large-scale manufacturing industries, the marginal capital-labour ratio in non-agricultural sectors is expected to rise. There is also considerable uncertainty as to the volume of net foreign assistance, and thus investible funds which will be available to the country in the coming plan periods. These considerations point to the conclusion that the agricultural sector must not only retain the current level of its labour force but also absorb additional numbers which are unable to find employment in the non-agricultural sectors.

The second major underlying factor is concerned with the sources of productivity growth in agriculture. In analysing the basic production relationships and in assessing the productivity performance of agriculture, it is convenient to consider the following identity:

$$\frac{O}{L} = \frac{A}{L} \times \frac{O}{A}$$

where O denotes output in agriculture, A planted hectarage, and L represents workers gainfully employed in agriculture. The identity states that, in incremental terms, an increase in output per worker (labour productivity) in agriculture results from, and can be partitioned into, an increase in land area per worker and an increase in output per hectare.

Capital is subsumed in this relationship, and its effects on production, either through the application of working capital such as fertilisers and improved seeds, or through the use of fixed assets such as irrigation facilities or farm machines (or through both) will appear either as a rise in yields per hectare or in a rise of the land area per man (or as both). It is clear that there are many ways in which factors of production can be combined to achieve a certain level of output. For any given set of factor prices, different techniques of production (i.e., different combinations of factors) can be arranged in order of increasing unit costs of production. This is also the order of decreasing profitability for any given price of the product. Given factor and product prices it is most profitable and efficient[2] to adopt and use the

[2] There are two assumptions underlying this statement. One is that the factor and product prices are given and the other is that there are no economies of scale. The statement remains valid if either one of these assumptions is dropped, provided that the other is retained.

method of production which minimises unit cost of production. From the economist's viewpoint, the fact that a certain operation is performed more 'efficiently' in an engineering sense, be it in fuel-energy conversion, body-weight-energy conversion, or in the power-draft-speed relationship, does not necessarily mean it is an economically more efficient operation than other alternatives. Economic discussions of the choice of techniques are couched crucially in terms of the relative prices (opportunity costs) of the substitutable factors of production.

The standard objection to the use of the 'static factor proportions argument' as given here is that factor costs may change markedly over time as a result of economic development, so that an advantage based on, say, cheap labour may prove to be quite limited in duration. The relevant consideration for the case of Pakistan's agriculture in this respect is that the agricultural labour force will continue to grow absolutely for several decades to come. Thus, the opportunity cost of labour is expected to remain low until the structural transformation of the economy takes place several decades hence. Another standard argument agains the line of reasoning given here emphasises the possibilities of reinvesting surpluses of large (e.g., mechanised) farms. I shall deal with this point in conjunction with the fourth element in the basic framework, the consideration as to the choice between the 'bi-modal' and 'uni-modal' patterns of agricultural development.

If land is in ample supply and labour is scarce, the primary objective of agricultural development may be to increase hectarage per worker, thus raising the output of each worker in the sector. On the other hand, if labour is abundant and land is scarce, the basic theme for growth of agricultural productivity will be the increase in yields per hectare to enhance the output per unit of available land. Thus, for example, given the differences in the prices of land and labour in the United States and Japan we would expect that growth of output per worker (O/L) in the United States would be closely correlated with changes in land area per worker (A/L) and in Japan with changes in output per hectare (O/A). These expectations are indeed confirmed by comparative studies of agricultural development in these countries.

Although it is often impossible to distinguish clearly the effect of an innovation in agriculture, it is convenient to associate loosely mechanical-engineering technology (such as tractor/combine harvester mechanisation) with an increase in area per worker and biological-chemical technology (such as the recent seed-fertiliser revolution) with a rise in yields.

In terms of particular inputs, additional supplies of irrigation water act on both (A/L) and (O/A), as will be made clear later, by increasing hectarage or the cropping intensity and by increasing crop yields. The new seeds and increased application of fertilisers in recent years have had the effect primarily of increasing yields per hectare (per crop). Inputs of herbicides and pesticides, however, have their beneficial effects on yields (per crop) and also save labour required for weeding and pest control, and therefore, to that extent open up the possibility of expanding hectarage per worker. By the same token, machines and implements of various types and sizes will affect the expansion of area per worker, to the extent that they save human labour required for various phases of agricultural operations, and increase yields per hectare so far as they perform agricultural operations 'better' than the manual labour alone.

It is common knowledge that machines and implements save labour required for any specific operation. Proponents of tractor (and combine-harvester) mechanisation, however, would emphasise the yield effect rather than the labour displacement effect when faced with a situation characterised by an abundant supply of labour. One of the most important of the favourable effects of mechanical power and 'efficient' implements on better yields is timely preparation of seed beds. Another is of course 'efficient', timely harvesting and post-harvest processing of crops.

Timeliness in seed bed preparation is less critical on irrigated land than on rain-fed areas, which have to rely on the moisture retained in the soil for the growth of crops. However, it is often argued that the time element in seed bed preparation becomes of crucial importance on irrigated lands if there is a pressure to achieve multiple cropping. Introduction of a time dimension into the picture does not unduly complicate the analysis. From an economic point of view, a high level of cropping intensity represents a situation where land is scarce relative to other inputs. Taken within each cycle of crop growth (determined by the growing period of the crops and the climatic characteristics of the area), therefore an increased input of any one factor should be judged on the same basis as before. Because intensified cropping means simply that a greater amount of output is obtained in a given period of time due to an increased application of factors to land, presumably it is possible to intensify cropping by an increase in any factor of production other than land (say, labour, capital, or water) at some configuration of input prices. Furthermore, it is important to point out that in Pakistan, the most binding constraint on increasing the intensity of cropping is the availability of additional supplies of water, rather than mechanical power *per se*. Then, we may state that emphasis in the agricultural development of a land-rich, labour-poor economy will be given primarily to mechanical-engineering innovations, while the basic theme for agricultural development of a land-poor, labour-abundant economy will be biological-chemical innovations. In the event that both land and labour become constraints on the growth of agricultural production, both hectarage per worker and yields are proper targets for improvement. It is the contention here that, given the circumstances where (except in the relatively small rainfed area in the northern portion of the country) the agriculture of Pakistan has been directly dependent on irrigation water, the beneficial effects of extensive mechanical-engineering technology (tractors and combines, in particular) on productivity growth in agriculture are marginal relative to those to be realised by alternative innovations.

One cannot overemphasise the need for an integrative, comprehensive approach to economic development, in which the positive interaction between agricultural development and industrial growth can be exploited. Industry established on this basis will not only create the 'forward' and 'backward linkages' to the agricultural sector (e.g., industrial workers create a consumers' market for farm products; farms need industrial products; raw material processing, etc.), but it also opens up the scope for mobilisation of indigenous resources, development of skills, diffusion of technology, and employment and income creation. This is the third element in the framework underlying this chapter: a particular concern with the 'backward linkage' of the agricultural sector as it interacts and influences the development of indigenous industry.

A concomitant of rapid growth of the agricultural sector in Pakistan in the 1960s has been the burgeoning of a small-scale engineering industry which supplies key durable good inputs: mainly diesel engines, pumps, and strainers, but also various farm implements. This growth industry, located at various points around the Punjab, has attracted the attention of a number of economists for at least three important reasons:

1 It is a clear example of specific industrial/agricultural interaction, i.e., agricultural growth has generated demand for output of the domestic manufacturing sector, while conversely, the supply of agricultural inputs has been the *sine qua non* of the green revolution in Pakistan.

2 The industry is truly small-scale and, hence, has been a vehicle for marshalling indigenous 'minor' savings/investible funds, for development of entrepreneurial talent and management skills, for training of skilled and semi-skilled labour and for application of new (for Pakistan) technology.

3 This development has occurred spontaneously without subsidies, tax concessions, special credit arrangements, technical assistance or even recognition by official agencies.

An estimated 600 firms of this type in the Punjab, employing some

6-7000 production workers and producing some 75-85 million rupees worth of output in 1969, have helped to reduce the pool of partially and fully unemployed, have raised output levels and have mobilised savings that would not otherwise be forthcoming.

These firms have demonstrated their viability and their positive contributions to the economy. A conscious effort to develop agricultural technology with the strategy of emphasising biological-chemical technology in agriculture, will foster domestic industries, which in turn, will have beneficial effects on the agricultural sector by providing it with appropriate inputs on increasingly favourable terms. It would be a clear mistake to minimise this type of positive interaction between agricultural and industrial development.

Finally, the fourth element in the basic framework concerns the appropriate relative emphasis to be placed on the 'bi-modal' and the 'uni-modal' patterns of agricultural development. According to the agricultural census of 1960, only 8 per cent of the farms in Pakistan cultivated land in excess of 10 hectares, while there was a total of 15 million hectares of cultivated area and 4·9 million farms, which had an average cultivated area per farm of only 3 hectares.[3] Some foreign engineering experts have advised the authorities to pursue a program of rapid tractor-mechanisation, especially on farms of more than 10 hectares. Government policy has in fact favoured tractor-mechanisation. Agricultural capital goods have been substantially exempted from tariffs and taxes and imports have been licensed at artificially low official exchange rates. These measures, coupled with favourable output prices, have led to a considerable increase in the number of privately owned tractors in recent years, particularly in the Punjab. Upon the purchase of these tractors, a substantial and growing number of large landowners have set up management units of 60 hectares or more, and the numbers of hired workers on these farms have been drastically curtailed as a result.

The concentration of the government assisted program of mechanisation on the farms with more than 10 hectares under cultivation was essentially aimed at maximising the rate of growth in the economy. The government was successful in achieving high growth rates during the 1960s. The policy has, however, stimulated the emergence and growth of a subsector of agriculture which is 'progressive', large-scale and capital-intensive, and is quite distinct from the 'traditional' small-scale farming in which the overwhelming majority of the farmers is engaged. In the process, economic power has become concentrated in the hands of an exceedingly small number of people (the so-called

[3] Even if we add the reported cultivable waste of 10 million hectares and assume that water and other essential inputs were forthcoming as required, total land area available per farm would be only a few hectares greater.

twenty families) and the vast majority of farmers have failed to benefit from the growth, or worse have been left to suffer from its adverse consequences (at least until the recent fall of the Ayub regime).

With such a 'bi-modal' pattern of agricultural development, commercial sales of output expand faster than the growth of total output, and unless effective demand increases as rapidly as the farmer, farm commodity prices (mainly foodgrains) must decline below the current favourable level. The consequences of such a price decline are not difficult to imagine. The great bulk of the farmers excluded from the program will face a decline in net income due to the worsened terms of trade for their small marketable surpluses and landless workers will be driven off the land without prospect of absorption in other gainful employment. Large farmers will find that, in order to cut the costs of operation per hectare, their operating area has to be further expanded and their operations made more commercially oriented. Such expansion would take place, in the absence of the effective legal restrictions, so long as mechanical-engineering economies of scale outweigh managerial diseconomies of large-scale operation. The economic logic of this process is simple but hard. Once investments are made on fixed assets (such as tractors), the short-run cost function becomes 'lower' than the long-run cost function. Since fixed costs are foregone in the short-run, they do not affect the short-run level of output. Product prices can fall to the levels that cover only variable costs and not fixed costs, though the impact of such a situation can be disastrous for small farmers whose total costs are largely variable.

The extent to which the pressure for maximum growth creates and enhances the problem of the 'misery rate' (measured by the proportion of the population excluded from and suffering the consequences of the high growth in certain areas and sectors) has become clearer with recent events in Pakistan. In the agricultural sector, the trends are plainly for the large to get larger and the small to get smaller, with the prospect of the rural population becoming polarised into the large, rich farmers, and small, poor farmers. The tragedy will not only be that the small will suffer, but that they will have no positive prospect if the present trend continues, despite an apparently promising future with the development of biological-chemical technology. The undesirable effects of massive grain imports from developed countries (e.g., under P.L. 480 programs) will be reproduced in the domestic scene. The only differences will be a change in the actor who plays the role of surplus exporter and, more important, in the use of increased earnings in the hands of the new large supplier. In Pakistan, the agricultural sector is very lightly taxed, which means that the mobilisation of increased cash incomes of large farmers becomes difficult, if not impossible. Despite their unwelcome effects, the P.L. 480 foodgrain imports did provide Pakistan with counterpart funds which enabled the government to finance a substantial fraction of its development program as well as provide the hungry with needed foodgrains.

## PRODUCTIVITY GROWTH IN PAKISTAN'S AGRICULTURE

Although there have been important differences in the rates and the patterns of productivity growth among the various districts of the region (Table 9.1),[4] and despite the obvious difficulties pertaining to the available data, a general picture of the productivity growth in Pakistan's agriculture emerges rather clearly. During the early 1950s, the influx of refugees, the natural increase in population, and the state of urban/industrial development created a serious employment problem. In spite of increases in planted crop hectarage, the land/man ratio deteriorated during the decade. The annual increase in yield was extremely slow and failed to counterbalance the worsening land/man ratio. What was once the granary of the subcontinent had become a net importer of foodgrains by the mid-1950s.

In contrast to the virtually stagnant rural environment of the 1950s, the agricultural situation took a remarkable turn for the better in the following decade. For the first time the expansion of yields (real output per hectare per year) exceeded the growth in the agricultural labour force and more than compensated for the worsening land/man ratio during the Second Five Year Plan period, 1960-5. This striking reversal of events and the further even better performance recorded in the second half of the decade can be attributed to the following three major factors on the supply side of agricultural production: quantitative and qualitative improvements in water-resource availability; a change in the availability of fertiliser and the farmers' attitudes toward it; and the green revolution (sometimes called the seed-fertiliser revolution). These factors were themselves reinforced by the marked improvement in the cost/benefit ratio of production supported by the agricultural policy.[5] Nevertheless, the encouraging performance during the Second Plan period was essentially a surprise. It was largely unplanned and unnoticed until well under way, and resulted mainly from water-resource development in which private tubewells rather than government efforts were particularly important.

The key element in the agricultural development during the Third Five Year Plan period, 1965-70, lay in the shift of development

[4] It has been observed that the agricultural growth in Pakistan during the latter part of the 1950s and the mid-1960s was the result of rather remarkable rates of growth in a relatively few districts (Gotsch 1968).

[5] In the 1950s and in the early 1960s the terms of trade were very unfavourable for agriculture relative to manufacturing. See, for example, Lewis 1968.

TABLE 9.1

AGRICULTURAL LABOUR FORCE, PLANTED AREA AND GROSS CROP PRODUCTION, PAKISTAN, SELECTED YEARS 1951–69

| *Years* (*averages*) | *Agricultural Labour Force*[a] ('000) *L* | *Area Planted under Crops*[b] (*million hectares*) *A* | *Gross Crop Production*[a] (*Rs million*)[c] *O* | *Compound Annual Rate of Growth* *A/L* | *O/A* |
|---|---|---|---|---|---|
| 1950–2 | 6188 | 9·99 | 4105 | | |
| | | | | 0·2 | −0·2 |
| 1954–6 | 6620 | 10·76 | 4385 | | |
| | | | | −1·1 | 0·6 |
| 1960–2 | 7570 | 11·49 | 4870 | | |
| | | | | −1·2 | 2·9 |
| 1964–5 | 8148 | 11·78 | 5587 | | |
| | | | | 0·9 | 4·8 |
| 1968–9[d] | 8868 | 13·29 | 7600 | | |

[a] Agricultural labour force: computed residually assuming the following constants observed during the two census years (1950/1 and 1960/1): the share of agricultural labour force (1961), 59·3 per cent; compounded growth rate total labour force, 2·99 per cent; that of non-agricultural labour force, 4·58 per cent.

[b] Crops included are: rice, wheat, bajra, jowar, maize, barley, gram, sugarcane, rapeseed, mustard, sesamum, linseed, cotton and tobacco.

[c] In constant 1960–1 prices.

[d] Preliminary estimates.

*Sources:* area planted: Central Statistical Office 1966 and Pakistan, Economic Adviser to the Government, 1969; gross crop production: United States, Agency for International Development 1968 and Pakistan, Economic Adviser to the Government, 1969; the rest: Kaneda 1969.

priorities toward agriculture. Encouraged by the performance during the Second Plan period, and influenced profoundly by the events following the war with India in mid-1965,[6] the government introduced the so-called foodgrain self-sufficiency program in 1966/7 as one of the specific objectives for agriculture in the Third Plan. The government promoted new farming techniques by enhancing farm profits with its

[6] The war with India not only caused dislocation, but also entailed directly (because of an upward shift in defence outlays) and indirectly (because of the termination of United States grants for military imports) substantially higher claims on domestic and foreign resources for military purposes. Furthermore, the available foreign financial assistance fell far short of the expected amounts and its terms became harder, thus severely circumscribing flexibility in the use of resources. In the face of these difficulties, the government had to revise the initial Third Plan considerably. One key element in the revised strategy was the shift in the use of resources away from fixed investment towards current development outlays and other uses that enable higher output from existing productive capacities, particularly in agriculture.

policies for prices of inputs and outputs. For inputs, the most prominent were subsidies on fertilisers, pesticides and water rates. Tax and tariff exemptions for the import of agricultural investment goods and import licenses at an artificially official rate of exchange were also important. For output, the key features of policy were price-support schemes on such commodities as wheat, corn, rice and peanuts. In addition to this favourable market environment, large investments had previously been made in agriculture, particularly in the development of water resources by the public sector and by private enterprise, which offered an opportunity for substantial increases in output. It was in this situation that the availability of high yielding varieties of wheat and rice opened up the possibility of a dramatic breakthrough. Since 1965 the area under crops has itself increased faster than the agricultural labour force, while the annual rate of growth of real output per hectare rose to almost 5 per cent. Increased supplies of water, fertiliser, and the new seeds have combined to produce a dramatic rise in foodgrain production, particularly since the wheat harvest of 1968, with the important consequence that Pakistan has become independent of foodgrain imports and is moving rapidly towards a surplus of wheat and rice.

### TUBEWELL WATER AND CROPPING INTENSITIES

In order to provide additional water for irrigation and to reclaim waterlogged and saline soils, a number of Salinity Control and Reclamation Projects (SCARPs) had been underaken by the Pakistan Water and Power Development Authority well before the beginning of the Second Plan period. Actual construction of the first of these projects (SCARP I) began in 1959 and by 1962, a total of 1980 public tubewells had been completed. Subsequently, by fits and starts, the public tubewell program was extended to SCARP II, SCARP III, and SCARP IV. It was intended, by continuous operation of these tubewells:

1 to eliminate waterlogging and to permit draining by lowering the water-table;

2 to leach the salts by supplying continuity of excess water; and

3 to enable a much higher intensity of cropping by supplying a new source of water (in addition to the surface water).

The results of these public tubewell projects have yet to be fully evaluated. In much of the area covered the actual accomplishment with respect to reclamation has been disappointing, but the investigation leading to these projects did reveal that Pakistan has, in its groundwater, an extensive potential reservoir of additional water. Subject of course, to physical and economic limitations on its use, this

TABLE 9.2

IRRIGATION WATER (FIELD AVAILABILITY) BY SOURCES

| *Year* | *Canals*[7] | *Public Tubewells* | *Private Tubewells* | *Persian Wheel* | *Total* |
|---|---|---|---|---|---|
| 1960 | 55 | 2 | 0·3 | 1·7 | 59 |
| 1965 | 58 | 2·7 | 5·3 | 1·7 | 68 |

vast aquifer of non-saline water lies underneath some 8 million of the 12 million hectares of potentially irrigable land.

From a very modest beginning in the 1950s, the number of privately installed tubewells increased at a striking pace in the 1960s. By 1964, when the first survey of private tubewells was taken, there were 25,000; by 1965 the total installations exceeded 31,000, and in 1967, when the last count was taken, the number had risen to approximately 53,000 (Ghaffar and Clark 1968). On the basis of an estimated annual rate of installation of 8-10,000 tubewells in private hands, the total in 1970/1 may have reached 70,000.

Largely unplanned and unnoticed as it was until 1964, this development considerably increased the field availability of irrigation water. According to estimates made by the International Bank for Reconstruction and Development and International Land Development Consultants (in Bose 1969: 9), the whole Indus Basin received the discharges shown in Table 9.2 from all sources in 1960 and 1965. Private tubewells thus contributed 5 out of 9 million acre feet of additional irrigation water during the Second Plan period. It is safe to say that the relative contribution of private tubewells in supplying additional irrigation water has increased since 1965, for while private tubewells more than doubled their 1965 capacity other sources did not increase their capacities much.

Significant as this quantitative improvement in irrigation water was, the qualitative improvement in the use of water made possible by the tubewells in the farmers' hands was more important. The 'run of the rivers' and the existing canal system was such that the cropping intensity during the summer (*kharif*) months had been kept low, since there was not enough water in the canals during the sowing and maturing periods of additional *kharif* crops.[8] Now the tubewells could

[7] According to Ghulam Mohammad (1964a: 360), an average of 74 million acre feet of water was annually diverted into canals during the five years ending 1956/7. Because of conveyance and application losses, only 34-41 million acre feet of water were available for crop use.

[8] The British engineers who designed the canal system fixed the capacity of most perennial canals (those carrying water the year round) to provide for a cropping intensity of 25-30 per cent in the summer (*kharif*) season and 50-5 per cent in the winter (*rabi*) season. The additional non-perennial canals were designed to carry water only during the summer season to permit about 70 per cent of the gross area to be irrigated solely during the season (Mohammad 1965: 44).

provide additional water from mid-June to mid-September, at the time and in the volume that the farmer required.

It is not surprising, therefore, to find that the area under *kharif* crops increased much more rapidly than the area under *rabi* crops in the perennial canal areas where the installation of tubewells had assured an adequate water supply, and that the largest increase took place in the area under cotton and rice. According to the pioneering study by the late Ghulam Mohammad (1965), additional water made available by the tubewells enabled the farmers to increase cropping intensity (both in *rabi* and *kharif*, but particularly in the latter season), use more fertilisers, change cropping patterns in favour of higher value crops and adopt other practices to raise yields. It was estimated, furthermore, that the output increase resulting from these improvements enabled the farmers to recover the full cost of tubewell installation in a period of two to three years (Mohammad 1965: 44), thus implying extremely high rates of return.

It is important to note, however, that the major impact of additional water from tubewells on annual yields was in its favourable effect on cropping intensity. On the basis of a simple Cobb-Douglas production function study, incorporating a variance-covariance analysis with respect to regional and size variations of the farms in the sample, the author analysed the gross annual outputs of some 125 farms before and after the installation of private tubewells. When the increased area under crops after tubewell installation was explicitly accounted for, the yield effects attributable residually to additional inputs of fertilisers (rather modest per hectare), changes in the cropping pattern, and other increases in the efficiency of production, were quite small, falling in the range of 1-5 per cent of the gross output depending on the farm size (Kaneda and Ghaffar 1970). In the absence of adequate quantitative information on the inputs of capital assets and current inputs for the sampled farms, the conclusion is by no means precise. Nonetheless, the presumption is strong that during the Second Plan period and before the spread of new seeds and greatly increased application of fertilisers, the benefits of supplementary irrigation from tubewells were mainly in increasing the intensity of cropping.

## THE GREEN REVOLUTION

Since the latter half of the Third Plan period the short-stemmed varieties of wheat and rice imported from abroad and the increased use of fertilisers have dramatically increased the output of foodgrains in Pakistan. This recent breakthrough in the agriculture of the Indus Basin is yet another example of the green revolution taking place in various parts of Asia and other less developed tropical countries.

TABLE 9.3

ESTIMATED AREAS IN PAKISTAN COVERED BY IMPROVED VARIETIES, 1965/6 TO 1969/70

| | *1965/6* | *1966/7* | *1967/8* ('000 *hectares*) | *1968/9* | *1969/70* |
|---|---|---|---|---|---|
| Wheat (dwarf) | 5 (0·1)[a] | 101 (2) | 957 (16) | 2384 (39) | 2885 (48) |
| Rice (dwarf) | | less than 0·2 | 4 (0·3) | 350 (25) | 501 (31) |

[a] Figures in parentheses are percentages of total hectarage for each crop.
*Source:* Directorate of Statistics, Agriculture Department, Government of Pakistan in Eckert 1970.

TABLE 9.4

AREA AND OUTPUT OF WHEAT AND RICE, AND FERTILISER USE IN PAKISTAN, 1955/6 TO 1969/70

| *Year* | *Wheat Area* ('000 *hectares*) | *Output* ('000 *tons*) | *Rice Area* ('000 *hectares*) | *Output* ('000 *tons*) | *Fertiliser sales to farmers* ('000 *nutrient tons*) |
|---|---|---|---|---|---|
| Average 1955/6 to 59/60 | 4705 | 3620 | 1074 | 895 | n.a. |
| Average 1960/1 to 64/5 | 4984 | 4087 | 1244 | 1141 | 31[a] |
| 1965/6 | 5155 | 3854 | 1393 | 1296 | 70 |
| 1966/7 | 5344 | 4266 | 1410 | 1343 | 114 |
| 1967/8 | 5983 | 6317 | 1420 | 1475 | 193 |
| 1968/9 | 6160 | 6513 | 1555 | 2000 | 248 |
| 1969/70 | 6056 | 7123 | 1622 | 2372 | n.a. |

[a] The figure is for 1960/1.
*Sources:* wheat and rice: for 1968/9 and 1969/70 Agriculture Department, Government of Pakistan (personal communication). For other years Pakistan, Economic Adviser to the Government 1969. Fertilisers: estimates by Aresvik (1969). The figure for 1960/1 was obtained from Falcon and Gotsch 1968: 282.

Because of the generally favourable conditions in Pakistan in regard to irrigation water (improved since the Second Plan period) and solar energy, and owing to the unusually favourable weather in 1967/8 in particular, the green revolution spread very rapidly in the Indus Basin. It has been a highly successful demonstration of the process of international transmission of agricultural technology through the transfer of scientific knowledge embodied in key inputs such as fertilisers and new high yielding varieties of seeds.

The use of fertilisers in Pakistan had been increasing steadily during the 1960s (except for 1965/6 when there was a setback owing to import reduction following the 1965 war with India). The distribution of locally improved varieties of seeds had also increased through government and private (farmer to farmer) channels, though the quantities of these inputs used were still quite small during the early half of the decade, owing to their limited overall availability. After the major change in the development priorities with the inception of the Third Plan, adoption of the new technology accelerated. This was especially due to the substantial and rapidly rising amounts of foreign exchange made available for the import of fertilisers and other inputs. The statistics compiled by the agricultural economics adviser to the government of Pakistan reveal the adoption patterns of the dwarf varieties of wheat and rice in Pakistan (Table 9.3).

Wheat production reflected this dramatic trend and rose from the previous peak of 4·5 million tons to 6·3 million tons in 1967/8 and further to over 7 million tons in 1969/70. The output of rice (including IRRI varieties) increased from 1·5 million tons in 1967/8 to over 2 million tons in 1969/70 (Table 9.4).

In an illuminating analysis of 'leading inputs' (irrigation and fertiliser) Professor Shigeru Ishikawa (1967: 90-2) observed that for irrigation to facilitate an increase of the cropping intensity and the introduction of new technology (including increased application of fertiliser and use of better seeds), the quality of the existing irrigation facilities must be improved. We find in the development of private tubewells and the green revolution in Pakistan a practical demonstration of this apt observation. There is indeed strong evidence that additional supplies of water from tubewells, and the fact that the farmers have personal control over the pattern of their application, have played the role of catalyst in introducing the new technology. In the first place, the rapid growth in agricultural output during the Second Plan period was, and the spread of the recent green revolution appears to be, limited largely to a few of the agricultural districts of Pakistan where private tubewells are concentrated. In the second place, the recent development has highlighted the problems of the wheat farmers in the *barani* (rainfed) areas, who cannot easily adopt the innovation. These wheat farmers, normally receiving unreliable and poorly distributed rainfall, are left behind by the seed-fertiliser-tubewell combination. Similarly, the farmers in (then) East Pakistan dominated by the constraining physical characteristics of their part of the country, lagged far behind their fortunate counterparts in the West wing before the split (Gotsch 1968).

While the recent increases in the number of tubewells, use of fertilisers and in the area under new varieties of seeds are indeed remarkable, water supply is still the major constraint upon full realisation of the agricultural potential for a large part of Pakistan. In many parts of it fertiliser is still not available at the right time and in the quantity required for the optimal performance of the new varieties. In short, the well-known problems of the traditional type still remain over the majority of Pakistan's territory. In addition, the dawn of the green revolution and the speed with which it has progressed have created a new layer of problems.[9] The problems of a stagnant agriculture were primarily on the supply side; the new problems include an increasingly difficult one of expanding effective demand, investment in storage and marketing facilities, and reformulation of the incentive programs as increased output begins to seek commercial outlets.

### THE RISE OF THE TUBEWELL INDUSTRY

The economic relation between the agricultural and non-agricultural sectors involves an exchange of products, flows of productive factors, and diffusion of ideas. Typically, a less developed economy is fragmented and lacking in the cohesive forces of adequate transportation and communication. Although they are not obvious under these circumstances, the intersectoral flows of products, productive factors, and ideas characterise a two-way relationship between agriculture and industry.

Albert Hirschman alleges that, in a traditional setting of less developed countries, there is no backward linkage from the agricultural sector to manufacturing (1958: 109). Observers of recent developments in Pakistan acknowledge that important relationships exist between agricultural development strategy and industrial growth (Falcon 1967; Johnston and Cownie 1969). A concomitant of rapid growth of the agricultural sector in Pakistan has been the burgeoning of a small-scale engineering industry which supplies key durable good inputs, mainly diesel engines, pumps and strainers but also various farm implements. While the public tubewell program has relied principally on imported equipment, private tubewells have used pumps, diesel engines, coir string strainers and other items of a simple design, produced in machine shops located in various towns in the Punjab. The spectacular growth of this industry offers an important illustration of the interaction between agriculture's demand for inputs and the generation of output and employment in domestic manufacturing.[10]

[9] See for example Kaneda 1969.

[10] In collaboration with a colleague, the author is now in the process of completing a detailed analysis of this industry on the basis of a sample survey conducted in 1969. It is hoped that the study of this particular example will provide further understanding of intersectoral relationships and develop new insights into the subject of linkages. What follows is but a preliminary report of the unfinished analysis.

As with the case of private tubewell installations during the Second Plan period, the rise of this industry was a surprise. It has developed spontaneously without subsidies, tax concessions, special credit arrangements, or even recognition by official agencies. The industry, furthermore, is truly small-scale: well over half of the estimated 600 firms which make it up employ fewer than ten production workers.[11] Finally, this industry has been a vehicle for marshalling the indigenous minor savings and investible funds, for development of entrepreneurial talent and management skills, for training of skilled and semi-skilled labour, and for application of new technology.

Lahore, Daska, and Gujranwala are the centres of tubewell related products, whereas Lyallpur is the largest manufacturing centre of agricultural implements other than tubewell equipment. Lahore has a relatively large number of diesel engine producers of comparatively large size, while the smaller firms there are mostly strainer manufacturers. Daska is another diesel engine manufacturing centre but the size of firm is smaller than in Lahore. All these towns share the characteristic of having shops of a similar size existing side by side, producing the same products, and competing for farm customers. There are however some notable geographical variations among them, attributable partly to historical accident (e.g., to patterns of migration of the forerunners of the industry from India) and partly to the cropping patterns of the farm areas in proximity. In this sense, Lyallpur, with its

[11] At the time of the survey the interviewers were asked to prepare separately a list of an estimated number of establishments classified by major products, and by size of the firm (in terms of the total number of production workers), for each town they visited. The interviewers made physical counts of the establishments and verified their counts with inquiries from the owners and managers they interviewed. Classified with respect to the major products, the estimated number of firms in the seven towns was in April 1969:

| City | Diesel Engines | Pumps | Strainers | Others |
|---|---|---|---|---|
| Lahore | 115 | 18 | 15 | 22 |
| Daska | 150 | 3 | 7 | 7 |
| Multan | 1 | 24 | 30 | 5 |
| Lyallpur | — | — | — | 57 |
| Gujranwala | 29 | 11 | 5 | 4 |
| Sahiwal | 7 | — | 5 | 4 |
| Gujrat | 3 | — | — | 11 |

The total estimated number was 533 in the cities where the interviewers visited. Of course, it was not an attempt at enumerating all the firms in all the localities in the Punjab, nor was it a complete enumeration of all the establishments in the industry in each of the cities visited.

apparent lack of tubewell equipment producers, is a good example.[12] Overall, the observation made by Falcon and Gotsch (1968: 276) in the mid-1960s remains appropriate: 'Whole streets in such cities as Multan, Lyallpur, Lahore, Gujranwala, Sialkot, and Daska have been devoted to the manufacture of pumps and engines, and the skill, ingenuity, and training demonstrated in these shops have been impressive.'

Regardless of product specialisation and geographic location, the firms in this industry were invariably privately owned, by individuals, families and partnerships. There was no public ownership, nor were there any incorporated enterprises among the firms surveyed.[13] Twenty five per cent of the survey firms were individually owned; 63 per cent owned by families; 12 per cent owned by partnerships. This is not surprising when it is realised that virtually all of the initial capital was derived from personal and/or family savings. Some twenty eight firms, i.e. only 16 per cent of those surveyed, reported loans in their initial capital, of which only eleven relied exclusively on loans. The overwhelming majority (twenty five) of the loans of these twenty eight firms came from 'family' sources. Moreover, the average loan amounted to about Rs2000, which in most cases comprised only about 25 per cent of the initial capital investment. It should also be noted that *bona fide* loans, with interest and terms specified were rare, and there was no borrowing from the government or government agencies.

Analysis of the age distribution of firms in the industry by major product categories revealed that single-product diesel engine firms were in general on the younger end of the age distribution, with over half of them established after 1965. In contrast, more than half of the firms producing other tubewell equipment and/or agricultural implements as well as diesel engines (i.e., multi-product diesel engine firms) were in operation before 1960. Nonetheless, the unprecedented, rapid increase in the number of these diesel engine firms and the firms producing other tubewell equipment during the late 1950s and the 1960s was the *sine qua non* of the rapid growth in private tubewell installations during the last decade.

The modal value of total fixed capital (excluding land and buildings) of small single-product diesel engine firms was in the neighbour-

[12] The largest concentration of shops producing fodder choppers, sugarcane crushers, and other relatively simple agricultural implements, is found in Lyallpur, which is also the seat of the Pakistan Agricultural University and the Ayub Agricultural Research Institute. It appears to the casual observer that the engineering sophistication required (and demonstrated) by diesel engine producers (such as those in Daska and in Lahore) is somewhat greater than that exhibited by simpler implement producers in Lyallpur.

[13] Some 180 of the estimated 600 firms were surveyed and efforts were made to contact almost all of the largest firms in the industry.

hood of Rs15,000 with three to four employed production workers. Typically they had two lathes, a drill press and perhaps a planer. These firms, for instance in Daska, might buy castings from larger neighbouring producers with fully integrated facilities (from casting to assembly). More often however, castings and especially forged steel crank shafts, along with other essential material inputs such as copper tubes, sheet metal, ball bearings and other precision parts, would come from suppliers in Lahore.

One interesting aspect of this industry was its capacity to train its workers from the available local stock. The overwhelming majority of workers were recruited from the immediate local area. They were trained gradually through the process of learning-by-doing on the job. There was no input observed of graduates from the government sponsored technical institutes. While most firms trained their own workers or recruited skilled workers from other firms in the industry, some appeared to rely on inter-industry shifts of workers skilled in operating machine tools. Despite the lack of formal technical training, some skilled workers seemed to attain a quite impressive level of engineering proficiency, e.g. in redesigning diesel engines and developing new products. With regard to the latter, there is a strong presumption that the training and the engineering skill developed in the industry has 'spilled over' to the expanding consumer goods industry.

In relation to diesel engines the 'engineers' in this industry have demonstrated not only an ability to copy anything that can be dismantled and studied in detail but also a capacity to adapt design to local requirements. The engines commonly produced in the Punjab are said to have been derived from the horizontal Rustin engine of the 1920 vintage produced in England. Several modifications have been performed on the original model in order to adapt it to the local conditions, of which the most important was the reduction in the engine speed to permit long hours of operation in the hot climate. The engine itself remained bulky and heavy owing in part to the need to allow for the low quality of castings produced by the rather primitive backyard furnaces. Nonetheless, under the pressure of increasing raw material prices and the sheer shortage of iron and cokes for castings, some firms have begun to experiment with lighter, high-speed, vertical engines derived from West German prototypes. Furthermore, some firms have recently branched into the production of stationary wheat threshers (which have incorporated engineering improvements each year), and even tractor parts, on the basis essentially of the part-time efforts of skilled workers and foremen.[14]

[14] From the point of view of the farmers installing tubewells, the diesel engine tubewells cost more in initial costs and operating expenses than electrical installa-

Import liberalisation and relatively liberal foreign aid on the commodity aid basis during the Second Plan period, increased the availability of raw materials for tubewells at lower prices, and encouraged the production of tubewell equipment by private entrepreneurs. However, restrictions on imports and the foreign aid have been tightened since the early Third Plan period. It was not surprising, therefore, to hear an almost unanimous complaint of the shortage of raw materials in recent years. With such critical inputs as pig iron and coking coal in short supply (secured mainly on barter deals), under-utilisation of capacity appears to be serious for small and medium sized diesel engine firms that lack import licenses. Faced by this raw material bottleneck, the industry seems to have arrived at the point where a certain change in structure is inevitable. Although there is no indication of mergers having taken place among the firms surveyed, those with favoured access to raw materials (if nothing else) would have a definite advantage in consolidating their position, while the remainder must continue to struggle for survival in increasingly adverse conditions, or disappear from the scene. It is in the light of this recent reality that the question of an appropriate strategy for agricultural development becomes relevant to the small-scale, engineering industry in the Punjab.

## SOME BASIC CONSIDERATIONS FOR THE ADOPTION OF AGRICULTURAL TECHNOLOGY

Recent agricultural growth in Pakistan has been the result of improvements in the production relationship in the sector. However, in discussing the ways in which the country may sustain the recent growth performance one cannot neglect the relationship between land and man in agriculture and that between agricultural production and other economic activities for these have a determining influence upon the impact of agricultural growth. The contrast between the two directions of mechanisation in Pakistan's recent agricultural history amply illustrates their signficance. While both power irrigation and tractor mechanisation provided opportunities for strengthening the linkage between agriculture and industry, their respective influences on labour

---

tions. Nonetheless the farmers expressed preference for diesel tubewells because: availability of power was still limited and quite unreliable; it was difficult to obtain a sanction for an electrical connection; and there were a variety of administrative constraints to installation of private tubewells, though less onerous than in the case of diesel tubewells. One question in the survey of firms was prompted by the possibility that these phenomena were temporary. When asked what they would do if electricity became more abundant and reliable, almost all of the respondents knew the facts, some of them disputed the assumption, and almost all replied that they would be able to produce electric motors.

employment in agriculture were opposite. By enabling an increase in cropping intensity, the growth of power irrigation contributed to a fuller utilisation of labour resources in agriculture and required little socio-economic adjustment in the sector. Tractor mechanisation, on the other hand, led to significant labour displacement and thus added to the already critical problem of employment in agriculture and elsewhere in the economy. It is as well to note too that neither choice of techniques nor the phasing in of new technologies can be determined solely upon engineering logic, however attractive it might be. Technical data must be subjected to economic analysis which takes into account the production relationship, the man/land relation in agriculture and the linkage between agriculture and industry.

Mechanical-engineering technology in agriculture can embrace a wide range of different configurations involving motive power, machines and implements, and there are many ways of classifying it. From our point of view however, it is important to recognise that mechanisation can be applied to each specific farm operation and to several or all of the farm operations; and that mechanisation as introduced into a given socio-economic organisation of agriculture may call for minor adjustments or it may require highly sophisticated levels of organisation and co-operation not easily attained in a given situation.

The first distinction recognises the many different operations related to growing and harvesting of crops, such as levelling of land, irrigation/drainage, ploughing, seed bed preparation, planting and fertilising, crop protection, harvesting, and preparation of products. Machines and implements can be designed specifically for a limited number, for many or for all of these operations. Aside from implements relying on human or animal power, tubewells, low-lift pumps and power threshers, Japanese power tillers, dusters and sprayers are examples of machines designed specifically for limited tasks. On the other hand, the development of the tractor and so-called power take-off, together with the development of various implements, make possible the more direct and widespread application of engine power to many different operations. In the United States, all cultivation processes came under the aegis of mechanical power; and harvesting processes and many post-harvest operations became thoroughly mechanised with the advent of the self-propelled combine harvester. Nevertheless, it should be born in mind that a tractor is by no means the only way to mechanise agriculture.

Second, the increase in size and specialisation of farm has been one of the most significant changes in the farm structure and organisation associated with the adoption of tractor mechanisation in the United States. Modern equipment and machines are so expensive in many instances that it is advantageous for the farmer to develop larger farms and enterprises (i.e., specialise) to make full use of the new resources and to hold down the unit costs. The typical relationship between area and unit costs is a curve that shows a steep decline in costs until utilisation reaches about one fourth to one half the maximum possible with the machine, and thereafter a very moderate reduction takes place in costs with greater use. High costs per hectare or per hour with limited use reflect of course, the fixed charges, unrelated to the actual use of the machine. Introduction of larger machines and implements, therefore, necessitates large management units, output standardisation, and uniform cultural practices, which call for highly sophisticated organisation and co-operation.

Biological and chemical innovations that have characterised the agricultural development of Japan, and which have brought forth the green revolution, are neutral to the scale of farm operation. They can, therefore, be incorporated into the existing institutional framework of agriculture in less developed countries without drastic adjustments. Small-scale peasant farms can adopt these innovations with relatively minor adjustments in contrast to technical innovations involving tractors and combines. Undoubtedly, this aspect of recent innovations has contributed to the very rapid diffusion of technology in the Punjab (as elsewhere), where the benefits of such innovations have not been limited to a few large farmers.

Tubewell water, new seeds and increased application of fertilisers, which dramatise the green revolution on the input side, are basically complementary to the farm resources of labour and land. By making it possible to grow more crops, more lucratively, per unit of cultivated area, these inputs have increased the use of labour on farms as well as the incomes of farm workers concerned. In contrast to investments in tractors and combines that are fundamentally labour-displacing, investments in inputs such as seeds, fertilisers and even tubewells are much more conducive to augmenting the incomes of the bulk of a nation's farmers. Emphasis on these inputs, furthermore, would retard the polarisation of the rural population and land holdings, and make it possible for the majority of farmers to achieve satisfactory increases in output at modest costs in terms of scarce resources of capital and foreign exchange.[15]

## CONCLUSION

It should be clear from the foregoing argument that this chapter emphasises the virtue of biological-chemical technology in the economic context of a country such as Pakistan, but not to the exclusion

[15] For a detailed analysis of these points see Kaneda 1969.

of all forms of mechanisation in agriculture. Capital investments in agriculture should most appropriately be in the form of small-scale, divisible, easily adaptable units to remove labour bottlenecks, to involve the bulk of the nation's farmers, and to generate employment opportunities in agriculture. It is argued, therefore, that serious attention should be paid to the design of machines for selected farm operations, such as threshers, winnowers and sprayers as well as tube-well equipment. The strategy of emphasising capital inputs of this nature will generate effective demand for the final products of the domestic, small-scale, engineering industry. As it is important to have consistency and compatibility in industrial planning, it is important to recognise and exploit positive intersectoral relationships of this kind that promote rapid economic development. As the growth of the tubewell industry illustrated, emphasis on selective mechanisation could nurture indigenous industries supplying the agricultural sector of the country. The development of such industries could in turn enable farmers to acquire the equipment on increasingly favourable terms, encouraging further use of such inputs. It would be a mistake to minimise this type of positive interaction between agricultural and industrial development in favour of using large amounts of capital and foreign exchange for the encouragement of tractor/combine mechanisation.

This chapter has traced recent developments in rural Pakistan, focusing particularly on the development of supplementary irrigation by tubewells since the Second Plan period, on the more recent green revolution, and on the demand for output of the domestic manufacturing sector arising from agricultural growth and the green revolution. With the benefit of hindsight, some successful results might be attributable to the conscious policy of the government, but much more has to be credited to the spontaneous, initially unnoticed developments begun and sustained by farmers and entrepreneurs. The perennial question is, here again, whether the government has the capacity to maintain the initiative, and the institutional flexibility to play a leading role in the crucial phases of agricultural development.

# 10

# The New Agricultural Strategy in Uttar Pradesh, India, 1968-70: Technical Problems

*Elizabeth Whitcombe*

> Most certainly the question, are we making the most out of the land, must be answered in the negative . . . because the land itself yields nothing like what it should (Hume 1879).

FORMING his conclusions thus in 1879 on the state of Indian agriculture, Allan Octavian Hume—the founder of the Indian National Congress—drew not merely on his nine years' experience as secretary to British India's newly-formed Department of Agriculture. Hume had also been in the revenue administration, as magistrate and collector for several districts, most notably Etawah, in the North-Western Provinces which, with Oudh, formed the United Provinces, now Uttar Pradesh (Figure 10.1). In the course of this career he had accumulated a store of observations on agrarian principles and practices of the Doab. It was clear to Hume that the primary fault of persistently, and increasingly, poor agricultural performance was not to be laid at the farmers' doors. On the contrary,

> . . . it is due to the patent, frugal and not unintelligent husbandmen of India to admit freely that, looking to the conditions under which they labour, their ignorance of scientific method, their want of capital (and all that capital enables a farmer to command), the crops they do produce are, on the whole, surprising . . . (Hume 1879: 4).

The root cause lay rather in the state to which the soil had been reduced, in the fact that 'agriculture in India has become, and becomes daily, more and more, what Liebig happily designated a system of spoliation' (Hume 1879: 24). Low yields were the index of a declining productivity, for which not only the cause, but also the remedy, was immediately clear:

> . . . with proper manuring and proper tillage, every acre, broadly speaking, of land in the country can be made to yield 30, 50, 70% (according to circumstances) more of every kind of crop than it at present produces, and this with a fully corresponding increase in the profits of cultivation . . . (Hume 1879: 20).

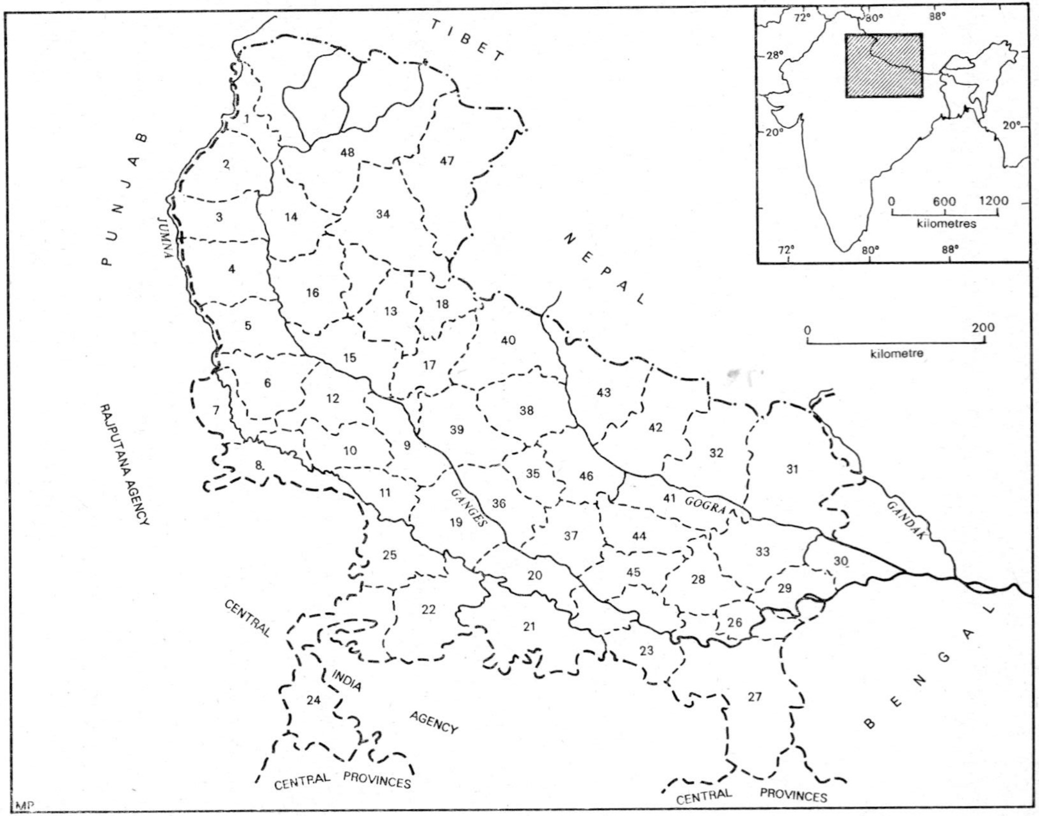


Figure 10.1 The United Provinces of Agra and Oudh. Districts are:

1 Dehra Dun
2 Saharanpur
3 Muzaffarangar
4 Meerut
5 Bulandshahr
6 Aligarh
7 Muttra
8 Agra
9 Farrukhabad
10 Mainpuri
11 Etawah
12 Etah
13 Bareilly
14 Bijnor
15 Budaun
16 Moradabad
17 Shahjahanpur
18 Pilibhit
19 Cawnpore
20 Fatehpur
21 Banda
22 Hamirpur
23 Allahabad
24 Jhansi
25 Jalaun
26 Benares
27 Mirzapur
28 Jaunpur
29 Ghazipur
30 Ballia
31 Gorakhpur
32 Basti
33 Azamgarh
34 Naini Tal
35 Lucknow
36 Unao
37 Rae Bareli
38 Sitapur
39 Hardoi
40 Kheri
41 Fyzabad
42 Gonda
43 Bahraich
44 Sultanpur
45 Partabgarh
46 Bara Banki
47 Almora
48 Garhwal

This was the essential beginning. Other improvements were necessarily to follow in co-ordination: the arrest of the processes of salination and the reclamation of the already vast tracts of salt land; judicious, small-scale mechanisation; the relief of peasant indebtedness and the provision of comprehensive public credit for agricultural purposes.

Hume's recommendations for the urgent improvement of agrarian conditions had been repeatedly shelved because of the opposition and because of the prior concerns of the Department of Agriculture's two senior partners, Revenue and Commerce. In the North-Western Provinces, large-scale irrigation schemes progressed throughout the second half of the nineteenth century to cover all but the eastern Doab with a network of canals, stimulating the cultivation of soil-exhausting commercial crops, indigo, cotton, sugarcane, wheat. The provision of nutriment for the soil, in the form of organic or inorganic fertiliser was, however, nowhere organised in co-ordination.

Irrigation was the preserve of the Canal Department, the senior technical service of Government; Agriculture, a newcomer to the bureaucratic scene, never had the means to carry out schemes for comprehensive improvement which fell within its purview. Proposals for the relief of scarcity in organic manures by the selection of areas for fuel and fodder reserves were rejected by the revenue administration as impractical. The manufacture of chemical fertilisers from the saline-alkali deposits in central and east districts of both the North-Western Provinces and Oudh (a symptom of severe ecological malaise) was effectively prevented by the Salt Tax.

During the first decades of the twentieth century, agricultural developments tended to follow rigidly the lines laid down by their nineteenth century antecedents: the building of the Sarda Canal, the formation of the sugarcane agro-industry around the staple commercial crop of the Provinces, small-scale improvements in agricultural machinery and varietal experiments (chiefly on export staples). By the early 1930s, a significant advance had been achieved in what is now known as minor irrigation with Sir William Stamp's pioneering tube-wells, first developed in the field in parts of the (highly productive) Meerut Division. The application of chemical fertiliser, in the form of (imported) ammonium sulphate, had barely passed the experimental stage. It was not until 1946/7, when A. B. Stewart conducted his balanced dressing experiments with phosphatic and potassic as well as nitrogenous fertiliser on selected cultivators' fields in Kanpur district, that the essential chemical components for the improvement of the fertility status of the soil were established.

It remained to co-ordinate these with irrigation and to develop at the same time crop strains which would ensure higher and better quality yields. The 1950s saw considerable activity along these lines in the established agricultural institutions, most notably in Kanpur College, now the Institute of Agricultural Sciences. But, as in Hume's day so in the days of his successors in the Congress: the selection of priorities reinforced by budgetary restrictions prevented both large-scale experimentation with, and adequate diffusion of, improved techniques.

This position was dramatically changed by the end of the decade. Intensive Agricultural District and Area Development Programmes for the 1960s, heralded by the establishment of agricultural universities on the model of United States land-grant colleges with sister relationships to their United States counterparts, marked the reversal of priorities for planning for industry in favour of agriculture. These schemes, based on an all-round 'package' of practices (fertiliser, irrigation, plant protection) have culminated in the High Yielding Varieties Programme, in operation in Uttar Pradesh since 1966/7, and based on the dissemination of exotic dwarf wheats and rices. Here at last were the co-ordinated improvements for soil and for seed required for planned agricultural developments.

The selection of the Tarai and, more precisely, of the 1214-hectare government farm near Rudrapur as the site for the Uttar Pradesh Agricultural University and as the focal point for the dissemination of the new intensive agriculture programs in co-ordination with the Indian Council for Agricultural Research and the State Department of Agriculture was apposite. The fertility status of Tarai soil, newly reclaimed (between 1946 and 1952) from jungle and malarial swamp, is the highest in the State. The groundwater reservoir of the region, fed directly through the gravel beds of the Bhabar tract to the north from run-off from the Siwaliks and ultimately the Himalayas, has been assumed to be virtually inexhaustible. The high water-retention capacity of the soil is a function of its rich clay-mineral component. These conditions have been fully exploited by the new agricultural strategy, in a consistent drive to maximise high yielding hectarage and high yields in the region.

Investment in agriculture has tended, predictably enough, to flow most freely where risks are least. The typical Tarai farm, of a minimum of more than a hundred hectares under co-operative registration, represents in its combination of abundant natural resources and private capital the ideal in creditworthiness. This is appreciated not merely by the Reserve Bank's Agricultural Refinance Corporation, the State Bank of India, and numerous commercial banks but also by the World Bank, as exemplified by its assistance to the newly-formed Tarai Development Corporation. Agricultural performance in the Tarai to date has justified expectations. By following recommended monocultural practices to maximise potential productivity for wheat, viz. regular applications of 157-179 kilograms of nitrogenous fertiliser per hectare, plus 90-112 kilograms per hectare each of superphosphate and muriate of potash, and five, six or even seven regular irrigations at precise, optimum timings in the crop life cycle, objectives of aggregate rice and wheat yields of 10 tons per hectare over a single, two-harvest agricultural year are not unreasonable.

How far can this pattern be followed in the greater part of the 19·4 million hectares which make up the cultivated area of Uttar Pradesh? How adaptable is the new pattern of monoculture to the complex disparities which form the present-day agro-geomorphology of the alluvial plains? A more modest response to the stimulus of the new agricultural strategy is to be found in the northwestern uplands of the Meerut and upper Rohilkhand divisions: the home of Uttar Pradesh's agricultural aristocracy (as distinct from the pioneering nouveaux riches of the Tarai). Uplands-Meerut cultivation and Jat cultivators were a byword for agricultural prosperity in Moghul times; their sugarcane and wheat production aroused the admiration of the first British administrators.

Large-scale canal irrigation, an innovation which became the basis of modern Indian agriculture, was first introduced into the Meerut districts in the early 1830s (the East Jumna canal system). The object of irrigation, then as now, was to maximise production and its profits; the stimulus it provided to the cultivation of sugarcane and wheat in the region satisfied that objective. The tradition of such cultivation, in which proportions under each crop vary in accordance with market conditions, has persisted to the present day, with concomitant facilities: abundant irrigation from both canals and tubewells; public credit distributed through 'cane' as well as regular co-operative societies; a competitive market for sugarcane ranging upwards from village-level refineries to sugar factories; a proportionately greater mileage of tarred and concrete-track roads than in other Uttar Pradesh districts; rubber-tyred 'Dunlop' carts. Fundamentally, these uplands regions have a distinct geomorphological advantage: natural drainage is for the most part adequate, and the permeability coefficient of the soil is such as to ensure downward percolation of moisture at a fairly uniform rate.

To the south of the Meerut division, where the uplands proper merge with the central and lower Doab drainage basin, the dominant problems of the Uttar Pradesh plains become increasingly evident. In south Meerut and southwest Bulandshahr the general gradient of the land slopes off, in the central Doab, to less than 0·3 metres per mile, with long shallow depressions, at irregular intervals. During the *rabi* (winter-spring) season, dark green Mexican wheat fields stand out against ragged patches of grey-white, waterlogged and saline soil.

These latter, the product of defective and impeded drainage, vary in size from a few square metres to vast tracts of 400 or more hectares, barely covered at intervals with stunted, vestigial monsoon grasses. From the northwest through to the southeast, the proportions of green to grey-white indicate the extremes in fertility: the ratio of highly productive to severely deteriorated land. In the mid-Doab basin, the central tracts of Aligarh, Etah, Mathura, Mainpuri, Farukhabad, Etawah, Kanpur, Unao and Fatehpur districts, the indices of soil disease intensify (Etah and Mainpuri alone have each between 90,000 and 122,000 saline-alkali hectares) and the contrasts of rich against poor land become even more pronounced.

Farukhabad, for example, boasts of the most intensively cultivated area in India. It has a market gardening tract, on the lip of the Ganges uplands some 3·2 kilometres in radius from the cities of Farukhabad-Fategarh, which produces a minimum of three, frequently four, and in some fields five garden crops per year (rotations of potatoes, tobacco, vegetables, maize and wheat). Beyond it lie vast grey expanses dotted with clumps of cereal crops where irrigation by private tubewells, installed by mortgage of land on the security or with the profits of intensively cultivated parts of a holding, is making it increasingly possible to wash down the toxic accumulations of salts from the root zone to an extent where paddy tolerant of moderate salinity can be cultivated. Such marginal reclamation is characteristic of land on the fringe of the central drainage basin. The vast saline tracts in the heart of the basin, moving east and south from Farukhabad, present a more complex problem. Vertical drainage is frequently impeded by the presence of an indurated layer or hardpan at varying depths (sometimes as close as 0·5-1 metre from the surface). The groundwater of upper aquifers may itself be alkaline, which prevents its use as a reclamative agent (for leaching) and restricts its irrigation utility to alkaline-tolerant crops: marginally, tobacco and certain varieties of paddy. Mathura district, for example, particularly in the west is severely afflicted by brackish groundwater.

These problems proliferate in the eastern regions of the State, between the Ganges and the Gogra rivers, where in addition, the supply of irrigation water is severely limited both by the intermittent occurrence of impervious strata in the soil profile, which can make well and tubewell boring difficult and expensive, or even impossible, and by the fact that these areas are served only by the tail-ends of the Sarda Canal system and small-scale pump canals from the Ganges. Such canals barely cope with the demands of the prosperous pockets, particularly as these rise steadily with the progress of the new agricultural strategy. In the central portion of this Ganges-Gogra Doab, alkalinity is manifest both in vast surface expanses and in intensity. For parts of Azamgarh and Ghazipur Districts pH readings of 11-12 (9-9·5 is already severely toxic) are recorded.

Beyond the Gogra, to the north, salinity is virtually absent. The problem of waterlogging occurs there as a consequence of heavy clay soils of low permeability and high water-retention capacity (one-crop-per-year soils), and of inundations in areas adjacent to the great shallow bed, braided rivers of the north.

The south of the State—the other extreme—is as dry as the northeast is wet. The characteristic light soils and surface rock formations of Bundelkhand are accompanied by impervious strata at intervals through the profile, and by irregularly distributed, often confined, aquifers which, if they can be tapped, yield very limited amounts of groundwater. Traditionally, agriculture in the region was a form of shifting cultivation for the most part, fed in the cool season off rainwater stored in tanks or in pockets of thick black clay soil. Land was sown once a year with a mixture of winter cereals and pulses when the rainwater had seeped below surface level.

In this environmental pattern of both large-scale and minute disparities in fundamental soil and water conditions, ecological problems increase in frequency and magnitude as one descends from the northwest to the southeast. At the furthest extremities to the east and south of Uttar Pradesh, these problems are seen in their most severe form: in sterile soil and unusable or inaccessible water. Vast tracts are affected; approximately two thirds of district Azamgarh, for example, is significantly to severely alkaline. These areas are ringed by patchy cultivation, indicating that the condition is spreading. At the same time scarce resources for agricultural improvement are necessarily being concentrated on relatively confined pockets of prosperity, in keeping with the State level irrigation-intensive agricultural strategy. These are principally sugarcane and wheat areas, and secondly good, irrigated uplands soils sustaining millets or paddy and wheat rotations.

Other, predominantly institutional problems have tended to follow the geomorphological pattern. The impact of population growth, for example, is most severely evident in the declining areas of east Uttar Pradesh. Densities of more than 370 persons per square kilometre were recorded in Ghazipur, Azamgarh and Ballia tehsils by the 1961 census. The combination of land reforms (in reality redefinition rather than redistribution of proprietary title) and mortgage laws have channelled recent investment in agriculture into immediately productive areas, from which a tiny minority of the land and the population dependent on it can benefit. Disproportionate attention and benefit in favour of northwest as against southeast on the one hand, and, on the other, of productive as against problem areas within individual districts is striking even from the most casual observation.

This observation itself becomes more meaningful if we inquire into the cause of the physical disparities. They are not so much the indices of enterprise as against backwardness, but the result of geomorphological processes exacerbated by the introduction of public works in the nineteenth century. They are not therefore a condition of stagnation, betraying a lack of capacity for change, but the dynamic consequences of modernisation. For the processes set more rapidly in train in the nineteenth century persist with their intensified momentum.

The deforestation which accompanied the construction of the railway networks throughout Uttar Pradesh over the greater part of the century (intensifying from *c.*1860), stripped off greater quantities of alkaline-tolerant, water-absorbent *dhak* jungle more rapidly than the slow expansion of agriculture achieved. The resultant increase in rates of evaporation and subsoil capillarity provided sufficient preconditions for the accumulation of alkali salts in the upper layers of the soil. Erosion (leading to loss of topsoil and exposure of nobbly calcareous rubble) started and became particularly severe in undulating Bundelkhand and along the myriads of river catchments of minor tributaries of the Ganges which intersect the eastern districts. Obstruction to natural drainage lines by road and railway embankments insufficiently provided with culverts increased the susceptibility of low-lying tracts to waterlogging. But the most complex and far-reaching distortions arose from the introduction of the essence of agricultural modernisation, large-scale irrigation by canals.

The construction of the great canal systems tapping the Jumna and Ganges (on the model of smaller systems developed in Italy, France and, to a much lesser extent, England) from the 1830s and 1840s was promoted according to the principle of maximisation of production. Areas of guaranteed and potential high productivity (i.e. the northwest) suggested, by reason of their geographical advantages and celebrated agriculture, their selection as the recipients of large-scale investment which promised an immediate return. The introduction of canal irrigation marks the beginning of modern Indian agriculture. It brought in its wake northern India's first green revolution.

In the canal tracts of the Meerut and Agra divisions, crop patterns were substantially changed in the years immediately following the opening of canal branches. The old patterns of small-scale, well irrigated 'polyculture' gave way to block cropping of *kharif* (autumn) commercial crops (indigo and cotton), sugarcane and the *rabi* export staple, wheat. Millets and pulses were partially displaced and the wells themselves which had regulated, by labour intensive methods, the supply of water to the soil, were superseded and in many cases demolished by the inroads of the canals. The best tracts of central northwest uplands could withstand most of the pressures exerted by the sudden change, though the danger of overcropping was early manifest (in the 1860s) and the lack of restoration of nutriment to the hungry soil became a matter of public debate in consequence.

Further to the southeast, in the great central region of the Doab described above, the consequences of canal irrigation in terms of a radical distortion of the delicate mechanism of the hydrological cycle were early observed. Embankment obstruction in low-lying areas combined with lateral seepage through the (unlined) canal walls, while flush irrigation at the surface flooded the soil with more water than could be drained horizontally or vertically. Evaporation rates, heightened by deforestation, drew up sub-soil water in which alkali salts, previously dispersed through the profile, rose in solution to accumulate in toxic quantities in the root zone and, in their dehydrated crystalline forms, in the saline efflorescence on the surface (*reh*). In 1879, a year after the first governmental inquiry into salinity in the North-Western Provinces and Punjab and twenty years after the first detailed reports accompanied by chemical analyses of saline deteriorations on lands irrigated by the Western Jumna Canal, A. O. Hume spoke in no uncertain terms of the progressive problems of canal tracts:

> At first the result may be good, and marvellous are the crops that have been raised in the Doab on the first introduction of canal irrigation, owing to the first slender doses of potash and chloride of sodium . . . Time passes on, some crops begin to be unprofitable; in the hottest time of the year, a glimmer as though of a hoar frost overspreads the land. The land grows worse and worse, but every night and day nature works slowly on, and the time comes when, abandoned by the cultivator, the land glitters white and waste as though thickly strewn with crisp, new-fallen snow; never alas! to melt away, except under the rays of science.
>
> Along the little old Western Jumna Canal, thousands of fields are to be seen thus sterilized. Along the course of the mighty Ganges Canal—a work as it were but of yesterday (opened, 1854)—the dreary wintry-looking rime is already in many places creeping over the soil.
>
> Come it quickly or come it slowly, the ultimate result here is also certain; and, unless a radical change is effected in existing arrangements, we know, as definitely as we know that the sun will rise tomorrow, that the time must come when some of the richest arable tracts in Northern India will have become howling saline deserts . . . (1879: 25).

By 1891, the first agricultural chemist to tour India reported some 0·8-1·2 million hectares of saline soil in the northwest alone. Only the most peripheral experiments were done in the interests of organised reclamation. Canal development, however, continued. In 1926, the first branches of the Sarda Canal, slightly modified from the original 1872-8 design were opened to serve eastern Rohilkhand, central, southern and fringe-eastern Oudh. Saline-alkali tracts throughout modern Uttar Pradesh may be conservatively estimated to cover between 2·4-2·8 million hectares (Whitcombe, 1971).

Uttar Pradesh's second green revolution has proceeded on principles not dissimilar to the first: maximisation of guaranteed and immediate productivity by means of irrigation-intensive techniques. One might interpolate here that the fertiliser factor, so prominent in the new agricultural strategy, surely marks a fundamental break with the past. Is this really so, however, if we consider that the recommended fertiliser inputs are those demanded by the new, high yielding varieties for immediate growth? One may query any long-term benefit to the soil, especially when it is remembered that soils in high temperature zones have a chronic incapacity to store nitrogen. Substantial amendment of declining fertility in the soil, for example, in areas bordering the intensely productive tracts and marginally affected by salt accumulation or erosion, can hardly be achieved by inorganic fertiliser. Abandonment and, in many instances known to the writer, official discouragement of the use of organic manures raises the question, how the mechanical properties of alluvium are to be sustained, and their water retentiveness strengthened. Further, the range of chemical treatments available through public supply channels include no neutralising agents, e.g. calcium sulphate (gypsum), for the rectification of salinity and alkalinity.

The crucial element of the new planned strategy, as of the old pattern of piecemeal modernisation, is irrigation. The new, high yielding varieties, exotic paddy and wheat, and the hybrid maizes and millets which form subsidiary programs, are thirsty. The cost of their adaptation to the soils of the high temperature zones of India is not merely to be measured in multiples of the recommended doses of N/P/K fertiliser (at the current rate of 134/90/90 kilograms per hectare), but in the amount of irrigation water required both to dissolve fertiliser for plant uptake and to sustain, by regular waterings at precise points in the life cycle, the amount of vegetative growth of the high yielding varieties. This, by virtue of the fact that their capacity to absorb nitrogen is at least double that of local and improved indigenous strains, means a water demand at least double that of indigenous and local improved cereals. In the case of wheat, five to six irrigations are recommended, as against two to three for local varieties.

The first fertiliser recommendations, calculated not on the basis of optimal requirement but of estimated available supply (in 1965/66), were pitched at the modest level of 67/45/45 kilograms per hectare

N/P/K. By 1969/70 this had doubled, with the expansion of fertiliser manufacture to allow more precise tailoring of recommendations. But whilst it is generally admitted that the optimal doses in relation to varying soil conditions are still not established,[1] it is clear that only the top income group of 'progressive farmers'[2] apply fertiliser according to recommendation. A minority of large-scale farmers amongst these do in fact regularly exceed the recommendations, to the extent of applying 202 or even 224 kilograms of nitrogen per hectare (with proportionately increased P/K basal dressings), in the interests especially of high yielding seed production. Suppose, then, that top productivity demands remain at this level, that the numbers of participants in the High Yielding Varieties Programme increase even marginally, thus bringing a few thousand more hectares under increasingly intensive cultivation, and that the general level of fertiliser application rises whilst a greater area is brought under the program as irrigation facilities are developed. The question then arises, whether the water demands can be met.

The net irrigated area of Uttar Pradesh in 1966/7 is officially recorded at 6·2 million hectares (the net area sown in that year was 16·7 million hectares). The development of, particularly, minor irrigation (chiefly private 10-centimetre delivery tubewells), more recently has added several hundred thousand hectares to the total irrigated area, though exact figures are not yet available. The total area represents an all-source aggregate, calculated in terms of command areas of the various installations. In 1966/7, canals accounted for the largest irrigated hectarage, 2·4 million, with wells other than State or private tubewells a close second (2·1 million hectares). The tubewell area was recorded at 1·1 million hectares in 1966/7, and may well have passed the 1·2 million mark by 1971. Assuming these figures do give an approximate estimate of command (that is, irrigable) area, questions arise in the context of the new agricultural strategy, as to the hectarage actually irrigated.

In both the Ganges-Jumna and the Sarda canal systems, the water supply, drawn direct from the great rivers, is a function of annual monsoon precipitation: a shortfall means a reduction in irrigation water available for the following *rabi* season. As a consequence of the drought which coincided with the first two, and in the case of east Uttar Pradesh, three, years of the new agricultural strategy, supply decreased as demand increased in accordance with the propagation of

[1] Either by the all-India co-ordinated fertiliser experiments on research farms and cultivators' fields or by the routine soil sampling conducted by regional research stations on samples sent in by the National Extension Service village-level workers.

[2] Cultivators participating in the new strategy, selected initially on the basis of access to guaranteed irrigation facilities.

new techniques. A fundamental problem of regional distribution of irrigation water was thereupon aggravated. In the tail ends of the canals, commanding central and southwestern Doab districts, a persistently defective supply was reduced, in the drought, to nothing. In the upper reaches, sufficient water for a minimum of two, and a maximum of three, irrigations has been mostly available on main branches and distributaries. Complaints of water shortage in minors even in favoured Meerut and Bulandshahr Divisions, were widespread. The maximum supply legally available does not however come up to full High Yielding Varieties Programme recommendations, and cases were frequently reported locally of water thefts, which were said to have been committed by 'progressive farmers' who were anxious to ensure maximum yields and who were fully dependent upon canal irrigation.

Further, the traditional (since mid-nineteenth century) roster system of distribution is not geared to the precise requirements of the program schedule whereby the new varieties' sensitivity demands irrigation at specific points in the life cycle. Complaints, therefore, as to the uncertainty of canal supply are increasing as are, predictably, water thefts again on account of the inadequacies. Beyond these immediate problems, arising from the subjection of an old irrigation network to new demands, the now historic manifestations of irrigation induced hydrological imbalance, waterlogging and salinity, persist.

The groundwater reservoir seems to offer a more stable alternative for exploitation. Tubewells, first developed by Sir William Stamp in the Meerut area in the early 1930s, are synonymous with 'assured irrigation', the prime requirement for 'progressive farming'. State tubewells are constructed and administered by a subsidiary office of the Canal Department, still the senior technical service of the State. They are generally of 15-centimetre delivery, tapping aquifers at a depth of 55 metres and below with a capacity of *c.*36,000 gallons/hour. In accordance with the established (commercial) policy of the Canal Department, State tubewells are sited in level uplands, readily irrigable areas, commanding a maximum *rabi* acreage (ensuring maximum revenue in water rates). Initially, commands were fixed at 243 hectares per tubewell, a figure which early proved impracticable. Cultivators beyond the 81-hectare periphery were provided with perhaps one irrigation in three or even four weeks. In 1969/70, the command was finally reduced from a temporary 162-hectare area to 81 hectares.

Policy has already led to a heavy concentration of State tubewells in the immediately productive areas. As late as October 1969, Pratapgarh District, a predominantly *kharif* area, with at least 40,000 hectares of highly saline soil and much riverine erosion, had only one State tubewell; Farukhabad tubewell Division, on the other hand, had 311, and by February 1969, a single subdivision of sandy loam uplands accounted for seventy eight of these. The operation of State tubewells has, however, proved frequently irregular. Electricity cuts, mechanical faults and 'mischief' of the operators themselves have been the most general complaints. These difficulties, the excessive areas fixed as commands, the inadequacies or absence of canal irrigation facilities, in the face of the demands of and prospects for the intensive agricultural district and area development programs, and now the High Yielding Varieties Programme provided the necessary stimulus to exploit the upper aquifers (at varying depths of *c*.6-24 metres) by means of tubewells of 8 to 10 centimetre delivery, privately owned and operated.

The degree of control which might be exercised by government over private tubewell development in the interests of planned agriculture is, *a priori,* limited. No comprehensive hydrological survey of groundwater resources as yet exists. No detailed calculations of safe limit (for withdrawal) or of safe distance (for spacing between installations) to avoid excessive demand on the groundwater reservoir can therefore be made. Locally, approximate calculations are made on general analogies with neighbouring or apparently similar regions.[3]

Even where such calculations have been made on the basis of pumping tests and hydrogeological investigations, no legislation exists which might be used to enforce them to assist planning. The only restriction on private tubewell enterprise is the prohibition, amounting to an injunction against the infringement of departmental rights, against granting electric supply to tubewells installed within 183 metres of a canal line. This was a measure designed to prevent the increase of lateral percolation by tubewell suction within such proximity and to protect canal department rights to water sales. Diesel tubewells can, however, be operated with impunity, though at greater cost in maintenance to the owner.

General costs of tubewell installation are financed either by private capital (as predominantly in the Tarai) or, most commonly, by public credit distributed according to private creditworthiness. Hydroelectric power provides the chief source of motor power. In the interests of reducing costs of electrification on government account (0·2 kilometres of free line is allowed per tubewell), the Hydel Department, which controls energisation independently of the minor irrigation department of the food and agriculture ministry (co-ordinated at district planning office level), encourages clustering of tubewells within the smallest feasible radius. A vegetable, wheat and sugarcane producing village, for example, of 243 hectares cultivated area some 4 kilometres

[3] In Pilibhit Tarai, figures from the partly similar area of Najibabad in Bijnour were used as the basis for safe limit estimates.

from Etawah commands, in October 1969, no less than forty four minor irrigation installations, including twenty six private tubewells, each with an irrigating capacity of about 8 hectares.

Taking the existing developed areas, where clusters of tubewells have been in operation at least since the introduction of the High Yielding Varieties Programme, as a model for present and future development, and bearing in mind the absence of a hydrological survey, can we make any deductions as regards groundwater supply from present manifestations? As far as these developed pockets may tell us whether it is true, as a recent advertisement for an international drilling-rig firm would have it, that 'India has more water than she knows what to do with' (Atlas Copco, *Financial Times*, 26.10.70).

The most overt effect of clustering, leading to a regular withdrawal of more than 1-2 cusecs per sq. kilometre, in coincidence (a vitally important factor) with recent drought, has been to depress the water table to a depth indicating at least a temporary excess of withdrawal over recharge. In the Rudrapur region of the Naini Tal Tarai, the water table which was observed at depths of 0·6-0·9 metres in May-June as late as the early 1960s is now recorded at levels of 6-8 metres. The High Yielding Varieties Programme developments in the Bhabar catchment north of the Tarai, through which hill-stream waters percolate rapidly to feed the Tarai groundwater reservoir are at the same time reducing the available supply from one major source of recharge. Elsewhere, on the plains, extreme clustering is resulting in so rapid a depletion of upper aquifers that both old and new borings are being extended to deeper water-bearing strata. Reports from the Farukhabad-Fategarh strip suggest repeatedly that private tubewells are lowered by 0·9-1·5 metres per year, to maintain the level of supply. Nearby, in Kasganj block Etah district, a pocket of upland prosperity (traditionally derived from the cultivation of indigo, wheat and sugarcane), in stark contrast to the sprawling saline-alkali patches in the southwestern two thirds of the district, new tubewells are built now to tap the second aquifer, the first being already on the point of exhaustion.

Clustering, with such consequences for the water table, can be beneficial. In a few low-lying pockets in Meerut and upper Rohilkhand Divisions, later percolation from canals has led gradually to a rising water table inducing root zone waterlogging. Intensive tubewell development has achieved something similar to a hydrological equilibrium through the balance of withdrawal and recharge. This phenomenon was cursorily observed in tehsil Jansath District Muzaffarnagar (in the Upper Ganges Canal command) is of great significance, but again, is one for which no precise measurements from the Uttar Pradesh field are as yet available.

Where no such artificial source of recharge such as canal seepage supplements the regular (but fluctuating) supply from precipitation and from river seepage, the effect of clustering seems less beneficial, and again, the coincidence of drought heightens the effect. In Varanasi district, for example, widespread complaints have arisen against tubewell interference with the water supply, derived from upper aquifers, in masonry wells. In 1968/9, for example, the supply of drinking water from wells in areas with pockets of tubewell development was severely reduced.

The agro-geomorphology of Uttar Pradesh dictates the regions where tubewell development is practicable. Problems of insufficient and of inaccessible aquifers abound both in the eastern districts and in Bundelkhand and resources available for both State tubewells and for minor irrigation in these areas are necessarily concentrated on productive pockets, chiefly *rabi* wheat-growing areas. Other problems arise from the quality of irrigation water in upper aquifers in regions severely affected by waterlogging and salinity-alkalinity. Brackish, alkaline groundwater, with its high boron content, is inimical to plant growth. It is a problem which inhibits High Yielding Varieties Programme minor irrigation development in alkaline tracts of eastern districts (Sultanpur-Pratapgarh, Azamgarh-Ghazipur, for example) and, widely, in Mathura district, where prolonged waterlogging in trans-Jumna basins has led to a serious deterioration in groundwater quality, making it impossible for the benefits of the program to be extended to the major part of cultivation in the area.

Costs of intensified irrigation on the present lines of development are not merely to be measured in terms of patchy soilwater deterioration by waterlogging and salinity, or in the depletion of aquifers by excessive tubewell demand. All irrigation installations permit at present, through the general lack of lined distribution channels, significant percentages of water loss. Irrigation engineers generally admit of losses of up to 50, or even 60 per cent from canal to field, of 10-20 per cent from private tubewell to field, and 5-10 per cent from State tubewell to field.[4] It is clear from these estimates that a major economy is required in the form of a planned distribution network. Financing of private tubewells by public loans (at the rate of Rs6000-10,000 per tubewell, according to drilling costs) rarely covers the cost of channel lining at present. It is also clear that such a network should be developed by clay, or better, pvc or aluminium portable piping, since a substantial amount of the cost is recoverable in the form of reduced water loss, thus leading to reduced demand on aquifers.

[4] This last is regularly supplied, at government cost, with 1·6 kilometres of lined channel.

The all-important question of the economic exploitation of water resources, the consideration of old and new irrigation problems, and of the increasing demands of chemical treatment of the soil, dictates the need to assess the performance of the high yielding varieties to date. Measurement of actual performance is as yet elusive. Since gross output figures for the high yielding varieties (as against total agricultural production, hectarage expansion and area under improved local varieties), are not available, estimates are made on the basis of areas reported sown and of estimated yield capacity, with selected crop cutting experiments for corroboration.

According to these estimates, the yield of a new variety of exotic parentage reaches a minimum of twice, frequently three times, and occasionally four or even five times the average indigenous yield, in the first year, and if high inputs are sustained according to recommendation, in the second year of its life. Thereafter, genetic impurities have appeared, according to frequent reports, with the reversion of the dwarf characteristic, which inhibits high fertiliser uptake (owing to lodging), and in a decline in yield to a level, over two succeeding seasons, of 40 or even 50 per cent below its peak first year performance. 'No one talks of S64 (the breathtaking wonder wheat which began the program) any more', it was said as early as the *rabi* of 1968/9.

In the absence of progressive experimentation to establish yield trends for several varieties over a period of years,[5] these general estimates (largely derived from substantiated local complaints) must suffice. We may safely conclude that the principal new varieties of paddy and wheat have a clear superiority in yield, even under 'control' conditions. Against this must be discounted the quality of the grain, which is not consumed locally but marketed where possible and fetches a price which may, as in the case of Mexican wheat, reach a level Rs0·2/kilogram (20/quintal) below that of local improved wheats. This consideration becomes more significant when it is remembered that if the original seed is maintained through local multiplication, the yield level drops after the second season, perhaps to 3·2 to 3·7 tons per hectare, as against the average yields of local improved varieties of 2·7 to 3·2 tons/hectare.

Costs also must be considered. The minimum all-in production costs for Mexican wheat current in 1969/70 were generally quoted as Rs250-300 and for indigenous improved varieties, approximately one half to two thirds that figure. To this must be added the rising costs of irrigation exploitation (in the case of over-developed aquifers) of fertiliser, and of constantly renewed seed to maintain high yield levels.

[5] The all-India co-ordinated experiments are conducted on selected varieties with differential fertiliser treatments for one year only.

In view of the necessity for regular varietal changes, one must also estimate the wheat area—for example—regularly pre-empted for seed rather than foodgrain production. The official demand for top quality seed demands top productivity conditions for its multiplication.

In addition, there is the stimulus provided by the pre-release seed market for the production of the latest in high yielding strains, sold in the first instance (e.g. triple gene dwarf, in 1968/9) for Rs200-300/ kilograms. Regular seed sales, on a yield of say 4·6 tons/hectare of released varieties in their first season fetch a minimum of Rs20/kilograms. Since the programs began with next to no stock, the market is a growing one, and financing seed production a safe risk. One might almost say that the fortunes of the Tarai are founded on genetic instability.

Costs in technical and economic terms are, regrettably, matched by costs in disease. The high yielding varieties have a notoriously low threshold of resistance. Where soil and water conditions are poorest, disease inevitably spreads, assuming epidemic form on occasion. A classic example was provided by the paddy epidemic, chiefly due to tungru virus (known in the Philippines, the home of 'miracle' rice and in other areas of South-east Asia), in the eastern districts in the *kharif* of 1969/70. Some twenty two districts were reportedly affected. The present writer saw vast areas typically affected in the Ganges-Gogra Doab in that season, and estimates of 8094 hectares per district, for the most severely damaged, would seem reasonable. This becomes all the more serious when it is remembered that, because moderate to severe salinity and alkalinity are widespread in the east, paddy is for many areas the only crop which can be sown and, that during the recent drought, the last *kharif* saw areas sown for the first time for three or even four seasons. Losses of 60-100 per cent of the crop were generally reported from affected areas, and disease tended to spread from the exotic into local varieties.

Hybrid millets, the most costly of the new crop varieties in the State program, have proved similarly susceptible to diseases. These, however, are controllable—with heavy doses of chemical pesticides and fungicides. The paddy virus could not be controlled, and the suppression of lesser paddy diseases proved to be beyond the logistics of the plant protection service.

Unfortunately, it now seems probable that certain selections of Mexican wheat are not going to prove immune. Outbreaks of *alternaria* blight, most commonly in S227, were widely reported in Doab districts during *rabi* of 1969/70. Some part of the increase in susceptibility may be caused by a decline in available trace elements in the soil. In all probability, a consequence of intensive fertilisation with ammonium sulphate has led to a deficiency in some highly productive areas. Cases of this have been reported from the Government Rice Research Farm at Masodha, Faizabad District and, widely, in the Tarai. The paddy disease *khaira* is believed to be symptomatic of such a deficiency.

In all, the list of liabilities measured against the asset of yield superiority, seems formidable. What are the alternatives? In the interests of conservation of resources (especially water), of planning for their control, and of preventing the increase of disease risk on the basis of significant warnings now patent in the agricultural condition of the State, can a more comprehensively based strategy be formulated? The answer would seem to be, yes. Uttar Pradesh is a natural laboratory, an agglomerate of complex conditions, for which a unique technical documentation[6] can be used to plan more precisely for conditions as they are and as they developed on a detailed time scale of recent history. In addition, local agricultural botany offers hardy strains, not merely in both paddy and wheat but in millets, maize, and, more importantly, pulses which are the essential vegetable protein constituent in diet.

Certain major cereal strains have already been selected and multiplied. The enthusiastic response of cultivators, particularly on the grounds of fine grain quality, should not be ignored in this respect. But far more support could well be lent to the genetic development of local material for widespread dissemination, at lower cost in technical inputs, and therefore at reduced pressure on natural resources of which (pending the preparation of adequate surveys) the extent and propensity to depletion is unknown. Emphasis is important not merely on selection of local botany (north India provides one of the richest 'genetic banks' in embryo in the world), but also on judicious improvement of local methods of cultivation. The current emphasis required by exotic varieties on uniform monocultural practices implies heavy pressure on available nutriments in the soil.

Mixed cropping, for which the optimal patterns can be decided by experiment[7] relieves such pressure without compromising aggregate yield to the point of impracticability. Indeed, an all-round agricultural development based on the dissemination, over a far greater acreage than can ever be brought under exotic strains, of improved local strains of a general nutritional pattern (cereals, pulses, supplemented by vegetables) would seem possible, provided agencies for preparation and distribution both of local improved seed and its produce and the

6 Observations by scientists and engineers as well as revenue officers over one hundred and more years.

7 As has been done, for example, in the case of different paddy varieties and for wheat plus gram, by Professor S. K. Roy of the Indian Statistical Institute, Calcutta.

accompanying publicity required for such a program could be worked up to the standard attained currently in favour of exotic varieties.

It should surely prove possible also, in co-ordination with such a program, to begin reclamation work on the major problems affecting the State: most notably the increasing waterlogging and salinity-alkalinity menace. Such reclamation might well begin with low-cost techniques on areas as yet marginally affected where the required vertical drainage could well be achieved by co-ordinating tubewell development with reclamation and, simultaneously, by sowing salt tolerant varieties more widely on tracts taken up for reclamation. Only then would it be possible to begin to arrest the process of 'spoliation' by which Hume characterised the basic predicaments of modern Indian agriculture nearly one hundred years ago.

11

# The Green Revolution at the Village Level: A Philippine Case Study, 1963-70

*Gelia T. Castillo, Alice M. de Guzman, Soledad L. Pahud and Lorna Paje*

AMONG the latest additions to the development experts' vocabulary for international conferences and for the exploding development literature are such expressions as: the green revolution; the second-generation problems; the political costs of agrarian modernisation; and even pandora's box. In recent writings on the subject, the most salient issue is one of equity—who have benefited most from the recent advances in agricultural technology? Conclusive evidence to answer this question is not easy to establish, nevertheless attempts should be made to assess in some way what has happened to the small farmer.

In 1963 a rural development project was launched in seven very under-developed, predominantly rice growing *barrios* (villages) of small, mostly tenanted, rice farms in the municipalities of Santa Maria and Mabitac, Laguna, Philippines.[1] In these seven villages,

[1] This Pilot Study of a Co-operative Approach to Rural Development was undertaken originally by three agencies (de Guzman *et al.* 1969): the Bureau of Agricultural Extension (BAE), the Presidential Assistant on Community Development, and the University of the Philippines, College of Agriculture (UPCA), particularly through its Farm and Home Development Office (FHDO). Two development teams, each composed of four technicians in crops, livestock, home management, and community development, were organised to undertake agricultural extension and community development work in these villages. The basic features of the project were : teamwork among the agencies, a field of specialisation for each of the four workers in the team, and a *barrio*-centred approach. The four members as a team jointly assumed the responsibility of guiding the development program of the four pilot *barrios* covered by each team. Each member served as the representative of the team in the *barrio* where he or she was staying but each one was available to other team members to provide assistance in his particular field of specialisation. Without planning it, each team happened to be made up of two men and two women representing three co-operating agencies and four fields of specialisation. The project was originally planned for three years but was actually extended to five. However, before the end of the third year, changes in administrative leadership in the two government agencies and difficulties in implementing the original policies regarding assignment and promotion of personnel left the Farm and Home Development Office of the College of Agriculture to carry on alone. As a consequence most of the activities focused on the improvement of rice production.

TABLE 11.1

PROPORTIONS OF FARMERS ADOPTING SPECIFIC RICE PRACTICES OVER SIX CROPPING SEASONS, PHILIPPINES, 1964/5 TO 1967

| *Rice Practices* | *Dry Season (per cent)* 1964/5 | 1965/6 | 1966/7 | *Wet Season (per cent)* 1965 | 1966 | 1967 |
|---|---|---|---|---|---|---|
| Application of fertiliser at planting time (basal) | 10 | 8 | 15 | 7 | 9 | 12 |
| Application of fertiliser during the growing period of the rice plant (top dressing) | 26 | 20 | 46 | 16 | 15 | 54 |
| Planting in straight row | 25 | 31 | 53 | 28 | 42 | 59 |
| Use of rotary weeder | 18 | 28 | 45 | 25 | 35 | 58 |
| Use of weedicides | 32 | 34 | 45 | 34 | 50 | 58 |
| Spraying or soaking seedlings in chemicals before transplanting | 35 | 36 | 39 | 38 | 41 | 43 |
| Spraying against pests and diseases | 27 | 33 | 48 | 31 | 44 | 50 |
| Germination test | 39 | 40 | 38 | 33 | 47 | 39 |
| Making and application of compost | 6 | 4 | 17 | 14 | 4 | 2 |
| Seed selection before harvesting | | | | | | |
| (a) paddy | 45 | 56 | 57 | 54 | 50 | 54 |
| (b) panicle | 13 | 14 | 8 | 22 | 13 | 6 |
| Rat control | 12 | 10 | 11 | 9 | 17 | 12 |
| Planting of recommended varieties | 4 | 23 | 69 | 11 | 49 | 88 |

Total no. of farmers reporting = 179.

average rice yields in 1963 were 20-39 cavans/hectare in the wet season and 20-33 in the dry season. These were comparable to the low national average (27 cavans/hectare) in that year but were well below overall yields in Luzon (40 cavans/hectare average).

The project concentrated primarily upon the introduction of a variety of new techniques with the objective of modernising rice growing.[2] These techniques and the results obtained for the wet and dry seasons of three crop years are shown in Table 11.1.

It will be noted that there was an increasing trend of adoption for nine out of twelve innovations introduced but the most impressive performance was in the planting of new recommended varieties. Decreasing trends were observed for germination test, rat control, panicle method of seed selection, and application of compost. The latter two practices were considered extremely laborious. Fortunately for these *barrios,* rats were a problem in only two of the villages. But even this low acceptance of rat control was a substantial improvement over the days when rats were regarded as 'untouchables' because of the

[2] This section is based mainly on Pahud 1969.

superstition that they vent their wrath on those who try to destroy them.

Fertiliser application at planting time was less acceptable than fertilisation during the growing period. Furthermore, fertiliser application during the wet season was traditionally not practised because it produced lodging. However, in the wet season of 1967 more than half of the respondents had accepted the practice of applying fertiliser two weeks to a month after transplanting. This increase in acceptance could be attributed to the fact that most of the farmers had shifted to new varieties which were more responsive to nitrogen. The fertilisers commonly used were urea and complete fertiliser. Ammonium sulphate was seldom applied.

When the technicians first came to these *barrios*, the rice fields looked more like weed fields but planting in straight rows to facilitate the use of the rotary weeder was one of the easily accepted practices. To induce the farmers to plant in straight rows, the field technicans demonstrated the method with the use of the planting board. Farm and home visits were made. When farmers complained that local planters did not have the expertise on the method, the technician brought some experienced planters from another town and the local planters did the job side by side with them. A field trip to the College of Agriculture and to the International Rice Research Institute was also organised so the farmers saw for themselves the advantages in the method. The aesthetic value in seeing rice plants growing in straight rows was an additional inducement.

The farmers' recognition of the negative effects of weeds on yield and on the general stand of the rice crop had made weeding an important cultural practice. Those who did not use the rotary weeder applied weedicides, the most popular of which was 2,4-D. In the beginning there were problems in this regard because some farmers applied the chemical when the weeds were already tall and therefore control was not very effective. Some harmful effects observed were due to incorrect proportions used.

Preventive measures in controlling pests and diseases were adopted although the concept of prevention rather than cure had not been fully comprehended. Farmers who did not spray in the seedbed reasoned that there were no pests and diseases damaging their crops anyway. Others sprayed only when infestation had already reached an advanced stage.

New rice varieties were recommended in the seven *barrios* to increase yields. A number were introduced from 1965 onwards, including PETA, BPI-76, BPI-76-1, C-18, tainan, tjeremas. Some of these gained rapid popularity. However, from 1966 onwards, distribution of the IRRI varieties commenced and by 1970, these (IR8, IR20 and

IR22) together with C-4 (from the College of Agriculture, University of the Philippines) had virtually replaced the earlier improved varieties. Local varieties showed a definite decline in use both with respect to numbers of farmers planting them and in the area devoted to them. Table 11.1 shows that overall acceptance of high yielding varieties increased rapidly in both seasons over the three years to 1967.

Significant structural changes took place in three of the seven *barrios* after it was found early on that the prevailing cropping pattern involved planting during the months when the supply of irrigation water was inadequate. Seasonal planting times were shifted to correspond better with water availability and improved rice strains, which were suited to these seasons, were introduced. Acceptance of these changes took place with relatively little time lag.

Since the major reason for the shift in cropping pattern was to take maximum advantage of the periods when water was available for plant growth, and since the majority of the cultural practices recommended by the technician required water for their effective use, one might have expected that the new cropping pattern would be accompanied by more complete adoption of these practices. This proved to be so in practice. Between half and three quarters of those who shifted to the new planting schedule adopted nine of the twelve recommended practices. This was a considerably higher range than that recorded amongst those who did not shift their planting schedule. The difference was particularly marked for the adoption of new rice varieties.

Overall, the project and the technical changes it promoted, produced a substantial increase in yields, even in three years (Table 11.2). There was a notable shift in the frequency of yields of farmers from levels below 1·8 tons/hectare in 1964/5 to 2·6 tons/hectare and above in both dry and wet seasons.[3] Significantly, in the *barrio* where the water problem had been most critical and rice yields lowest and where the seasonal shift in planting had been first accepted, the yield response was greatest.

The foregoing provides an outline of the technical basis of the project and some appreciation of the impact it had on the seven *barrios*. It should be noted that the advent of high yielding varieties was not

[3] Yield reports have to be taken cautiously because of the pressures to understate them. Expectation that the area will be declared a land reform area militates not against achieving high yields but against accurately reporting them, because leasehold rentals will be computed on the basis of the average yields for the three crop years preceding the declaration of a land reform area. The higher the known yields, the higher the rentals. Furthermore, even area planted is based on very gross estimates since many of these farms do not really have accurate surveys. The size of the farms is usually estimated from the amount of seed they use for planting and not on actual measurement of the land.

TABLE 11.2

YIELD DISTRIBUTION, PHILIPPINES, 1965/7

| *Yield (cavans/hectare)* | *Dry Season* | | | | | | | | *Wet Season* | | | | | | | |
|---|---|---|---|---|---|---|---|---|---|---|---|---|---|---|---|---|
| | *1964/5* | | *1965/6* | | *1966/7* | | *Total* | | *1965* | | *1966* | | *1967* | | *Total* | |
| | (*No.*) | (*Percentage*) | (*No.*) | (*Percentage*) | (*No.*) | (*Percentage*) | (*No.*) | (*Percentage*) | (*No.*) | (*Percentage*) | (*No.*) | (*Percentage*) | (*No.*) | (*Percentage*) | (*No.*) | (*Percentage*) |
| 20 and below | 19 | 10 | 14 | 7 | 10 | 4 | 43 | 7 | 49 | 26 | 15 | 7 | 7 | 3 | 71 | 11 |
| 21–40 | 78 | 40 | 63 | 30 | 59 | 25 | 200 | 31 | 78 | 41 | 52 | 24 | 47 | 21 | 177 | 28 |
| 41–60 | 57 | 30 | 89 | 42 | 71 | 30 | 217 | 34 | 40 | 21 | 73 | 34 | 74 | 33 | 187 | 30 |
| 61–80 | 27 | 14 | 29 | 14 | 38 | 16 | 94 | 15 | 14 | 7 | 35 | 16 | 42 | 19 | 91 | 14 |
| 81–100 | 7 | 4 | 11 | 5 | 34 | 14 | 52 | 8 | 4 | 2 | 28 | 13 | 28 | 12 | 60 | 10 |
| 101 and above | 5 | 2 | 4 | 2 | 27 | 11 | 36 | 5 | 6 | 3 | 11 | 5 | 28 | 12 | 45 | 7 |
| Total parcels reported | 193 | 100 | 210 | 100 | 239 | 100 | 642 | 100 | 191 | 100 | 214 | 100 | 226 | 100 | 631 | 100 |
| Average yield | 45 | | 47 | | 61 | | | | 37 | | 55 | | 63 | | | |

anticipated by the project, but these villages nevertheless were participants in the events which contributed to changes collectively referred to as the green revolution.

Research was built into the project in a number of ways: through field technicians' diaries and periodic reports; in two intermediate surveys of rice farmers (1966 and 1967); in a benchmark study of social and political factors in 1963 and a restudy in 1969; and through special follow-up studies of different aspects of the project. Publications on various aspects of the project are already available.[4] This study focuses on the social and political factors in the development of these *barrios*, i.e. on the somewhat elusive and unpredictable social and political ripples which have accompanied the technological changes from the project. It is based on data from 370 (household head) respondents for both the 1963[5] and 1969 surveys. Personal interviews were made in each of the two years to help capture the attitudinal, perceptual, cognitive and social structural aspects in the pilot area. The analysis of change over a six-year period includes such variables as:

1 State of livelihood, income, level of living, and perceptions of social differentiation;

2 attitudes toward rice farming as an occupation;

3 family responsibilities, aspirations and expectations for the children;

4 expectations of different levels of government and of different government officials;

5 political participation and civic responsibilities;

6 community problems, leadership patterns and propensity to organise;

7 related developments.

### THE RESPONDENTS

The typical respondent and his wife had about four years of formal schooling. Only about 6 per cent of the husbands and 4 per cent of the wives had attended high school; none of them had graduated from the secondary school. The majority of husbands were more than 40 years of age while their wives were younger. More than 90 per cent earned their livelihood from farming; 74 per cent of them had some land to cultivate while 17 per cent were hired farm hands. The pro-

4 See Castillo, Dimaano, Calleja and Parcon 1964; Castillo 1964 and 1970 and de Guzman and Dimaano 1968.

5 The questions for the 1963 survey were developed by O. F. Sison, G. D. Feliciano, B. M. Villanueva and G. T. Castillo.

portions with non-farm occupations[6] in 1963 and 1969 were 5 and 7 per cent respectively. The farmers had considerable farming experience, with an average of about fifteen years. Almost half had spent twenty years or more in their places of residence.

Although the average size of farm was about 2 hectares, more than 40 per cent of the farmers had only 0·5-1·5 hectares to cultivate. This is very significant in relation to the dependency burden on the household head. In 1963 the average household size was 5·6 and in 1969 it had risen to 6·4.[7] In the light of the estimated average income from all sources of only about ₱1200 ($US200) and the fact that some 40 per cent had less than that to support the household, poverty was an obviously common characteristic of the area studied. Over half of the respondents had other sources of income. About 50 per cent had hired farm jobs including land preparation, transporting, weeding and harvesting of rice and coconuts. The remainder mentioned sales of farm products such as bananas, chickens, sari-sari stores, etc.

Between 1963 and 1969 there was a slight fall from 81 to 75 per cent in the proportion of respondents who derived their income mainly from farming. Knowledgable individuals in the villages revealed five reasons for this. Small landowners, living near the area, decided to cultivate the land themselves; aged and ailing farmers transferred responsibilities for cultivation and chose to work as hired hands during the cropping season; urgent need for money had forced some to sell land; others had sold land when moving from the area and on their return were working as farm labourers; and, in one case, loss of a work animal had forced a farmer to accept work as a labourer.

Within the same time period, other changes took place. First, the proportion of lessees rose from 1 to 12 per cent, while the proportions of share tenants and part owners fell respectively from 71 to 66 and from 12 to 4 per cent. The shift to leasehold occurred despite the fact that the pilot project area was not covered by the Land Reform Program, so the changes apparently occurred by mutual consent of landlords and tenants. Of the 1963 part owners, by 1969 more than half had become share tenants and a few owners, labourers, leaseholders and traders. Actually the term part owner is something of a misnomer, for their land ownership claims had no legal basis, but referred to upland areas where the farmers practised 'slash and burn' agriculture and exercised usufructuary rights. With better opportunities in lowland rice farming, some chose to concentrate there as tenant farmers. The proportion of owners incidentally changed little in the six years (rose from 16 to 18 per cent).

6 These included store owners, rig drivers, small traders in farm crops and livestock and fruit, carpenters, labourers, etc.

7 Excluding children who had left home and resided elsewhere.

The proportion working on lowland rice alone doubled, from 19 to 37 per cent, over the period, but the majority (64 per cent in 1963 and 52 per cent in 1969) carried on cultivating a combination of lowland rice, upland rice and other upland crops, both to secure food supplies and to earn cash incomes.

For share tenants the prevailing arrangement for harvest sharing was 1 : 1 during the dry and 2 : 1 during the wet seasons. Under the former system, all production expenses are equally shared by tenant and landlord. In addition, the tenant furnished his own and his family's labour, work animals and farm implements. In the latter arrangement, two parts of the harvest go to the tenant and one part to the landlord but the tenant shoulders most if not all the expenses. The rental rates of leaseholders varied widely, 1-30 cavans/hectare. The majority of the lessees pay up to half a ton per hectare per crop.[8]

Since the major objective of any rural development program is to improve livelihood, especially where people are so obviously wanting in the material necessities of life, this study included several variables which reflect livelihood. In 1963 an attempt to obtain figures for cash and non-cash income proved to be very expensive in terms of interviewing time and even then it was difficult to judge the reliability of the data. As a consequence, the respondents were asked in 1969 to compare their current income level with that of 1963 and were also asked to state the reasons for their answer. A second easier way of gauging income used was direct observation of material items purchased. This was considered more meaningful than income figures because it represents the particular style of life which they could afford. Thirdly, respondents were asked to compare their state of livelihood with that of their parents, of their neighbours and to state their future expectations in a relative sense.

## INCOME AND STATE OF LIVELIHOOD

Even though *barrio* people do not keep cost accounts, debit-credit ledgers or farm and home budgets, they do have a fairly accurate idea of their level of income. Of course, the adequacy of income depends not so much on its absolute value but more on the extent to which it satisfies wants. When respondents were asked to assess how their 1969 income compared with that of 1963, 41 per cent said it had increased; 28 per cent reported a decrease, and 31 per cent said it was

[8] Discussions with a farmer *barrio* official who was a lessee showed that rentals had recently risen with a change in land ownership. Previously they had stood at 5 and 10 cavans/hectare for each of the two crop seasons. The new owner charged 10 and 15 cavans/hectare. The lessee admitted that the former rental was hardly fair, especially with the increased yields.

the same. Amongst those who reported a decrease, 53 per cent attributed it to poor yields from their upland crops and from rice owing to the typhoon and the drought. Higher costs of living (18 per cent) and a lack of dependable additional sources of income (14 per cent) were also mentioned. Loss of, or reduction in, cultivable area, mentioned by only 9 per cent, is of concern because of expectations that will increase with added population pressure on limited cultivable land.

From the viewpoint of the project's objective of raising agricultural productivity, it was encouraging to find that 64 per cent of those who said their income had increased attributed it to increased rice yield. A further 16 per cent mentioned additional sources of income from farm jobs. Conversion to leasehold was regarded as the key to increased income only by 3 per cent. The most noteworthy feature of these findings was summarised by one farmer who said: 'My income increased when my yield increased. Even if I am a share tenant, my share also increased.' The majority of those respondents who reported increased income from higher yields were share tenants though this did reflect the composition of the sample with respect to tenure status.

Despite these developments, there is much to be done to help the 59 per cent who felt their income levels had decreased or had remained the same. The market pricing system is one way. Only two farmers commented on prices in relation to their relative incomes. The reason for this may well be the procedure of creditors' pegging prices on expected harvests for advance loans. Under this system the creditor controls prices, and makes the market decisions. An alternative is needed to the present market credit tie-up to encourage greater market orientation amongst farmers.

In general, evaluations of their income and livelihood compared with that of parents, and of other *barrio* people had moved up from 1963. Almost one third of respondents claimed their 1969 livelihood was better than that of their parents. This was twice as many as had claimed that in 1963. There was a fall from 55 per cent to 40 per cent in respondents who said they were below the *barrio* average in income. Expectations of the future in 1969 did not look as bright. About 30 per cent were either uncertain or expected the future to be poorer than in 1969. In 1963, there were only 16 per cent in these categories. In the pessimistic group, 52 per cent felt their income had decreased from 1963. In the uncertain group, respondents were almost equally distributed in the three categories of increased, decreased or the same income in 1969 versus 1963. Apparently, even those whose life had improved since 1963 could not anticipate what the future would bring. Overall, the proportion of respondents expressing uncertainty has increased from 5 per cent in 1963 to 15 per cent in 1969.

There was in fact good reason for respondents to feel uncertain. A large proportion who said their income had decreased from 1963 levels had experienced low crop yield owing to typhoon and drought damage. This cycle of typhoon, flood and drought is an annual phenomenon and while one can make plans to try and avoid its maximum fury on the crops, no scheme has so far been devised for escaping such disasters altogether. However, they are not without their functions. Every typhoon which passes through seems to establish further the credibility of the new rice varieties which somehow manage to stand up while all others are lying down. As a consequence, some harvest can usually be salvaged where almost total losses would have resulted with the traditional varieties.

### *Material Levels of Living*

Where poverty is a dominant feature of everyday life, material possessions are quite meagre, simple, and difficult to acquire. In the pilot *barrios*, the material comforts and trappings of daily life are few.

> Houses in the project area are similar to those which can be found in most of the poorer *barrios* in the country. Each home is made of light materials like bamboo, buho, nipa or cogon with galvanized iron sheets being occasionally used for roofing materials. The house is ordinarily a one to two-room affair, sometimes with a makeshift division between the kitchen and the all important multipurpose receiving-dining-bedroom-sala-porch area. Here the farmer and his wife may receive their guests, while away their time, eat, and sleep. Generally what passes for furniture is a bench called a 'bangko' which if present would be offered to a visitor. Otherwise, everybody sat on the floor, ate on the floor and slept on the floor. 'Wall-paper' consisted of cut-outs of favourite screen personalities or faded formal photographs with the subjects peering at the camera uneasily and stiffly. When night came, the multi-purpose living room automatically became a bedroom by spreading a large mat which is shared in groups of threes or fours depending on the size of the household. If the visitor stayed overnight, the farmer would graciously offer this living room and he and his family retired in the kitchen even if the division between the kitchen and this room were more imaginary than real. This kitchen had no form of drainage whatsoever and so kitchen water remained under the house, creating a very muddy, smelly, unsanitary condition. Built with an open hearth of flimsy flammable structure, it also had few utensils to speak of. Pots, a basin, two ladles, a turner, frying pan, a kettle and a few utility plates for eating and other purposes cluttered the kitchen. A drinking glass or two for the entire family completed the set-up. Spoons and forks, if there were any, were reserved for very special occasions. Drinking water was usually stored in earthen jars and not always provided with a dipper so that one had to dip

TABLE 11.3

HOUSING MATERIALS AND VALUES, PHILIPPINES, 1963 AND 1969

| | *1963* (*per cent*) | *1969* |
|---|---|---|
| *Construction* | | |
| Light materials like nipa, bamboo, cogon, sawali and coconut leaves | 24 | 20 |
| More durable materials like wood, concrete, hollow blocks and galvanised iron roof | 5 | 26 |
| Combination of the two | 71 | 54 |
| | 100 | 100 |
| *Value* | | |
| ₱30 and below | 16 | 3 |
| ₱31–60 | 20 | 7 |
| ₱61–150 | 21 | 15 |
| ₱151–450 | 22 | 23 |
| ₱451 and above | 21 | 52 |
| | 100 | 100 |

deep into the jar with the common drinking glass. Toilets were rarely found and whatever existed were no more than dug holes. The majority of the people answered the call of nature under bamboo clumps, banana plants, and bushes. Stray animals always helped dispose of whatever waste there was (de Guzman 1969).

Improvements in housing were evident in the types of construction materials used in 1969 (Table 11.3). From a mere 5 per cent of houses in 1963 made of durable materials like wood, concrete, hollow blocks and galvanised iron roofs, the 1969 survey showed 26 per cent of the houses belonging to this category. Inevitably, the value of the houses increased as well because of the improvements or the new construction.

Amongst household possessions, there were marked increases in the proportion of households with specific items such as chairs, coffee tables, dining tables, benches, cabinets, bedrooms, beds, kitchen drains, kitchen sinks, stoves, radios, sewing machines, toilets, and sources of water for household consumption. There were more chairs purchased in sets rather than individually; more of the benches were made of wood than of bamboo; more cabinets had mirrors. Seven per cent of houses had gas stoves, 4 per cent had electricity and 3 per cent had spring beds (items which no one had in 1963). More houses were enjoying artesian well and piped water and fewer used shallow wells.

The only items with a lower incidence were the wall mirror, the trunk, and the stairs. However, the increase in cabinets with built-in mirrors took over the function of the individual wall mirror; some trunks disappeared in favour of cabinets; the stairs in some houses had to go because new construction or repairs made were of the bungalow style for which stairs were not needed. In the majority of houses where stairs were present, there was an improvement in materials. Predominantly bamboo in 1963, they had been changed mostly to wood and some to concrete. Toilets, although the incidence had doubled from 39 to 68 per cent, they were still predominantly the pit privy type. An overall comparison of household items, showed that households with ten and more items had doubled in 1969 (23 per cent to 49 per cent). Thus, life was still poor after six years but less miserably so, at least as far as material comforts were concerned.

### PERCEPTION OF STATUS DIFFERENCES WITHIN THE BARRIO

A third variable related to the preceding two, is perception of social differentiation. In what ways do the *barrio* people think their statuses in life differ? Have their criteria for evaluating status differences changed? To the question: 'In what ways do you think people in your *barrio* differ with respect to their status in life?', the most significant response in 1963 given by one fifth of respondents was: 'There are no differences here in status; we are all poor'. Thirteen per cent did not know how people differed in status; 12 per cent mentioned differences in tenure status; 21 per cent gave responses pertaining to relative wealth or poverty. An analysis of the responses belonging to this category revealed that 'rich' and 'poor' did not have the same social class meaning in the sense of upper and lower class because there were no really rich people in these *barrios.* This was evident in expressions they used such as 'some are poor; some are a little bit better off' or 'some have sufficient means for everyday needs; others don't'. Many respondents emphasised that more people were poor than were better off. The notion of sufficiency/insufficiency is related to the subsistence character of life in these *barrios* and sufficiency in the most essential necessities of life was already regarded as a better off situation.

In 1969 only 2 per cent of the respondents said there were no perceived differences in *barrio* peoples' status in life. Apparently, there was either more consciousness as to what produces status differences or more actual basis for differentiation. Perhaps the perception of being all poor had become less descriptive of the situation in the *barrios* after six years. Other changes included a drop from 21 to 14 per cent among those mentioning notions of relative wealth or poverty and an increase in those who were aware of differences in source of livelihood (occupation) and in those who saw differences in individual drive to improve livelihood.

Source of livelihood differed in terms of farm/non-farm but the distinction was stronger with respect to whether one was a farmer or a mere farm labourer. The distinction made in responses was greater in this respect than that between share tenant and lessee. What was crucial for being better off in the *barrio* therefore was having land to cultivate, whether or not one owned it. From this study, informal conversations and observations, the 'lowest guy in the totem pole' was one who had no land to cultivate and was therefore dependent on available but unstable farm jobs.

The increased attention paid to differences observed in individuals' drive to improve their livelihoods (a rise from 7 to 14 per cent) also deserves comment. The responses included definite statements such as 'some are lazy, some are industrious'; 'some work hard even when they already have, while others don't have and yet do not work hard'. This change might be explained by the fact that the introduction of the project brought certain opportunities for improvement in livelihood (particularly in farming) and that some had been more alert to such advantages than others. But this is mere speculation untested against actual direct evidence.

The third type of change was the emergence of criteria which were not or were hardly mentioned in 1963. They included differences in land area being cultivated, differences in yield, and availability of capital and other resources. Again here, the developments in rice production seem to be relevant. For example, farmers with bigger areas to cultivate would have had potentially greater total harvests. Yield differentials were also emphasised. Finally, those who had available capital were better able to take advantage of new farming methods and business opportunities than those who were lacking in such resources.

One should not ignore, however, the 4 per cent who believed that some are luckier than others in their farming and livelihood. In 1963 there was only one respondent who mentioned this. When one has seen the natural calamities these farmers go through from year to year and how some manage not only to survive but also to recover successfully, it is not difficult to appreciate the significance of being lucky or unlucky. But this is to be philosophical rather than empirical, for even in 1969 only fifteen respondents or 4 per cent cited the role of luck.

Insight into the nature of status differences can be given in terms of an income, sufficiency/insufficiency dimension. Survey data showed at least four levels: sufficiency of earnings even for unusual expenses; sufficiency for ordinary expenses; insufficiency for extraordinary expenses; and insufficiency even for ordinary needs. This last, lowest level, implies that there were times when even three square meals a day were not easily available. Survey data indicated an increase from 33 in 1963 to 44 per cent in 1969 of household heads who considered their earnings sufficient for ordinary expenses. However, there was an increase from 68 to 74 per cent of those who said they did not earn enough for unusual expenses. Any emergencies would undoubtedly put their families in a most precarious position; and even more so for those who did not have even enough for ordinary expenses.

## ATTITUDES TOWARDS FARMING

With the implementation of the rural development project in the *barrios*, it was hoped that the input of modern agriculture would result not only in better yields but also in a new outlook, new ideas and attitudes toward farming. These changes may be revealed in a number of ways.

### *Basis for Possible Decision to Leave Farming*

In 1963, a question posed to respondents was: 'If you were to leave your present occupation, what would be the most important basis for the decision you would make?' Sixty three per cent said prospects of higher income and 15 per cent mentioned an easier job. The remainder were split evenly between prestige and job satisfaction, a refusal to give up farming, and not knowing. The most significant change in 1969 was that 45 per cent of respondents felt they would not consider leaving their occupation and only 36 per cent would think of making that decision for reasons of higher income.

This could be interpreted as showing relatively much greater satisfaction with farming as a good source of livelihood in 1969. It was also significant that farmers placed a high value on income as a criterion for choosing an occupation. Prestige and satisfaction derived from the job which was mentioned by 7 per cent in 1963 was not mentioned in 1969. Economic returns rather than psychic returns appeared therefore to have the stronger attracting power.

### *Modernising Attitudes Towards Farming*

Over a six-year period the respondents' views on how to improve farming showed a remarkable shift towards recognition of the role of modern inputs such as chemicals, fertilisers, weeding, straight-row planting and use of new varieties and the hand tractor. Respondents who indicated one or more of these ways represented means of improvement increased from 10 per cent in 1963 to 65 per cent in 1969, with the first three inputs (57 per cent) being the most appreciated (Table 11.4). To those concerned with the impact of the new high yielding

TABLE 11.4

RESPONDENTS' VIEWS ON WAYS TO IMPROVE FARMING, PHILIPPINES, 1963 AND 1969

| *Ways to improve farming* | *1963 (percentage)* | *1969 (percentage)* |
|---|---|---|
| Use of: | | |
| (a) Chemicals (insecticide, pesticides) | 1 | 18 |
| (b) Fertilisers | 4 | 24 |
| (c) Weeding, clean culture, straight-row planting | 3 | 15 |
| (d) Use of new high yielding varities | 1 | 4 |
| (e) Tractor | 1 | 4 |
| Irrigation water | 2 | 6 |
| Capital | 1 | 2 |
| Higher price for palay | — | —[a] |
| Wait and see others' experiences before trying new farm practices | 5 | —[a] |
| Consult knowledgeable persons such as agriculturists and other government workers | 25 | 9 |
| Follow the advice of *barrio* captain | 5 | 0 |
| Be the first to adopt new practices | 14 | 0 |
| Experiment on new farming methods | 4 | 7 |
| Rely on own experience and knowledge | 5 | 0 |
| Shift to leasehold, ownership or increase size of area being cultivated | 1 | 1 |
| Unite and organise farmers | 1 | —[a] |
| It is hard to change farming methods | 1 | 0 |
| Don't know how | 26 | 10 |
| Total number of responses | 100 | 100 |

[a] A fraction only.

varieties, it might be a disappointment that only 4 per cent of the responses referred to them. This can readily be explained, however, by the fact that the use of these varieties was already taken for granted for practically all respondents had adopted one or more. Indeed as shown earlier (Table 11.1), 88 per cent of the farmers had accepted these new seeds by 1967, and seed adoption was the most highly accepted and readily applied of the project's innovations.

The decline from 25 to 9 per cent in the response on consulting knowledgeable persons such as agriculturalists, etc., was also notable and on first appearance denoted a diminished recognition of the role of technical personnel. Other data, however, suggested that this was not the case. In reality the use of technically trained personnel (as with the seed variety) was again assumed because this was already the major source of information on modern farming. In 1963, 14 per cent mentioned being first to adopt new practices as a way to improve farming.

In 1969, no one gave this response. Again, it would seem that new farming methods have made headway and therefore being the first to adopt them was no longer the critical factor in 1969. The farmers had moved to a higher level in their farming and the reference point had also moved. It was no longer a question simply of what ways could improve farming but rather what ways could further improve what they already had in 1969. For example, reliance on own experience and knowledge had faded out and so had 'following the advice of the *barrio* captain' and the conviction that 'it is hard to change farming methods', but the need to experiment on new farming methods had gained ground. This concept of 'experimenting' is quite different from being the 'first' to adopt because in the former, the farmer is trying new methods on his farm in an effort to choose from alternatives. This is perhaps traceable to the idea of field trials, using different combinations of inputs in the farmer's field, which was very much part of the technicians' *modus operandi* in the villages. Noticeable also was the drop from 26 to 10 per cent of the responses indicating that they did not know how to change farming methods.

*Information on Modern Farming and Homemaking*

Other questions revealed that over the six years there was an increase from 47 to 63 per cent in responses mentioning technicians from different government agencies as their source of information for modern farming. There was a decline from 15 to 2 per cent in those who relied on personal experiences and trials. Other farmers as a source increased a little (10 to 13 per cent) because of farmer leaders who were given training. Mass media such as radio, newspapers, magazines, etc., were not significant as sources of farming information; nor were parents, relatives, landlords and *barrio* officials.

*Criteria for Naming Innovators*

Since farming before 1963 was largely traditional and new agricultural practices were mainly introduced externally, an innovator among these farmers had some status. When, in 1963, respondents were asked to name persons whom they considered to be innovators in their *barrio* and to explain their choices, 45 per cent mentioned possession of farm implements as an important criterion, and 30 per cent mentioned the use of modern farm practices, new seeds and achievement of high yields. In 1969, technical training (acquired from the College of Agriculture or from the agriculturists with whom they had contact) became by far the most prominent reason for considering a person to be an innovator, showing an increase of 13 to 47 per cent. The use of modern farm practices had ceased to be a distinguishing characteristic by 1969 since the majority of farmers had already adopted one or more of them. The importance given to possession of farming implements in 1963 arose from the fact that the majority of the farmers then owned only the barest of essential equipment for farming. The implication was then that those who had the implements had a better chance of adopting farming innovations. By 1969, hand tractors, sprayers, planting boards were no longer rarities in the *barrio*.

Another significant development was the increased recognition of capital (from 2 to 12 per cent) as a factor enabling farmers to adopt innovations. This is not surprising since practically all the new farming methods (fertilisers, chemicals, weeding, etc.) required some capital outlay. In other words, responses were more in the nature of what enabled them to innovate, and judging from the 1969 data technical training provided the opportunity to keep up with new knowledge. When this was combined with sufficient capital, the conditions for innovation seemed to be favourable. Another interesting result was that 8 per cent of the respondents in 1969 did not mention any names of farmer innovators. They felt that almost everybody started using the new practices at the same time, so being first was no longer anybody's monopoly. There was some merit in this observation for, where there was a change in the cropping pattern, it required several adjacent farms simultaneously to adopt new seeds and new inputs to go with them.

### *Investment Inclinations*

Given a hypothetical savings of ₱200-1000 what would farmers be inclined to do with it? Responses showed a difference between the investment plans with two different hypothetical amounts in both years. With ₱200, there was more emphasis on buying farm animals, farm implements, household items, and food. With ₱1000 the focus was on buying land or the right to cultivate land, or business investment, and on depositing in the bank. For both amounts however, the percentage of respondents inclined to spend their savings on farm investments declined: from 42 to 31 per cent for the ₱200 savings and from 54 to 40 per cent for the ₱1000 savings. There was, on the other hand, an increase in the inclination to use the savings for business investment (from 19 to 27 per cent).[9] The desire to spend savings on consumption items increased in 1969 over 1963 for both amounts. Amongst these items were food, schooling, things needed by the children, and the improvement of housing. It should be pointed out however that inclination towards farm and business investments far outweighed the alternative consumption inclination with totals of 72

[9] Within the context of these *barrios*, 'business' investments largely mean buying and selling of agricultural produce, small stores, and lending money at a profitable rate of interest.

and 62 per cent for ₱200 in 1963 and 1969; and 87 and 82 per cent for ₱1000.

Amongst farm investments, the inclination to invest money in farm animals declined, while that in farm implements increased. For the latter, hand tractors, sprayers, etc., had already attracted attention. The inclination to buy land or the right to cultivate land declined from 35 to 24 per cent but this nevertheless remained the item with the highest concentration of responses in 1963, while in 1969 it was second only to investment in business.

A series of open-ended questions were asked in both years to ascertain what broad implications tenure status, land ownership and size of farm had for the respondents. The answers indicated that none of the three was regarded as a pressing problem in the area. The lack of importance of the land tenure and farm size questions might be explained first because the project villages were in a district not yet declared a Land Reform Area so farmers had not yet become politicised in this respect. Second, since the majority of *barrio* residents were share tenants, there was a great deal of homogeneity with respect to tenure status and farm size. Being a lessee or an owner and/or having a large farm were rare characteristics. While there was status attached to them, the majority who were not in these categories seemed to have reacted in terms of their 'given conditions', accepted the constraints and tried to make the most of their available opportunities. Even within the limitations imposed by share tenancy and small farms, they perceived that a farmer could be progressive, could increase income, could be influential, could be a leader and a farmer-innovator in the *barrio*. This phenomenon was illustrated by Pahud's study (1969). This showed that share tenancy had not inhibited farmers from adopting modern farm practices, while on the other hand, being a lessee or even an owner did not necessarily mean there was more incentive to adopt innovations (Table 11.5). If anything, the tendency was in the opposite direction, with more of the lessee and owner categories being found in the low adoption category.

The high incidence of land possession responses to the question of how they would invest a hypothetical savings of ₱1000 was due to the wishful thinking or aspirational nature of the question. However, even this 'dream' declined from 35 to 24 per cent; perhaps not so much a manifestation of growing disinterest in land as a realisation that even a hypothetical ₱1000 is nowhere near the amount required to purchase a piece of land where the current value is from ₱15-20,000/hectare of rice land. Neither was purchase of rights to cultivate land easy where actual and potential farmer-cultivators asserted an increasing demand on available land. The reality component of answers even to the wishful questions was also evident in the fact that only 3 per

TABLE 11.5

TENURE STATUS AND ADOPTION OF RECOMMENDED RICE PRACTICES, PHILIPPINES

| *Level of adoption* | *Share Tenants* | *Lessee* | *Part-Owners* | *Total* |
|---|---|---|---|---|
| | | *(per cent)* | | |
| Low practices (16 per cent and below) | 27 | 34 | 36 | 30 |
| Medium (17–28 per cent) | 38 | 26 | 46 | 35 |
| High (20 per cent and above) | 35 | 40 | 18 | 35 |
| Total | 100 | 100 | 100 | 100 |

cent in 1963 and 2 per cent in 1969 thought they might use ₱200 savings to buy land or the right to cultivate land. Even the latter is hardly available for that amount.

*Concept of a Progressive Farmer*

When asked their basis for saying that a farmer is progressive, about one half of the respondents in 1963 and more than two thirds in 1969 mentioned good harvest and good income from farming which enabled him to buy his family needs and to have some savings and capital. Thrift and hard work were cited by only 7 per cent of the respondents. It was particularly enlightening that adoption of modern farm practices did not by itself constitute the hallmark of a progressive farmer, as indicated by the low proportion of respondents (6 per cent in 1963 and 10 per cent in 1969) giving this answer. This means that the more important focus was on the harvest and the income resulting therefrom and not on the mere adoption of modern farm practices which at times can be irrational and uneconomic.

A more important feature of the responses was the dramatic decline in emphasis on absence of debts as a characteristic of a progressive farmer (from 26 to 6 per cent from 1963 to 1969). This was a substantial change in the outlook of people who have traditionally regarded borrowing money as something to be avoided. Farm expenses then were regarded as expenditures and not as investments from which pay-offs could be expected. The intent was to minimise expenses and thus to minimise risks. Apparently, the concept of credit as productive investment had become more acceptable by 1969. In 1963 borrowing was resorted to only under critical circumstances and loans were used mainly for subsistence and consumption items.

Pahud's study (1969) showed that, in 1967, 77 per cent of the 179 farmers studied had debts either in cash and/or in kind. Respondents said that 56 per cent had been used for planting expenses, the bulk of which went as payment to the planters. Although more expensive, straight-row planting was one of the practices readily accepted. In

TABLE 11.6

SOURCES OF LOAN IN CASH AND/OR KIND, PHILIPPINES, 1963 AND 1967

| | *1963* | *1967* |
|---|---|---|
| | (*per cent*) | |
| Private money lenders (including rice millers, businessmen, storeowners, buyers of rice and coconut) | 23 | 38 |
| Neighbours, friends, co-farmers and compadres | 22 | 12 |
| Relatives | 32 | 13 |
| Institutional sources such as banks, Agricultural Credit Administration, insurance companies | 12 | 29 |
| Landlords | 11 | 8 |
| Total | 100 | 100 |

addition to straight-row planting, loans were also spent on weeding (8 per cent) and chemicals (3 per cent). This was an indication of the extent to which the farmers had begun to recognise the role of clean culture in rice production. The rice fields certainly looked neater in 1967 than in 1963. Fertilisers (10 per cent) and land preparation (10 per cent) were other items of expense. Owing to the decline in exchange among farmers there was a heavier cash expenditure for labour. The purchase of seeds (2 per cent) was also a new development. The new rice seeds have to be bought and at a premium price, whereas with the traditional varieties, farmers just used seed saved from previous harvests.

Along with the new uses of credit, there was also a noticeable shift in sources (Table 11.6). Between 1963 and 1967 emphasis swung away from relatives to private money lenders and institutions. The increase in patronage of private money lenders was a reflection of the increased need for credit and the inability of banks and other government agencies to meet such needs, although they were in fact used more frequently in 1967. The slight decrease in the use of landlords as a source was probably because lessees could no longer obtain loans from them.

Pahud's study (1969) found a positive relationship between amount of money borrowed and adoption of recommended rice practices. This was to be expected, since the use of these cultural practices requires some investment. Among the farmers who did not borrow significantly, 28 per cent were low adopters and only 19 per cent were high adopters. The latter were farmers who had enough capital of their own. Farmers whose loans amounted to more than ₱300 were mostly high adopters. For these small farmers, ₱300 was a sizable amount in relation to their

Table 11.7

SUFFICIENCY OF INCOME OR SAVINGS FOR OLD AGE, PHILIPPINES, 1963 AND 1969

| *Do you have sufficient income or savings for old age?* | *1963* (*per cent*) | *1969* |
|---|---|---|
| Yes | 15 | 13 |
| No | 82 | 86 |
| Perhaps | 3 | 1 |
| Total | 100 | 100 |
| *Anticipated source of support in old age for those without savings* | | |
| Will borrow money | 12 | 3 |
| Will work hard and save | 18 | 15 |
| Depend on children and other relatives | 16 | 23 |
| Lands and inherited properties | 8 | 9 |
| Farm crops | 9 | 21 |
| Income from livestock | 6 | 3 |
| Bahala na (God will take care) | 8 | 5 |
| Don't know; not sure | 23 | 7 |
| Sweepstakes, jueteng | — | 14 |
| Total | 100 | 100 |

income levels and the value of the peso in 1967. One can therefore appreciate the risks they took in the interests of improving their farming.

## FAMILY RESPONSIBILITIES, ASPIRATIONS AND EXPECTATIONS FOR CHILDREN

### *Sufficiency of Earnings and Support for Old Age*

In 1963 and 1969 only small proportions of respondents believed they would have enough income or savings to meet their expenses during old age (Table 11.7). For the majority in both years who had inadequate savings, what sources of support did they anticipate in their old age? There was in fact a shift in responses to this question from 1963 to 1969. Those who would borrow money declined to a low proportion. Perhaps this reflected a realisation that money is not easy to borrow in one's old age. Rather amusing but nevertheless indicative of their 'optimistic fatalism' was the emergence of a sizable proportion in 1969 whose hopes lay with sweepstakes and other types of local gambling and games of chance. A more accurate interpretation of this response is that they did not know where their support would come from. The proportion of respondents who admitted not knowing how they would manage in their old age declined substantially, but there was a noticeable increase of those who expect to depend on their children and other relatives. More encouraging was the increase of those who expect support from their farm crops such as rice, coconuts, bananas, and coffee (the latter three being planted in upland *kaingin* area). By and large, however, probably very few of them could really look forward to a life of comfort in their old age. The family serves as the one reliable institution to fall back on, even though living with their married children was favoured by only 33 per cent of the household heads in 1969. Staying with unmarried children was more acceptable to 76 per cent.

### *Management of Family Finances*

The view that the Filipino woman is the family treasurer was again borne out in these *barrios*. Interviews revealed that normative sanction for this custom was even stronger than actual practice with slightly more respondents endorsing it in answer to the question 'who should actually handle family money?' than to the question 'who actually handles it?' The proportions were 93 and 91 per cent respectively in 1963 and 90 and 83 per cent in 1969. An indication that this practice may be altering in some small degree toward a little more joint husband-wife responsibility and a little more control by the husband was apparent in the slight increase in the proportion of respondents who indicated husbands (4 to 7 per cent) and joint husband-wife responsibility (5 to 8 per cent). The normative definition was also in the same direction. One might speculate at this point that with the changes in the nature of farming in the area and the adoption of modern inputs requiring cash and/or credit expenditures, family money has to be managed with greater consideration for production needs. It is possible therefore that some husbands recognise the need to participate or even take over the control of family finances, particularly if their wives are not heavily involved in making farm decisions. Nevertheless the fact that 83 per cent of the households in 1969 still have the wife as family treasurer and that 90 per cent endorsed this role for the wife, provides evidence that her role in this respect has been only minimally challenged.

### *Parental Aspirations and Expectations*

One generalisation can be made about Filipinos: they place a high value on education, especially college education. Data from these seven *barrios* substantiate this generalisation and allow the following observations on the nature of the changes from 1963 to 1969:

1 There was a general rise in levels of parental aspirations for their children with respect to both education and occupation. The proportion of household heads wanting college education for their children

almost doubled from 36 to 69 per cent with a corresponding reduction of those wanting only elementary schooling (29 to 8 per cent). Those desiring white-collar occupations increased from 31 to 53 per cent. The reduction in those whose original aspirations were for their children to become farmers was from 22 to only 4 per cent.

2 Heightened parental aspirations were also accompanied by greater expectations. Those expecting high school education increased from 14 to 30 per cent and from 6 to 14 per cent for those expecting college education. Expectations for white-collar occupations increased likewise from 6 to 10 per cent.

3 There were more 'don't knows' for expectations than for aspirations. Apparently these parents knew what they wanted for their children but there was a high degree of uncertainty as to whether they could get what they want.

4 Within the six-year period, the shifts in aspirations and expectations were toward college education and white-collar occupations. The dreams and expectations of parents for their children were definitely away from farming in spite of some improvement in material levels of living, increased income attributed to increased yields by a majority of the respondents, and despite their modernizing outlook in farming. Apparently, however, the changes in farming have not made it attractive enough for them to want their own children to pursue it. College education, which they aspire to, is not associated with farming. Indeed, *barrio* people could not understand why people with college degrees (like the field technicians assigned to these *barrios*) would work in the *barrios* and get their feet muddy. However, it cannot be simply concluded that this college degree, white-collar orientation necessarily means a dislike for manual labour. A more plausible explanation seems to lie in the attraction to an occupational syndrome marked by less physical hardship, greater income stability, less exposure to the risks of natural calamities, and in many instances, higher income and greater security for the future. Nevertheless among some of these farmers studied, there were those (with the recent rises in yields) who definitely earned more than teachers and government employees. Perhaps it was precisely because of these gains made in farming that respondents aspire more and expect more because they can afford to offer a little more to their children.

5 While there were upward shifts in both aspirations and expectations for higher education and white-collar work, the aspirations expanded considerably faster, leaving a wider differential than before. Furthermore, when faced with the realities, 25 per cent of the parents thought their children would also be farmers although only 4 per cent wanted them to be such. The other type differential which can occur,

but which is not shown by the data at hand, is that between expectation and actual fulfilment. How many of those who expected their children to be in white-collar jobs would eventually have the satisfaction of seeing them get there? In terms of the larger social issues frequently being raised nowadays, which contributes to the build-up of social frustrations, the aspiration-expectation gap or the expectation-fulfilment gap? Does the existence of these gaps necessarily lead to the accumulation of 'steam' which could be potentially explosive? Unfortunately, the best answer this study has to offer is the general observation that within a six-year period the discrepancy between the proportions of respondents 'aspiring' and 'expecting' has widened from 30 to 55 per cent for college education and 25 to 43 per cent for white-collar occupation.

## EXPECTATIONS OF DIFFERENT LEVELS OF GOVERNMENT AND GOVERNMENT OFFICIALS

The contact between these seven *barrios* and government has been mainly through some minimal service from assorted development programs. Every two years of alternating local and national elections, however, *barrio* people find themselves the object of courtship, promise, and some level of delivery of material benefits: a stretch of road, a little bridge, a multi-purpose pavement, short-term employment on a public works project, a schoolhouse, and even a few pesos on voting day. On the other hand, the same *barrio* people have been accused of being too dependent on government. What is it that they are supposed to be dependent on government for? What do they expect of elected government officials (more commonly referred to as politicians) and of the different units of government which they represent?

The major expectation of the *barrio* captain was that he would bring *barrio* problems to the higher authorities and follow up whatever requests and petitions they had. Depending on the level of help available, this communicator-referral role was pushed vertically upward to the Municipal Mayor, the Provicial Governor, the Congressman, the Senators and, the President, the highest man in the hierarchy, who was expected to grant or approve their requests and petitions for assistance of all kinds.

The task of obtaining peoples' co-operation to discuss problems which affect the *barrio* so that they could be brought to the attention of higher level officials was assigned also to the *barrio* captain. There was a minimal expectation that government officials would come to the *barrio* for first-hand fact-finding and listening to complaints, and this was the only reason for wanting them to come to the *barrio*. If problems were 'appropriately referred' to the 'authorities concerned' and if they were acted upon, there seemed to be no pressing requirement that officials come down to the grass roots. The facilitator-intercessor and political dispenser roles were not assigned to the *barrio* captain. The power and resources were supposed to be at the higher levels from which financial, material, technical and other forms of assistance are supposed to emanate. In other words, to these *barrio* people, government officials represented a hierarchy of channels for pushing *barrio* problems upward till solutions materialised. Government officials were expected to facilitate this process by direct action on their part; by intercession or co-opting colleagues and personnel at their particular level of government; or by vertical referral to those in the next rung of the ladder.

Unfortunate though it is, legislation was rarely perceived as the role of the Congressman, and no one ever thought of the Senator in this role. Instead, these two officials emerge mainly as political dispensers. They also received the highest percentage of 'don't know' and 'can't do anything about our problems' responses in 1969 (34 and 46 per cent respectively). Their law-making function was observed by the consequences of the patronage system by which they got elected, hence they were perceived more as material providers than as law makers.

Respondents' expectations of different units of government with respect to possible help in the solution of *barrio* problems were parallel to their expectations of elected government officials. This finding comes as no surprise since the incumbent official represents the particular unit of government and is often seen as the personification of that level of government.

What was most significant about the data was the pattern of 'don't know' and 'none' or 'there is nothing they can do' responses. The percentage indicating they did not know what different government officials and government units could do to help solve problems of the *barrio* tended to increase with higher levels of government, i.e. towards the national level. In comparing 1963 and 1969, the proportion of respondents with 'don't know' answers went down for every government official and every government unit, but those saying 'there is nothing they can do' to help solve *barrio* problems increased. With respect to the President, in 1963, 15 per cent said their last hope was in him as the highest authority of the land, while in 1969, only 2 per cent gave this reply. The 'can't do anything' response increased from 10 to 33 per cent. The increase in this type of response for all officials and all units of government could be interpreted as:

1 A coming to terms with reality and with actual experience in dealing with government units and government officials who were not able to help solve their problems.

2 A disillusionment with government and elected officials manifested by cynicism side by side with opportunities. They knew only too well about unfulfilled promises and plans that never came through or came through badly for them. On the other hand, they could also point to projects (albeit unfinished or barely started) which had been supported by their Mayor, their Governor, or their Congressman and Senator. The President, in this regard, was rather remote from these *barrios* except for the ₱2000 cheques he distributed as part of the *Barrio* Development Funds before the 1969 elections. This may partly explain a rise in proportion of respondents (19 to 28 per cent, 1963 to 1969) who expected the President to play Santa Claus. Although the President ran on a 'rice, roads and schools' platform, there were no direct responses indicating that farmers in these *barrios* credited him with the improvements in rice production. At best, the indirect evidence was their increasing acknowledgment of government technicians as sources of information on modern farming.

From the experiences and data in this rural development project, the so-called dependence of *barrio* people on government was not a real dependence but more of an opportunism brought about by their voting power. Thus if they could obtain help from government, why not? One wishes at this point that they really did depend on government. If this were so, they would hold government and its officials to their duties and promises and perhaps eventually, government would be more responsive to their needs. However, the way things are, life continued for them whether or not promises were fulfilled. If the assistance came, it was welcome manna from heaven and adjustments to the better life brought about by windfalls were promptly made. The paved road is enjoyed and so is the new multi-purpose pavement. Because help is asked for on a catch-as-catch can basis, if it does not materialise, it was not seriously expected anyway and the verdict of 'nothing' may be disappointing but not hard to accept. Cynically, although optimistically, one waits for the next election for whatever blessings it may bring to the *barrio* whose occupants have learned to live with the political system and try to manipulate it to obtain whatever they can.

## POLITICAL PARTICIPATION AND CIVIC RESPONSIBILITIES

In relation to the preceding discussion on *barrio* peoples' expectation of elected government officials, it is significant to note the importance they attached to their voting responsibilities. There was a high turnout amongst respondents in both local and national elections (80 per cent or more in 1963 and 1969). Furthermore they indicated, at least hypothetically that they were prepared to overcome obstacles such as long distances from election precincts and avoid social distractions on election day in order to vote, and that they would vote for a candidate against their elders' wishes. For people who had only few grades of formal schooling and had minimal contact with government services, the respondents took voting seriously. Apparently they realised this was not only a duty but was a source of power, the only one they could use in dealing with a politician.

Further evidence of civic responsibility was shown by the fact that 90 per cent of respondents paid taxes. Nevertheless a third did not know how tax collections were spent, an indication that much needs to be done to bring the government and its functions closer to the *barrio* people.

## COMMUNITY PROBLEMS, LEADERSHIP PATTERNS AND PROPENSITY TO ORGANISE

### *Community Problems*

In assessing changes in these *barrios* which have had an input of agricultural technicians and Community Development workers, it is important to look at the *barrio* itself as a community, as a reference point for the individual respondents in the study. The most pressing problems in the *barrio* as perceived by the respondents were those of basic physical infrastructure such as roads, irrigation, water supply, schools, etc. These made up 47 per cent of those cited in 1963 and 54 per cent in 1969. The proportion of responses expressing need for irrigation, household water supply and electricity increased in 1969. This focus on basic infrastructure problems was a reflection of the under-developed state and sometimes non-existent facilities in the *barrio*. There was a decline in salience of production and income problems from 6 and 9 per cent to 1 and 3 per cent respectively, indicating that some other aspects of *barrio* life had assumed prominence besides the obvious infrastructure needs. Besides, the problem of production and increasing income was usually defined as the problem of the individual or the family rather than of the *barrio*.

Also of concern was the proportion (20 per cent) of respondents who said they did not know what the problems were or that there were no problems in the *barrio*. This should probably be taken as an index of indifference or a lack of a sense of community rather than of ignorance and absence of problems in the *barrio*.

When asked what is the best way to improve the *barrio* situation, respondents gave unity and co-operation among the *barrio* people at the top of the list for both years (33 and 40 per cent). Other responses in this direction are that on the need for good administration by the *barrio* council officials; the marked reduction in those who thought of asking help from government officials and authorities (from 20 to 2 per cent); the expressed need to construct infrastructure facilities such as roads, irrigation, etc., in order to improve the *barrio* situation; and the previously cited findings about respondents' increasing realisation that government officials and different units of government can do nothing toward the solution of their *barrio* problems. These responses together could well be interpreted as a sign that *barrio* people might be thinking of working out solutions by themselves or by means other than through political officials.

*Leadership Patterns*

For an insight into the leadership patterns in these *barrios,* three questions were asked: Who do you think are the most influential persons in your *barrio*? If there is a project to improve farming in your *barrio,* who do you think should lead in this project? Who are the farmer-innovators in your *barrio*? The respondents were also asked to indicate their reasons for selecting the individuals they named.

Leadership, influence, and innovativeness in farming revolved around only three or four persons in each *barrio* and in each, the three qualities were mentioned for all three individuals in 1963 and 1969. In one *barrio* there was some beginning of specificity in leadership with the *barrio* captain mentioned for general influence; another man was mentioned prominently for leadership in a farming project and still another one stands out as the recognised farmer-innovator. Leadership had been continuously in the hands of these few men for more than ten years.

There was a pronounced overlap of perceived general influence in the *barrio* and nomination as possible leader for a project to improve farming in the *barrio.* The *barrio* captains of all seven *barrios* were also nominated for leadership in the farming project. This was to be expected since the majority of these named leaders were farmers themselves.

The man named most frequently as a farmer-innovator in the seven *barrios* was indeed always first in trying anything new. He sought out innovations; was quite sophisticated not only in the 'whats' and 'hows' but also in the 'whys' of modern farming practices. Next to him in frequency of mention, was a woman farmer. She was an active decision-maker in the farm and underwent training both in crops and livestock production at the College of Agriculture.

Other features of respondents' answers were the sharp decreases in numbers who said nobody was influential or pioneered farming innovations in 1969 as against 1963, suggesting a greater recognition of these types. The numbers named as influential decreased, suggesting some crystallisation of views on the subject. In the case of leadership in a possible (*barrio*) farming project, respondents tended to assign it to the elected *barrio* official but this idea was vague to them and names were cited less frequently for this. The most noticeable change was in the area of farm innovativeness. In 1963, 61 per cent of respondents said nobody in their *barrio* was adopting farm innovations and in all, only 17 per cent of the total number of persons were cited by others as doing this. In 1969, only 17 per cent cited nobody, while others cited 27 per cent as innovators. Thus in the latter year there were more people recognised as farm innovators and more respondents had become aware of their existence and identity.

Those influential in the *barrio* were considered so because they were *barrio* heads, had good public relations, could be consulted on needs, were progressive, financially better off, responsible and dependable. All these qualifications, given prominence by respondents were related to *barrio* people's expectations of their officials.

Reasons given for considering certain persons as possible leaders for a farming project were almost identical with those for citing *barrio* influentials except that they were expected in addition to have had training at the College of Agriculture or from its staff and/or farm experience. The latter criterion lost weight between 1963 and 1969 and became of negligible significance, while the former gained prominence.

*Propensity to Organise*

One approach frequently used in agricultural extension and community development and repeatedly applied in this project is to work through and with organised community groups. It was found in 1963 that only 13 per cent of household heads were members of organisations. With the formation of a Farmers' Association and the Federation of Free Farmers, this had increased to 29 per cent in 1969. Furthermore, answers to hypothetical situations where group action on community problems was possible and their propensity to undertake such action showed positive change. Against these pointers, a high percentage stated they would not accept the position of *barrio* captain. However, in view of the hard work and lack of rewards monetary or prestige, such attitudes are understandable.

In assessing the implications for development of existing leadership patterns and the propensity of these *barrios* to organise themselves for problem solving, it must be borne in mind that, for a great majority of respondents (84 per cent), the *barrio* was to be the intended place of permanent residence. Developments at the national level will have to reach and be felt by these people in the *barrios* if their lives are to be affected. In turn, developments at this level have their repercussions on the world outside the *barrio*.

## RELATED DEVELOPMENTS

While the so-called green revolution was taking place in the rice paddies, other urban elements of the institutional framework were also responding to the situation. A 1970 survey[10] of these developments revealed the activation of latent entrepreneural potentials and the alert opportunism of existing entrepreneurs manifested in different ventures.

### *Credit*

Prior to 1967, private moneylenders were the only source of credit for the community. These comprised landlords, store owners, millers and employers, and in 1970 there were around thirty of them. Rates of interest were as high as 100 per cent and business was apparently lucrative. The familiar problems of borrowers and lenders were evident in the survey. Farmers were aware of the usurious interest rates but continued their patronage out of necessity. The moneylenders took a big risk since promise of payment at the next harvest was purely verbal and sometimes was not honoured.

In 1967, the Farmers' Association for Commodity and Exchange Services (FACES) was organised. This provides cash and commodity loans at 6 per cent per season or 12 per cent per year. Credit was, however, extended only to its members, and although there were plans for town coverage, in 1970 it was operating in only one *barrio*. There were only twenty one members of whom only thirteen had paid fixed deposits, thus reducing the availability of credit. Despite the advantages of FACES—low interest rates on loans, lower priced production inputs, storage facilities for rice and patronage refunds, membership had, in 1970, risen by only three since mid-1969. Reasons most commonly given for not joining were the difficulty of meeting the membership fee and fixed deposit requirements (₱50), and a preference to wait and see if it developed. Previous experiences with failures of farmer co-operatives have left a social scar which is hard to erase. Clearly, though, the Association needed more capital for its lending operations in order to meet farmers' credit needs (Dimaano and Guzman 1969).

The Rural Bank, established in 1968, offered loans at 12 per cent per annum but few were able to take advantage of the services because collateral such as house and/or lot together with the landlord's signature as co-maker was required.

Besides the Farmers' Association there were five stores selling production inputs such as fertilisers and chemicals. The first opened in 1965, two years after the start of the Rural Development Project in the area. Three others opened in 1968 and 1969. Among the fertilisers

10 Conducted by A. M. de Guzman and Lorna Paje.

and chemicals sold were urea, ammonium sulphate, granulated and liquid chemicals. On the average, these stores sold 150 bags of chemicals in granular form per season in 1970. Sales of small farm equipment such as rotary weeders and sprayers were handled by two stores. Sprayers, however, were no longer in great demand as they had been rendered more or less obsolete with the advent of granular systemic insecticides.

*Rice Processing and Marketing Outlets*

The existing rice mills provided milling and warehousing services. A farmer could deposit his *palay* and sell it to the warehouse at a later date or withdraw it to be sold to other dealers. Two warehouse millers provide the storage service and charged a deposit fee which varied with storage time. The rest provide only milling services.

There were four *cono* type and three *kiskis* type of rice mills in 1970. The *conos* were located in the town proper and the *kiskis* ones were found in the *barrios*. Three of the *conos* had been in operation since the mid-1940s while the fourth was set up in 1969 in response to the increased rice production.

There were eight known wholesale rice dealers in town. Five of them had their own transportation facilities for hauling rice. The smaller rice dealers hire transportation as needed. The common outlet for rice coming from this area was Rizal Province and San Pablo City in Laguna. There were other wholesale rice buyers from Tanay, Rizal who generally offered higher prices than the local dealers. The rice they purchased was purportedly being used as feed for duck raising, which was an important industry in Rizal.

*Hand Tractors for Hire and for Transportation*

The formerly indispensable *carabao* is gradually being replaced by hand tractors. The first tractor in the area was acquired in 1963. At the end of that year there were only about five tractors operated for hire. In 1970, there were 100 tractors in the area, owned mostly by tenant farmers and used for preparing the land for transplanting.

There were three repair shops servicing hand tractors in town. The first one was established in 1966. Before this, farmers had to go back to the assembly plant for major repairs or to the dealers from whom they bought these tractors. In most instances, the tractor owners have learned to take care of simple problems themselves and go to the mechanic only for major repairs.

The cost of one tractor was ₱6000 in 1970. Because most owners were small farmers, they bought on an instalment basis. Under this arrangement there was a down payment of ₱700 and payments of ₱700 every sixth months thereafter for a period of six years. Since there were no tractor dealers in the town, farmers have had to go as far as

Bulacan, Manila, Santa Cruz (the capital of Laguna) and San Pablo City to obtain these tractors.

During the off-season months, tractors were used for transportation and hauling of rice from the *barrio* to the town. In *barrio* Coralan, eleven *kubota* with improvised trailers attached to the hand tractor have been the only means of transportation since 1966. The net earnings from this function have been used to pay the instalments of the tractor or to repay loans. To service all these vehicles, there were about twenty jeepneys in 1970 (which have made regular trips to the neighbouring towns of Siniloan and Paete since 1965) and tricycles (which became the most common means of transportation in the town after 1968, when a gasoline station was constructed).

## CONCLUSION

Over a six-year period and with the input of development manpower (agricultural technicians and community development workers) and agricultural technology into relatively virgin, isolated and underdeveloped *barrios*, traditional rice farmers have shown encouraging trends in modernisation, not only in their response to agricultural innovations but also in their outlook and attitudes toward farming. Having started at a very low subsistence level they have now been introduced to the rudiments of 'scientific' agriculture and to the novel experience of having some surplus harvest over the normal requirements for the family's food. However, they have a long way to go before their production can be described as being commercial and market oriented, and which economists can conveniently computerise in an input/output analysis.

*Barrio* people participated in the green revolution, their lives have improved materially and they have gained a measure of linkage with the world outside their *barrios*. They are no longer as physically and socially isolated as before, for just as modern technology is an input from the outside so is surplus production an output to the market outside.

At the same time, aspirations and expectations have risen and dreams for the futures of their children are away from farming and markedly toward college education and white-collar occupations. *Barrio* residents are active political participants in terms of voting and fulfilment of citizen responsibilities such as payment of taxes, etc. Expressed needs of the *barrio* are largely for basic infrastructure facilities which are usually beyond the purview of *barrio* resources. Although there are some signs of an increasing propensity toward group action and organised efforts to solve community problems, the position of *barrio* captain remains unattractive. A new criterion of technical expertise in agriculture has emerged as a basis for leadership in the *barrio*. Government and government officials are still defined as political dispensers and Santa Claus but there is a growing realisation that different government units and elected government officials cannot do anything to help solve their *barrio* problems.

As far as issues of equity and growing income disparities are concerned, it is difficult to avoid the suspicion that some well off townspeople have become wealthier but this was inevitable without any effort on their part. From this study, one can at least take comfort in the observation that the small, poor farmers have become a little bit richer too. To those preoccupied with problems of man-land relations, available data and inferences from these data indicate that the critical factor in the near or distant future may not be the issue of share tenancy versus leasehold but rather the starker reality of farmer-cultivator versus farm labourer. The problem is whether or not there will be adequate arable land to cultivate.

Despite the obvious material gains made in these *barrios,* it is misleading to think that affluence has set in. Earnings are far from adequate even for ordinary everyday living expenses let alone comfort, emergencies and old age. It suffices to say that these *barrios* have begun to be part of a *developing* society. Sustaining the gains made by moving on from this stage of development is by no means a trivial undertaking.

12

# Agrarian Reform in Asia: the Green Revolution and its Reform Effects

*Wolf Ladejinsky*

A quarter of a century has passed since Japan inaugurated the first agrarian reform in modern Asia. In the years following, 'land to the landless' became a critical issue throughout rural Asia. It no longer called for banner headlines when heads of State after State proclaimed their devotion to agrarian reformism in search for an answer to the poverty of their agriculturists, the low productivity of their agriculture, and the political instabilities these bring in their wake. Agrarian reform and the wrenching readjustments it calls for are as old as recorded history. With due allowance for the passage of time and changing conditions, the problems now, as then, are fundamentally the same: how to root the peasant securely and beneficially on the land he cultivates for a pittance, and how to ensure a country's political stability. The answer to these questions constitutes what is broadly known as agrarian reform. Agrarian reform in the sense considered here encompasses all or most of the following elements: distribution of land among the landless and favourable financial arrangements for tenant land purchases; security of tenure and fair rents; better methods of cultivation through technical assistance, adequate credit, co-operative marketing facilities, etc. Yet, while agrarian reform is a combination of a great many things, not all of them are of equal importance. In particular, unless those who work the land own it, or are secure of it as tenants, all the rest is likely to be writ in water.

Very germane to this problem, and exacerbating it, is the rise of the green revolution with its new and innovating package of practices and technology. It can be said here in passing that while the green revolution is the most promising portent for a sharp rise in agricultural productivity, at the moment it is anything but a boon for the unfinished business of agrarian reform in Asia. In India, at the moment, there are ample signs that the burden of the change falls squarely on the tenants. Numbers of them had been separated from the land before the advent of the green revolution, and more are in danger of facing the same process in consequence of it. The reasons for this will be dealt with in the second part of this chapter. All that needs to be added here is this: it is a fair surmise that the above consequences hold true of any Asian country where the new technology —only as technology—is beginning to make its appearance and where agrarian reforms are still in the doldrums.

## AGRARIAN REFORM

### *Realities of Asian Reforms*

Today, nearly two decades after the start of the rural Asian reform movement, the reforms have certain things in common, and not only with regard to the condition of the peasantry. Their purpose is the same whatever the wording of the official pronouncement. On the issues of substance, however, the reforms speak in a variety of voices, which more often than not are a far cry from the exalted pronouncements made on the supposedly historic occasions.

They reflect inevitably the political climate in the country concerned. This determines the will or lack of will to proceed with the task; the kind of specifics with which the general measures are or are not endowed; the care or lack of care with which the enabling legislation is formulated; the preparation or lack of preparation of the pertinent technical and administrative services; the presence or absence of technical agricultural services with their bearing upon success or failure; and, finally and most importantly, the drive or lack of drive behind the enforcement of the provisions of the law. For these reasons, the reform movement in Asia runs the full gamut, from Japan and Taiwan where fulfilment in the widest sense has been achieved but, as will be noted elsewhere, under very special circumstances; to Korea where much land has actually been redistributed but with less satisfactory results; to Iran where the peasants have been endowed with land but not much else; to South Vietnam where land reform measures have been engulfed by a civil war before they came to fruition. Then there are the Philippines, Nepal, Pakistan and Indonesia— all knowledgable in the writing of reform laws—where the record to date ranges between poor performance and non-performance. Finally, there is India, significant and encouraging for what it has attained in unprecedentedly difficult and bewildering conditions, and just as significant and discouraging for what it has failed to attain, and why it has failed. The case of India is sufficiently important to deserve separate treatment.

The picture, clearly, is mixed, and in order to assess it this question must be raised: why have some reforms succeeded, others fallen short of their goals, and still others failed to get off the ground?

*Japan and Taiwan*

Just as Soviet Russia was the progenitor of the communist type of reform, Japan and Taiwan are the progenitors of non-communist reforms in Asia. Japan, the leader, and Taiwan, the follower and innovator, provided between them all the pertinent elements of leadership, content and implementation which made for a successful reform.

In Taiwan as in Japan, reforms were not designed to satisfy the claims of both contending parties: the tenant was to gain at the expense of the landlord. Security of tenure is one of the cases in point. The sharp reduction of rents in Taiwan was an important move; but more important were the provisions that allowed, for all practical purposes, the tenants to remain on the leased land undisturbed even after the expiration of a contract. This virtually ensured the enforcement of rental provisions. Less carefully worded stipulations might have undermined this part of the reform program, as has been the case in a great number of other countries.

Provisions for security of tenure in overcrowded Asian villages, where tenants compete fiercely for the privilege of cultivating somebody else's land, are notoriously difficult to enforce; yet in Taiwan a land redistribution program for the benefit of tenants who cultivated 40 per cent of the land became much more feasible as a result of such provision. With security of tenure and rents effectively reduced, the price of land subject to reform declined sharply. The result was that the traditional approach to land as a prime investment became unattractive. This in turn created a favourable condition for the last phase of the reform—the 'land-to-the-tiller' program—a condition which did not exist in any other countries of Asia concerned with reform.

The idea that the reforms were meant to benefit the tenants is also apparent in the principal provisions about land redistribution and the creation of peasant proprietorships among tenants in these two countries. Absentee landlords had to sell all of their land at fixed government prices; resident landlords were compelled to sell land in excess of the permissible retention or ceiling. Neither Japan nor Taiwan aimed to do away with tenancy as an institution; but with determination to enlarge the area of individual, private ownership, the low ceiling on land proprietorship made it possible to extract a great deal of surplus land for redistribution. The result was that, whereas before the reform 54 per cent of Japan's land was owner-operated, after the reform the figure rose to 92 per cent. The respective figures for Taiwan are 60 and 85 per cent.

What matters is not only at what level the ceiling is set or the size of the rentals, or the character of the security of tenure provisions, but that all these and other measures are effective. The ceiling problem of

Pakistan,[1] to cite but one example, is not merely that 200 irrigated hectares and 400 unirrigated hectares are altogether too high in a situation where two thirds of the owners average 2 hectares each; it is also the exceptions and subdivisions of large holdings among members of a family which were made on the eve of the reform and which have combined to divest the ceiling, and, by the same token, the entire land distribution program of any meaning. It is not surprising, therefore, that even if Pakistan had implemented this part of the reform, only an estimated 7·5 per cent of the country's 2 million tenants might have obtained land. And Pakistan is not an exceptional case. In Japan and Taiwan, on the other hand, the area that could be retained by a landlord was fixed retroactively, both on the basis of the household as a unit and of land owned by that household. The same care was exercised about every measure intended to attain the reform goals and the results surpassed expectations. There was a sharp rise in peasant proprietorship, a reduction in tenancy and an increase in security of tenure, investment in land improvement, improvement in agricultural techniques, expansion of agricultural productivity, an increase in the standard of living, and a welcome reshuffling of the power structure in the village community. It also created a new and active concern on the part of the politicians about the state of agriculture and the welfare of agriculturists. Most important also, is the fact that in both countries the unprecedented economic growth in subsequent years has served to solidify the gains already made.

### *The Special Case of India*

It is safe to say that all the disabilities which peasant and land can suffer are to be found amongst the 600,000 villages of India. An observer will find many striking and promising manifestations of a resurgent agriculture; but these are still only islands in the vastness of the debilitating conditions under which large numbers of tenants, share-croppers and agricultural labourers cultivate somebody else's land. Of a recently estimated rural population of 434 million, about 103 million owned no land at all and another 185 million operated less than 2 hectares per family. Taken together they represented 67 per cent of the total rural population, and of these 110 to 154 million lived in abject poverty, or at a level of Rs200 ($US21) per capita per year. This situation, though involving fewer people, prevailed in India twenty years ago when the Indian Government decided to ease the lot of the peasantry by a drastic overhaul of the land system, the complexity of which almost defies description. This would-be agrarian revolution was to have been attained through the familiar pattern

[1] Refers to West Pakistan before the recent split.

already described, and through the elimination of the *zamindari* system, a peculiarly Indian institution.

The *zamindari* system was a by-product of the early British rule under which a *zamindar* or intermediary was given the right to collect land taxes and undertook to pay the British administration a fixed revenue. In return, he was permitted to keep a portion of the revenue and came to be recognised as the proprietor of the revenue-bearing land. In time the system covered much more than 40 per cent of the cultivated land of India, and it created some of the worst abuses that can be perpetrated upon a peasantry, including a long chain of non-cultivating sub-lessees all of whom received a share of the highly inflated rent from the same piece of land and the same cultivator.

Despite opposition and administrative problems, the *zamindari* tenures have been virtually abolished. Not all have benefited equally and not all the 20 million cultivators affected have received permanent, heritable and transferable rights without any strings attached. Nevertheless, the effort was undeniably a first and major step. It cleared the ground for a reconstruction of Indian agriculture. Why this measure succeeded is not difficult to answer. The *zamindari* system, with its absurdities and injustices, was the weakest enemy to attack, because it was imposed by a foreign power which handed out property rights to which neither the British nor most of the recipients had any claim. Thus abolition of the system became one of the symbols of freedom from the British rule, and it is not surprising that the abolitionists largely succeeded in eliminating it.

Getting rid of this system did not improve tenancy conditions in other parts of India. There was the multitude of tenants—not to mention the millions of agricultural wage labourers—cultivating at least a fourth of the country's arable land in *ryotwari* areas (as distinguished from the former *zamindari* areas) where owner-proprietorship predominated. The Government of India decided to provide tenants with security of tenure and reduction of rents, and to confer land ownership upon them through the familiar ceiling device and officially fixed land prices. Under the guidance and continuous prodding of the Planning Commission, the States have enacted a voluminous body of legislation presumably designed to meet these aims.

From the point of view of both the content of the legislation and the enforcement of it, these reforms are in serious difficulties. Here and there tenants have secured better tenure conditions or have acquired land. But as against such cases there are widespread evictions of tenants that can be traced to the legislation itself, the prevalent failure of the new rental regulations, and the wholesale evasion of the ceiling provisions upon which an increase in peasant-proprietorship was to rest.

Administrative problems are a formidable obstacle to the implementation of the reforms, but more to the point is the faulty content of many legislative enactments. In India, the most glaring manifestation of this is the seemingly reasonable but ill defined right of the landlord to resume tenanted land for what is euphemistically called 'personal cultivation'. This clause has led to mass evictions of tenants; to 'voluntary surrenders' of land by tenants in order to salvage some relationship to the land, even if it be as a hired hand; to augmentation of the ranks of agricultural workers; and, inevitably, to the failure of security of tenure and the new rent regulations.

The ceiling provisions did not fare well at all. For the moment, the question of how much land might have been available for redistribution is academic; of India's 28-32 million hectares of tenanted land fewer than a million hectares have been redistributed. In anticipation of ceiling provisions, the landlords divided up the land among members of their families so as to make certain that holdings were under the ceiling. Needless to say, such developments do not produce the incentives which lead to better living conditions, investments in land improvements, a rise in agricultural productivity, and thus to political stability. Yet these were the goals of the tenancy reforms.

It should be noted that while the need for reform in India is surely as great as in any country in Asia, the difficulties in the way are incomparably greater. Among them are the sheer size of the subcontinent; the administrative decentralisation, with each State a law unto itself; the paucity of good land records; the fact that a third of the tenanted land belongs to owners with 2 hectares or less; the fierce competition for any tillable plot of land on almost any terms. The peasant lacks initiative and is unable to comprehend the complex laws; there are poor prospects for alternative occupations despite the country's progress of industrialisation; and millions more people are added annually to the already overcrowded land, as illustrated by the fact that between 1961 and 1976 the population of India will have increased by 190 million.

All these are sufficient to give one pause before rendering any hasty judgment about the tortuous and far from successful path of Indian reforms (other, that is, than the elimination of the *zamindari*). And yet the handicaps do not quite explain why so much of the intent of the reform is still unrealised. What is significant is that most of the handicaps, poor content and poor enforcement or non-enforcement, are not always causes, but in a large measure consequences of attitudes displayed by state politicians and legislatures. This anti-reform sentiment has proved to be a crucial element in thwarting India's expectations.

*The Minimum Reform Program*

Despite the past record the reform issue may yet get off the dead centre. Future events may compel such a move. If so, what might be the minimum program to which the States of India should address themselves while there is still time?

First and foremost is the preparation of a record of tenancies without which no tenant's claim could be legally established and no tenant could obtain security of tenures. Written lease agreements should replace oral ones, and receipts for rents should be made mandatory. Crop share rents are extremely difficult to regulate and should be replaced by cash rents. To facilitate this, either the rents should be fixed as a multiple of the land revenue (as in Maharashtra and Gujarat), or else the State Governments should divide land in each area into a few broad categories and determine the average produce of each such category of land and the average price at which it may be converted into cash.

As a rule of thumb no rental should exceed a third of the crop, particularly where the tenant is responsible for all the inputs. Under Indian land/man ratio conditions, no rental below the 50 and more per cent of the crop can be enforced unless the tenant has the assurance that, barring misuse of the land or non-payment of rent, he can remain on the land undisturbed and with a right of lease renewal. This is the basic part of security of tenure and is crucial to the entire scheme of land legislation. To make this effective the most debilitating part of land reform in India, the ill defined right of an owner to resume or take back the tenant's holding for personal cultivation, should be abolished. In practice no tenant voluntarily surrenders land, so that this is accomplished under the threat of total separation from the land. If land is surrendered voluntarily then other tenants should be settled on the land. In all other instances, i.e. the majority of them, other tenants should not be permitted. Most of the surrenders are past history and restoring the tenanted on their formerly cultivated land is about as difficult as undoing the mischief of ceiling provisions. Nevertheless, no program of this kind can overlook it altogether. Finally, in any revised tenancy legislation, access to water where available should be included, especially to encourage the use of the new farm practices.

If some such measures are enacted and enforced, the familiar discrimination of the credit co-operative vis-à-vis tenants would undergo a drastic change. It is important too, that tenants are given a sense of participation in the implementation of a land reform program. With that in mind, and as is established practice in Japan and Taiwan, village Reform Committees should be established with a suitable representation of tenants, owner-lessors, and owner-cultivators. Under the name of Tribunals, Kerala is attempting a version of this in its revised legislation. The Committee would surely know anything the officials do not know about who is who in the village, ensuring the correctness of the procedures.

The ineffectiveness of land ceilings has been so complete that any corrective measures are probably futile. Nevertheless, the current attempts of one or two States to lower the size of the permissible retention is indeed welcome, as is prevention of land accumulation in the hands of big owners, notably in Punjab. There, between 1955/6 and 1967/8 holdings in the size group 40-60 hectares expanded by 40 per cent. How successful such measures might be is another matter. Considering the past record it is very doubtful that what has been so thoroughly scrambled up with impunity can be unscrambled, especially at a time when the role played by the bigger owners in the new technology is deemed so important. Besides, an effective implementation of lowered ceiling provisions would impose an administratively and politically difficult task of declaring null and void almost all former ceiling claims. Last but not least the ceilings would have to apply to holdings below 8 hectares, in order to create a good sized pool of land for distribution among the tenants; a proposition which would be doomed to defeat. For all these reasons, it is this author's considered view that now as in the past the crux of the issue is not land redistribution but a sound security of tenure system and its enforcement. This is where the emphasis should be, if and when action, as against verbalism, should come into its own.

### *The Lessons*

From the experience of Japan and Taiwan and from that of India one may learn why so few reforms in Asia have succeeded and so many have not. The primary lesson is that if agrarian reforms are to succeed, the reform idea, to begin with, must be accepted in such a manner that technical problems are not an excuse for inaction but are something to be resolved. The second and equally important lesson is that the decision making politicians must accept the reform idea with all its implications, for it is they who make or unmake agrarian reforms, and who as already mentioned, decide between reform and 'reform'.

There is no gainsaying the fact that the economic environment which includes the rate of overall economic growth, population pressure on the land, and customary relationships sanctioned by a a long history of social and religious traditions, exerts great influence on what happens to legislation designed to break old institutional moulds. But this does not invalidate the main premise, that the content and implementation of agrarian reform are a reflection of a particular balance of political forces in a country. This premise assumes even greater significance in Asia because the peasants themselves, while discontented, have not developed a movement, either in the form of tenant unions as in Japan before the reforms, or as peasant political parties like those of East Europe after World War I. Hence the third lesson, that by and large national and state legislatures in Asia do not represent the interests of the peasantry. If they did, reform might have taken on a different character altogether.

The fourth lesson is that in Asia, and beyond, it takes very special circumstances to ensure the writing and the implementation of a meaningful reform and this is illustrated by Japan and Taiwan. In Japan, the defeat by the United States and the American influence as an occupying power were crucial in the timing of the reform. But other factors were also important: the memories of peasant rebellions; the numerous, if unsuccessful, prewar reform measures; the strong tenant unions; the emergence of large groups of Japanese who were disillusioned with the old oligarchy; and an eager and active pro-reform leadership in the Ministry of Agriculture which drafted the enabling legislation. The sum total of these was expressed in clearly defined provisions, a minimum of half measures, a minimum of loopholes, and strict enforcement. Under 'normal' circumstances the same kind of reform would surely not have taken place, if at all. The situation in Taiwan on the eve of the reforms was not the same as in Japan, but here too, special circumstances, primarily non-economic or sociological in nature, created the setting for action. The Communist victory on the mainland and the subsequent prevalent belief among the Nationalist politicians that the Communists won because of the promise of land to the tillers played a crucial role in creating the favourable climate for reform in Taiwan. Certain elements in the Nationalist ideology worked to the same end especially when the beleaguered government recognised its need for greater social stability as a means to military security.

The final and most important lesson is that the key to successful reform in Asia is the degree to which the controlling political forces of a country are willing to support reform and their readiness to use all instruments of government to attain their goals. The fact is that those against whom reforms are directed are never ready to divest themselves of their property and of political and economic power simply because a government writes out a decree to that effect. The conclusion is inescapable, at least for the present. If the peasantry is to get what is promised, peaceful and democratically managed reforms are not going to suffice. Government coercion whether practised or clearly threatened is virtually unavoidable. There is ample evidence to substantiate this.

It is generally supposed that the Japanese and Taiwanese reforms and the abolition of the *zamindari* system in India were peaceful affairs. A closer look reveals that they were peaceful because the landlords in the first two countries knew that overt opposition would have met with drastic punishment. In India the *zamindars* knew that public opinion was overwhelmingly against them and as the British went, they went too. The Taiwanese reform took place under the aegis of a military authoritarian government. It is unlikely that General Chen-Cheng, the activist of the reform, had ever read Theodore Mommsen's comment on how the Roman politicians, having killed Tiberius Gracchus for his land reform efforts proceeded later on to enforce his program. 'Loud and often well-founded as were the complaints', Mommsen (1958 : 54) wrote, 'the Senate let the [land] commission have its way; for it was clear that if the land question was to be settled at all, some such unceremonious vigour was necessary.' General Chen-Cheng left no doubt in the mind of the opposition that he was ready to proceed with a good deal of unceremonious vigour.

One of the by-products of all these lessons is that, in Asia as elsewhere, only the countries in question can deal with their reform issues. No outside institutions or individuals, however well-meaning, have much to contribute to problems which spell passionate adherence to property, social status and political power on the one hand and attack on the other. It has all been tried, ranging from studies in depth to interested advice, and it has all proved that their efforts yield insignificant results. There is, to the author's knowledge, no case where in ordinary conditions even so involved an organisation like FAO has succeeded where others have failed. In short, in the kind of a confrontation reform gives rise to, it is the better part of wisdom for outsiders to stay out of it.

The story of why few agrarian reforms in Asia have succeeded and why many more have failed is not a cheerful one, but the end is not yet in sight. Landlords know, perhaps more so than tenants, that the issue will not fade away, not for example in a country like India where 20 per cent of the cultivators owned 75 per cent of all the land while 50 per cent owned less than 10 per cent, and multitudes none at all. The August 1970 nationwide attempts at land grabbing were a failure. Yet, it was a telling reminder to the bigger land-owners that while the cards are still stacked in their favour this situation is not forever. The extreme use of police force in the form of some 20,000 or more preventive arrests left the basics of the problem unchanged, while at the same time underscoring that this burst of official vigour only served to heighten the existing tensions. Certain effects of the green revolution lead to the same conclusion, because they materially strengthen the top crust of the owner class and undermine still further the already depressed position of the tenantry. But this, too, is not a closed chapter, for the very process of this polarisation (about which more later) creates its own imperatives for tenurial changes. A review of certain aspects of the green revolution based on the Indian experience throws considerable light on the process discussed here and on its possible outcome.

## THE GREEN REVOLUTION AND ITS EFFECTS ON AGRARIAN REFORM

### *The Plus Side*

The green revolution, or perhaps more correctly the 'green evolution',[2] is barely four or five years old. To date, its scope in India is limited to about 8 million hectares of irrigated land concentrated largely in North India, or about 6 per cent of the country's total cultivated land. The greater part of this land is under high yielding wheat varieties. It is not within our purview to explain how this new development came about, nor is it our task to recount in any detail the workings of the new technology as a means of raising productivity, or to speculate about problems that mounting output might bring in its wake. For India, with the new strategy in its infancy, this is premature. Our preference is to treat only those problems which have already made their appearance.

As background, this much is worth noting in summary fashion. What is taking place is a change in the sense that the combination of an adequate and controlled water supply, high yielding varieties and concentrated doses of fertiliser—to mention only the three principal ingredients—has brought about a two-fold increase in wheat yields, a rise in double cropping, and, in consequence, a sharp increase in output. A breakthrough in rice productivity, at least in India, is still a few years off, but even so, for the first time in many years, food shortages do not claim the headlines: with a record foodgrain output of 100 million metric tons, the food position of India is presently comfortable.

The most encouraging aspects of the shift from traditional to modern farming is not so much the physical output as the increasing use of inputs, and the willingness of the farmers to invest and take risks—and to be well rewarded for it, as is indeed the case. Aside from this and further improvements in store, the new widespread psychological attitude about the effectiveness of the new farm practices is of unquestionable importance. The desire for better farming and a better standard of living is evident not only among the relatively few 'revo-

[2] The term coined by Dr Morton Crossman of the Development Advisory Service, Harvard University, taken from Falcon (forthcoming MS:8, note 9).

lutionaries', but also among a great many farmers who, for a variety of reasons are still only observers of the new package of practices.

For those concerned with purely physical indicators of economic growth, a psychological change of this sort, not subject to quantification, is probably of no account as a development factor. But it cannot be denied that a new, if unquantified, factor of growth has been introduced. It is the recognition that these tangible and intangible elements are in the realm of the achievable that has given Indian agriculture a sense of direction and a promise that productivity is bound to grow in the hands of those who do practise the new technology. Its significance is not invalidated because this holds true only for a minority of India's 70 million farm families, and therein lies the current and potential contribution of the green revolution to agricultural productivity.

*The Seamy Side*

Not all of the prospects just noted may be realised, and yet it is just possible that in the forseeable future India's race between food and population will be won. If this comes about, it will not be without some flies in the ointment. Even though an increasing number of farmers—including small farmers—are engaged in the new technology, only a minority of the cultivators will have contributed to a state of food self-sufficiency. The drive for higher productivity does not, therefore, necessarily alleviate the condition of the rural have-nots. Many are left by the wayside, for the propitious circumstances in which the new technology thrives, like adequate resources to begin with, are not easily obtainable for multitudes of the small farmer-owners, and even less so for tenants with their additional problems. Constraints on the scope of productivity increase and the limited involvement of these groups of farmers are inevitable.

Even where the new technology is practised successfully, it is inducing a host of economic, social and political problems which are surely not a model for a new agricultural strategy. In the absence of appropriate counter measures, the benefits of development, whether governmentally sponsored or not, often accrue according to the law of increasing returns to the rich. In rural India, this is beginning to be demonstrated in the form of rising violence between the haves and the have-nots. It is no wonder that while food self-sufficiency is indeed a welcome anticipation for India, there is manifested concern that, for all the green revolution's technological feasibility, it might fall short in alleviating, indeed may add to, the crucial problems of unemployment and inequalities in income distribution (both relatively and absolutely), the rising sense of social injustice and the apparent perpetuation of an outmoded and exploitative tenurial structure.

This outcome explains why the extravagant anticipations of the new technology's performance have given way to a more sober appreciation of its limitations, and of the issues it has aggravated, even if unwittingly. It explains the fear that 'the red revolution goads the heels of the green' (Fleming 1970 : 99). It is the seamy side of the green revolution.

It will be pointed out elsewhere that dislocations are practically an unavoidable part of the process of growth and of the nature of other long-existing impediments, and cannot therefore, be placed solely at the door of the green revolution. On the other hand to the extent that the green revolution exacerbates them, they constitute the seamy side of that very revolution. In the meantime, it is well to detail the evidence that points to the disturbing side of India's new agricultural strategy.

*The Evidence*

*Small farms, income polarisation and the new technology:* whatever the statistical picture, an interested visitor to Punjab (the premier state of agricultural development), Western Uttar Pradesh, parts of Andhra Pradesh, or even a state like Bihar (which is still characterised by agricultural primitivism and pervasive poverty), can readily see the positive side of the technological changes. In Punjab for example, an innovation-minded group of farmers, mostly in the owner class of 4 and more of irrigated hectares have in a short span of years succeeded in translating promise into reality. More and more tubewells, more of all other inputs, and bigger crops, and making money is a new way of life. But this does not obscure the fact that in Punjab as elsewhere this picture of achievement is marred by the selective character of the changes, that is by the absence from the process of a majority of the cultivators. Good and bad reasons explain this, but all lead to the same thing: income polarisation on a widening scale.

Among the 'good' reasons, one should note that in typical Indian conditions of inequality of land ownership, resources and marketed surpluses, income inequality is the normal state of affairs. Another 'good' reason is the so far unavoidable fact that the entire new agricultural strategy is predicated on an assured supply of water whereas in India, rain fed farming predominates. If for no other reason, vast parts of the country have not been touched by the transformation at all and equally vast parts can boast only of small islands within. Rice, which accounts for 31 per cent of the foodgrain area, is only beginning to show signs of improvement here and there and must wait its turn for the development of suitable high yielding varieties. Of necessity therefore, and for the time being, these two facts of life leave the overwhelming number of farmers beyond the pale of the green revolution. In these circumstances the new technology has served to augment regional income imbalances by raising incomes in the technologically affected areas.

The same is true within regions and villages affected by the green revolution, for it is not only environmental conditions or lag in technology that causes the new package of practices to favour the few as against the many. In those very areas there are large numbers of cultivators who lack resources, or are institutionally precluded from taking advantage of the new agricultural trends. In both instances, for example, it is not so much the absence of water resources or unavailability of other marketed inputs as often the lack of means to dig a tubewell or buy the inputs. Because of these handicaps the correct assumption that the new technology is neutral to scale breaks down in practice. If the widening income gap between the large and small owners is to be reduced and the scope of the green revolution enlarged, the less privileged must be assisted to secure the costly and highly productive resources. It takes Rs10-12,000 to re-equip a 3-4 hectare holding, and more often than not the sum is out of the reach of the majority of farmers. Credit is not readily available, and is sometimes unavailable, to bridge this gap. This condition has little to do with the green revolution as such, but is rather a reflection of the village power structure which is very much biased in favour of the affluent. Whatever the causes, however, the argument is not against modernisation that makes the rich farmers richer still, but against its limited scope and the growing income disparity that leaves the poor peasant relatively poorer.

In more concrete terms, the Kosi irrigated district of Purnea (Bihar State) may serve as an illustration. There, even with water more readily available (together with the requisite resources) and the new package of practices no longer a stranger, the poor economic condition of the small farmer-owners has undergone relatively little change. If the small farmers are assumed to be those with 2 hectares and less, they would account for 75 per cent of all the households and a third of the cultivated land.[3] The enormity of the task that lies ahead of the new technology is apparent if the income of this group of farmers is to be raised and if the unevenness of its global distribution is to be lessened.

The new practices in Kosi are quite evident now that the irrigation project is gradually coming into its own and the dramatic expansion of the wheat acreage and yields illustrates the change. And yet, 68 per

[3] Not all the farmers with 2 hectares and less are necessarily small farmers—if they have water—just as not all the farmers with two to three times as much land are big farmers—if they have no assured water supply.

cent of households in Purnea do not use high yielding varieties; 80 per cent do not use locally improved; 70 per cent do not use chemical fertilisers; 91 per cent do not use improved implements; and 80 per cent of households do not use improved methods of cultivation. This cannot be explained only by backwardness and water shortages. Lack of owned resources and the failure of credit facilities to fill the gap contribute measurably to a situation where income per hectare on such holdings is less than Rs500, a level only one third to one quarter of that enjoyed by the progressive, well to-do farmers.

Purnea is not an isolated case, and nor is selectivity, which is the main prop of the green revolution. While these factors lead to the exclusion of large numbers of farmers from the new technology, by the same token they also lead to the growing disparities of income at a time when aspirations for betterment are widespread among all classes of farmers. It takes little imagination to surmise that denial of the opportunity to share in advancement with the more privileged farmers creates potentially disturbing problems. This is the other side of the coin in any assessment of the course of the green revolution.

The saving grace is that this situation is probably not immutable. It is true that the green revolution is accentuating the already existing sharp line of demarcation between the rich and poor farmers, but schemes such as the Kosi irrigation, and the package of new practices it generates, are diluting much that is undesirable in the Indian rural tradition, and are turning the minds of big and small farmers alike toward change. This, as noted earlier, is an important pre-condition of growth, and undoubtedly some small farmers capitalise on it. But if it is to acquire practical meaning for the greater number of disadvantaged ones, the current emphasis of the new technology on productivity above all else must somehow take into account the bothersome but vital consideration of the provision of the means to participate in the new technology. Until this crucial issue is paid the attention it deserves, regional and intraregional polarisation of the countryside, will continue apace as the green revolution expands along the old familiar lines. Let it be noted once again that, while this course may even assure India's self-sufficiency in food, it will hardly alleviate the problem of income disparities and much worse besides.

*Tenancy and the green revolution*: in the context of the new technology the story of the small farmers is not a pretty one, but the case of the tenants and share-croppers who cultivate somebody else's land is worse. The green revolution has not formally imposed new restrictions upon them; informally, however, they are there just the same. The very goal of raising output as quickly as possible has developed side effects which threaten their already unenviable status. Some of them are likely to be consigned to the lowest position of agricultural labourers, thereby undermining still further their capacity for making a livelihood. If it is recalled, that barring some exceptions, the lot of the tenant-farmer all over Asia leaves much to be desired, the subject discussed here has relevance beyond India.

Profitability is the hallmark of the new technology. Attractive support prices, in the virtual absence of land and income tax, makes modern farming in India an enticing proposition. Along with it is the inevitable rise in land values. Inflation has played its part of course, but it is the penetration of the green revolution that has pushed up land values to new peaks. There is not enough land changing hands to turn it into a boom-and-bust real estate operation, but in relation to the little that is being bought and sold, the farmers' refrain is the same: today's land prices are going to be higher tomorrow. Land sold five years ago at Rs5,000-12,500/hectare sells now at three, four or five times more. The positive aspect of profitability and rising land values is evident in the changing attitude of using savings for investment in the current phase of agricultural innovations. More than ever before, land subject to the new farm practices has become the 'real gold', and it is understandable that well-off farmers also invest in more land that successful farming with tax-free income can buy.

For tenants, rising land values are anything but a boon. In the absence of any effective restrictive rent measures, one of their immediate effects has been the rise in land rents. In Punjab, for instance, the Punjab Security of Land Tenures Act provided for a rental not exceeding one third of the crop. But this and other provisions like security of tenure, ceiling on holdings, occupancy rights, etc., have been so emasculated that the very title of the Act is one of the outstanding misnomers of the agrarian reforms of India. Following the enactment of the Act, there was large-scale ejection of tenants and gross evasion of the ceiling provisions, but rents remained at their customary levels of 50 per cent of the crop. Now rents have moved up from 50:50 to 70:30 in favour of the owner, or from Rs1,250-1,875/hectare where the arrangement is in cash. This holds roughly for all the areas where modernisation is in progress.

Since a hectare of irrigated wheat land on a well managed farm yields 3·7-4·6 tons per hectare or more compared with 1·8-2·3 tons per hectare formerly, in theory at least the tenant should receive more than before when he had been paying 50 per cent of the crop. This presupposes, however, that he uses the package of inputs in a manner applied by innovating farmers. Owner-farmers with sizable holdings who also lease in land are in a position to do that. For the tenants we are talking about, a 70:30 division is something else again. Under such terms they are in practice little better off than agricultural labourers. They submit to the terms, however, because even in the Punjab with the general pulverised pure tenant category, there are always others, whether share croppers or labourers, willing to take their place.

In sum, rising rents in consequence of rising land values may be an economically valid and unavoidable process, but they are a hardship for those who cannot extract from the land a return large enough to meet living requirements over and above the rental obligations and inputs which must be largely their own. The position of the tenants is further undermined by the recent trend for larger owners in the technologically affected areas to compete for leased land.

The already shaky situation as regards security of tenure is showing signs of deterioration. Under the impact of the green revolution owners would like to complete the enclosure process by which tenants are reduced to the lowly status of agricultural labourers. Owners contend that while land is indeed a prized asset, it would be more so if unencumbered by any claims of the tenants to any rights in the land, even if in practice they are minor. Once freed of these claims, owners can replace tenants with hired labour which has no claims on land whatsoever. The ostensible rationale underlying replacement is current farm policy which calls for efficient agricultural production. This in turn implies substantial resources, larger and compact holdings and centralised management in the hands of a single owner. Tenancy and share cropping are seen to have no place in it.

How far owners' wishes could be translated into reality is another matter. Statistical evidence one way or another does not exist, but observations on the spot indicate that covert separations from the land are not rare. On the other hand, the signs of the time are against a wholesale dispossession of the remaining tenants. Even the old familiar blatancy of not recording the tenants as such has lost some of its potency, though the problem is still there. Further action which would measurably augment the ranks of hired hands carries with it turmoil that might disrupt an area where the attempts were made. This is suggested partly in view of the recent wave of attempted land grabs. It is also suggested by the conditions of West Bengal, an example of politically charged atmosphere in which land owners are lucky these days to retain the status quo. Even there uneasiness exists among the tenants, and in theory at least, the threat to their position is heightened as more of the big owners become involved with the new technology, unless of course legal and forceful action is taken to remove it.

The merits and demerits of tenancy as a form of land usage are subject to reasonable differences of opinion, but there are virtually none about the system as practised in Purnea, where tenancy and share cropping account for 43 per cent of all the farm families and 26 per cent of the cultivated land, and where the green revolution has made its unmistakable appearance. Even supporters would admit, if pressed, that the system as practised is good neither for efficient production nor for the well-being of the tenants. The latter cannot add much to the progress of the green revolution, just as they cannot improve their own lot and it is not surprising in the existing conditions that only a small percentage (about 5 per cent) of the share croppers have adopted the new technology even to a limited degree. Purnea is of course only one of many examples.

The green revolution by itself has not been able to alter any of the handicaps under which the tenants labour, and the results are to match. Materially and psychologically many of them are worse off than before the advent of new agricultural strategy and its added rewards. The only increments are the grievances which are bound to swell the chorus of discontent among the small farmers, and, as we shall see in a moment, among the agricultural labourers as well. This discontent will be evident immediately in some countries, not immediately in others but it will be felt in all countries with the passage of time.

*Rural labour, mechanisation and the green revolution:* the landless agricultural labourers and mechanisation of agriculture are among the crucial ingredients of the green revolution and are subject to endless debate which runs the full gamut from most optimistic predictions to unrelieved gloom. It is understandable that these items claim so much attention, for nothing less is involved than the relief (or not) of unemployment and under-employment in the already bulging countryside of India. Carefully gathered empirical information on these ingredients and their relationship is far from abundant, and in the circumstances, reliance is mainly placed on personal observations, inadequate as they may seem. With this caveat on method, what is the position of labour and mechanisation in the new scheme of things?

The new type of farming with its land-saving and labour demanding practices and double cropping is labour-intensive at all levels. It is not surprising, therefore, that from the very outset of the drive for modernisation it has been hailed in certain quarters as a solution to under-employment among rural landless. The very same practices, however, also give rise to labour-saving devices which are already on the farms of many of the new practitioners of modern farming. These include tractors, rotary ploughs, harrows and tillers to go with them, electrically operated tubewells, sprayers and threshers.

Labour-absorbing and labour-displacing practices meet each other, and the resultant balance can be seen in the most advanced agricultural State of Punjab and in Pakistan. In the former, farm labourers who represent only 15 per cent of all the farm families can boast of a net gain. Daily wages have more than doubled, from Rs2-3 to Rs4-5 and in the critical harvesting season have reached Rs10-12.[4] In Punjab the increases in farm wage rates 1960-8 have been 60 per cent during the non-harvesting season and 130 per cent during the harvesting season, while the cost of living increased by 75 per cent.[5] The net gain was not obtained in Purnea. There, higher productivity and higher prices have caused a shift in wage payments from kind to cash, which is a distinct disadvantage to the farm labourer, whose wages in kind have assured him a minimum food supply. The best that can be said in this situation is that an increase in the number of labour days may have counterbalanced this particular disadvantage.

The question of days worked before and since green revolution days is very relevant to this discussion. Punjab shows that at this stage of development its fields are not a constant beehive of activity; about six months of the year are lean months. Farm employment varies accordingly, which explains why the numerous small state industrial undertakings find an abundance of labour in search of more steady employment at Rs3-4 a day. Nevertheless, if agriculture now provides a labourer with up to (an estimated) 200 days of employment he works 30-50 days more than was customary in the past, which is a distinct improvement.

This improvement does not justify the enthusiastic assumptions that the labour-absorptive capacity of the green revolution can solve rural unemployment or make a sizable dent in it. An actual case of a farmer with 24 double cropped acres illustrates the point. Not many years ago he single cropped 24 acres instead of the 48 acres he is now cultivating with the aid of tractors and other labour-saving devices. He also hires more farm labour, both permanent and casual, but his total labour force has increased by only 10-15 per cent, and this includes the peak harvesting season.

A detailed study of tractor mechanisation in Pakistan reveals not only no increase in labour utilisation at all but a dramatic displacement of the same. It appears

> that the labour force per acre had been reduced about 50 per cent from the pre-mechanization period. Assuming an increase in the

[4] Though rising costs of living have whittled down some of these gains.

[5] A study based on an exhaustive statistical analysis of official data disputes any gains made by agricultural labour in areas of the green revolution's greatest concentration. The author, however, does not take into consideration the increase in the number of days employed. His argument throughout is that increase in the cost of living has either neutralised the amount of real wages or in some instances has reduced them. Practically the sole exception is the State of Kerala. See Bardhan 1970: 1,239-46.

> intensity of farming, they (the authors) estimated that 600,000-700,000 labourers would be displaced in West Pakistan by 1985 if all farms over 25 acres were to be mechanized by that date (Bose and Clark 1969).

In a summary of the study the authors concluded that:

> The direct social benefits of mechanization appear to be less than the direct social costs. There is no straight-forward method of taking all the indirect costs and benefits into account in such an analysis, but it is apparent that if they were included the net benefit of mechanization to society would be significantly negative (Bose and Clark 1969).

Regardless of the causes of more or less displacement of labour, both lead to the same conclusion: the employment outlook gets darker as the technology and farm practices grow in sophistication and, as is occurring, the entree into the labour market rises.

Mechanisation in India will never reach the science fiction stage dreamt of by an enterprising Indian farmer, where tractors are guided by remote control. Most well to-do farmers are more modest in their expectations, but farming with them is not so much a way of life but a business in which economy of operations and returns are uppermost in their minds. They would like more modern, more powerful tractors and specialised equipment to go with them, and some would like to exchange their crudely made and none-too-efficient threshers for a version of a modern combine that would harvest, thresh, fill and tie the sacks ready for shipment. Not all of the hustle for mechanisation is economically justified; India and Pakistan provide evidence that the new equipment is often underutilised, but the idea that this is 'the thing to do' has its own values, which in some instances, are not unlike those held by the affluent with just one more spare car.

In practice, not all the pent-up desire for mechanisation will be met, but replacement of labour is likely to take place. A prediction that the employment of farm labour will eventually decline to vanishing point, as has happened in most agriculturally developed Western countries can be discounted, even for the largest and most well irrigated farms of Punjab. The maintenance of the irrigation canals alone will preclude this. Yet it is very doubtful that even if the green revolution extends its scope into other areas and into crops other than wheat, the farm labour employment picture will undergo the change expected by many of the supporters of the new technology.

The advocates of a 'go slow' policy for mechanisation are indeed fearful of this outcome, but they may not have the last word. The chances are that they would not win in a contest between further refinement of technology on one hand and a social policy of increasing rural employment on the other. Most of India's mechanisation policy leads to more imported or domestically assembled tractors, and more of other types of equipment. Between 1966 and 1970 the number of tractors in India increased from 54,000 to 90-100,000. In the eyes of the promoters of the new strategy this is not nearly enough; the annual demand is estimated at 40,000 and total backlog at 250,000.[6] The rate of expansion or the power component of the equipment may fluctuate, but it gives support to the proposition that farm mechanisation is as irreversible as the green revolution which fathered it.

One may bemoan the fact that in a country surcharged with millions of underemployed or unemployed ruralists, with more to come, certain sections of agriculture are assisted in adopting labour-saving devices, particularly when the many less prosperous and more densely populated states other than Punjab are taken into account. But an onlooker's concern does not determine policy and this policy and ensuing events are beginning to catch up with Nehru's and Gandhi's laments against farm mechanisation, or almost any form of mechanisation in the village, as a threat to peasant welfare. The drive for more equipment and fewer hands will go on so long as the new technology enjoys top rating on the ground of greater productivity, lower cost, profitability and savings for investment in and outside of agriculture, and while all the plentiful and valid arguments against mechanisation in a country like India are deemed to be of secondary importance. The estimated 35-40 million landless labourers are bound to grow in numbers and their rate of employment in any field of activity is not promising. Thus the outlook is for an overcrowded, low-wage farm market regardless of the scope of the green revolution.

## SUMMARY

The green revolution has much to its credit. Where it has penetrated, particularly in wheat areas, productivity has risen and, while it has a long way to go to demonstrate its prowess in rice and other crops, the application of the new package of practices has justified the strategy behind it. Depending above all on a breakthrough in rice technology, food self-sufficiency may well come to pass. The maligning of the green revolution on the grounds that it is not yet a do all for all types of farms and farmers, is not justified. In the light of India's resource position, the new agricultural policy was aimed at farmers with more than sufficient cash resources, whether owned or borrowed, and ample and irrigated land. Judged, therefore, by the letter and intent of the policy, the green revolution has lived up to expectations of rising productivity given certain pre-conditions. More equitable distribution of income was not part of the strategy, except that as

[6] From a statement by Union Minister of Food and Agriculture, Mr Fakhruddin Ali Ahmed, *The Indian Express* 24 September 1970.

production soared to those on top of the agricultural ladder, those lower down might have been expected to benefit from the 'trickle down' effect. To a degree, this is happening, but on the whole this implied part of the strategy has not worked well. But as an instrument of productivity the green revolution with all its limitations has left far behind all previous grow more food schemes and is pointing the way to a technological renewal of some of the country's farm economy.

This assessment is only part of the story, for in the application of the new strategy all the worrisome aspects of India's rural structure are coming into a sharper focus. A normally bad state of affairs is showing signs of becoming abnormally so. Briefly, the line of demarcation between the rich and poor farmers is more sharply drawn and income disparities are more obvious. The affluence and influence in the distribution of inputs are greater than before and, as part of it, are apparent in the allocation of credit. The farmer is indeed entitled to good prices for his produce, and this is one of the prime incentives of the green revolution. But the recipients are mostly the big farmers with a marketable surplus while at the same time an estimated 50 per cent of the farmers still buy food grains.

The small farmer has a harder row to hoe, certainly psychologically. All the uncertainties and bedevilments of the tenure system are there in still greater force. At a time when employment is fast becoming a more serious problem than food, a hastening process of mechanisation can lead only to a still greater surplus of the rural labour force, with no visible options outside of agriculture. Paradoxically, and for now, all these are undesirable side effects in the pursuit of a policy to raise agricultural production. At the same time, the old and the new provocations are widening the tide of discontent which may prove difficult to contain.

In one way or another these outcroppings or imbalances impinge upon the land structure of the country. Inevitably one returns back to square one: the familiar and seemingly intractable issue of agrarian reform. But before this chapter ends on this note, it should be made clear that the tendency to charge the green revolution with much that is wrong in the countryside is not warranted. It is true that it has exacerbated socioeconomic conditions but, on the other hand, it is equally true that all the causes necessitating reform attempts have been there all along. This is basic to any appreciation of what the green revolution can and cannot do. As it stands now, it reflects the Indian rural setting in all its manifold forms, the reasons why reform has not been a success, and the many other reasons why rural poverty in India is so widespread.

When all is said and done therefore, it is not the fault of the new technology that the credit service does not serve those for whom it was originally intended; that the extension service is more shadow than substance; that the village councils (*panchayats*) are mostly political rather than the developmental bodies they are supposed to be; that security of tenure is a luxury of the few rather than the many; that rentals are exorbitant; that ceilings on farm holdings are notional; that for the greater part tenurial legislation in general is deliberately miscarried, or that wage scales (Punjab aside) are hardly sufficient to keep body and soul together.

The point of this litany of woes is that to a large degree it is man-made. Even the impersonal forces of an over-expanding population on a limited cultivable area do not quite explain it. Ideally, the implementation of a reform in India could never approach anything resembling egalitarianism. There simply is not enough land to go around, and one cannot find in agriculture a way out of problems which in the final analysis are tied in with drastic and effective population controls and an equally drastic rise in general economic development. Barring these two pre-conditions, there is no final solution to India's agricultural ailments, however speedily the green revolution forges ahead. For this reason, final solutions of the problems touched upon are not now the issue while amelioration of the worst of conditions is. Only then can the reform become something of a resource builder.

If equity and welfare were combined with productivities, the efficacy of the new technology, the latter would leave its mark on a much larger segment of the farm community than it has to date. The minimum program for its attainment is a modest one and can be enforced if the states demonstrated the desire to do so. One might expect this in present day India, where there is a great deal of concern about removing the blocks to the realisable, thereby buying time to stave off a situation when the underprivileged might take the law into their own hands.

The possibility of this last situation is not excluded because articulating a problem and acting upon it are efforts which do not often ride in tandem. If in the days of nation-wide land grab attempts the consequences result in nothing more than the usual verbal reform affirmations, all the police force which placed the land grabbers behind the bars will not avail Indian rural peace and stability. And nor will the green revolution for all its productivity records, for growth and a measure of distributive justice cannot be treated in isolation from each other. But if it continues to be so treated, the notion that poverty in India can be tolerated indefinitely will prove a delusion. The festering tensions will not abate, and an agricultural revolution geared mainly to new techniques and higher productivity, a model extolled by most students of economic development, will unwittingly contribute to the same end.

Eventually, the current stirrings are likely to compel some meaningful concessions by the decision makers. Interestingly enough, the green revolution and what it stands for could just possibly contribute to that, having aroused interest in land-holding rights to a new pitch not only among the owners but among the non-owner cultivators as well. In the final reckoning some weight may have to be assigned to it. The question is one of timing. If in good time, the efficacy of the new technology might yet find the company of agrarian and other related reforms long overdue, the agricultural strategy would no longer be divided against itself. The benefits of such an alignment are quite obvious, even though not every farmer or landless labourer would share in them.

If things are left as they are, there are the predictions of the Ministry of Home Affairs, the country's law-and-order upholder, that an 'explosion' in the countryside cannot be ruled out, or that the 'green' revolution might turn 'red'. To this observer these dire anticipations seem premature, for the significant aspect of the land grab episode is that the peasantry displayed great reluctance to turn it into a mass movement. This said, it is worth recalling an exchange between Louis Fischer and Gandhi. Asked Fischer: 'What would happen in a free India? What is your programme for the improvement of the lot of the peasantry?' Replied Gandhi: 'The peasants would take the land. We wouldn't have to tell them to take it. They would take it.' (1944 : 43). As with all good prophets, it had not happened in his lifetime, but who can say that without a stitch in time India might not reap a version of Gandhi's prophecy when the patience of Job runs out and traditional submissiveness no longer holds?

13

# From Stagnation to Growth: Relative Roles of Technology, Economic Policy and Agrarian Institutions

*M. L. Dantwala*

THE post-war era has been very rewarding for agricultural economics. The tardy growth in agricultural production in most of the developing economies, right up to 1965, provided a fertile soil for a rich crop of literature. Inspiration and the new material for this paper have been derived from this literature.

Within this literature agriculture has been assigned a critical development role, although in the post-war period, it has more often than not proved to be a drag on economic development. Economists have been perplexed by this problem and have attacked policy makers in the developing countries as being responsible for the failure of agriculture to play its assigned role. This pursuit yielded ample literature—often impatient and angry—on post-war agriculture in developing countries. As one critic observed: 'The less-developed world is clearly losing the capacity to feed itself; stated otherwise, a growing share of the increase in population is being sustained by food shipments from the developed regions, largely from the United States, under the Food for Peace Program' (Brown 1965: vi). It appeared that the white man's burden would not ease even after he shed the political dead weight of colonial rule! In fact, the new burden seemed so great that, according to the celebrated authors of *Famine 1975!* it would become necessary 'to apply the classical medical *triage* method' to the starving world of less developed countries. 'Like doctors on the battlefield trying to make best out of the limited resources, the affluent countries will have to decide which starving countries to save and which to sacrifice!' (Paddock 1968). Fortunately no countries have been sacrificed. The scene changed so dramatically that the critics began to chant the song of the green miracle and Paul Paddock graciously extended the reprieve to 1980.[1]

[1] Paddock 1970.

## STAGNATION OR GROWTH IN INDIAN AGRICULTURE

The main scheme of this chapter is not the green revolution *per se.* I am concerned with the literature on the subject, with analysing the reactions of the economists who have keenly watched the drama of agriculture's 'failure' and 'success', and with attempting to draw some meaningful general conclusions regarding the factors which account for the stagnation and/or growth of agriculture in the context of the developing economies. For this purpose it may become necessary to examine the nature and magnitude of the alleged 'failure' as well as 'success' in agriculture; but only to the extent that this may be relevant to the identification of causal factors.

Though my interest is in the larger problem of agricultural stagnation and growth, much of the empirical data pertains to India. Beginning with the period with which the 'failure' of agriculture is associated, 1950-65, I suggest that it should be assessed by reference to the state of affairs before 1950, say from 1901. S. R. Sen (1967) showed that during the first twenty four years of this century, foodgrain production in India increased at a meagre annual rate of 0·3 per cent. During the next twenty four years, it actually declined, at an average annual rate of 0·02 per cent. For Undivided India in 1947/8, estimated production (66 million tonnes) was slightly less than that in 1900/1 (67 million tonnes). Estimates of foodgrain production in the Indian Union from 1936/7 onwards have a firmer basis. For the first fifteen years of the thirty year period from 1936/7 to 1965/6, production declined at the rate of 0·68 per cent per annum, but in the next fifteen years, it increased at an average rate of 2·75 per cent per annum. Thus, the most striking fact which emerges is that for nearly fifty years after the turn of the century, foodgrain production was almost stagnant.

George Blyn's study (1966) of agricultural output, availability and productivity in India from 1891-1947 confirms Sen's results. To ensure comparability of data, Blyn confined his study to British India and its constituent regions. It revealed that during the reference period[2] of fifty six years, aggregate foodgrain production increased at a meagre average rate of 0·11 per cent per year. Rice production, which then accounted for half of foodgrain output, showed a declining trend. The best performance was that of wheat whose production increased at an average annual rate of 0·84 per cent. It is significant to note that the annual rate of aggregate foodgrain increase in the last four Reference Decades was just 0·03 per cent. Rice output in this latter period was declining at the annual rate of 0·12 per cent. The performance of non-

2 The time span has been divided into ten equally spaced overlapping 'reference decades'. Five of the reference decades span the Census years, e.g. 1891-1901, and the other five are centred on the Census years, e.g. 1896-1906.

foodgrain crops was no better. During the entire reference period population increased by 0·67 per cent annually, and during the last four Reference Decades, by 1·12 per cent. As a result, the per capita all-crop output declined sharply. The Index Number, with 1891 as the base year, dropped to 84·2 in 1946. Inter-regional differences were considerable, with the largest decline in per capita availability in 'Greater Bengal' where the Index dropped to 54·1.

In judging the agricutural performance of developing countries in the period following their independence, either from colonial rule or from domination, some idea regarding the time required to bring about a yield breakthrough in agriculture can be derived from the early experience of the developed countries. In the United Kingdom, 'from the time national yield data first became available in 1884 until the advent of World War II, the average rate of increase (in grain yields) was only 0·2 per cent per year'. For the United States, the annual compound rate of increase in grain yields per acre was 0·3 per cent between 1866-70 and 1901-5, nil between 1901-5 and 1936-40, 1·5 per cent between 1936-40 and 1951-3, and a breakthrough of 4·8 per cent only in the post-war period from 1951-3 to 1961-3. Only in Japan was the growth rate consistently above 1 per cent per annum during the corresponding first two periods (Brown 1965 : 15).

It is in the light of this record of early performances that India's 'stagnation' in both agricultural production and productivity during corresponding stages of economic development should be viewed. In India, between 1949/50 and 1964/5, the compound rate of growth (1949/50 to 1951/2 = 100) of production was 2·98 per cent for foodgrains and 3·19 per cent for all crops. Productivity growth rates were 1·60 per cent for both (India, Ministry of Food and Agriculture 1966). If, however, the base is shifted to 1952/3-1954/5, the compound rate of foodgrains production drops to 2·50 per cent, and for productivity to 1·51. This was certainly not stagnation, but because of the annual growth of population of 2·5 per cent, the balance between food and population became precarious. The two successive disastrous droughts which brought down foodgrain production from 89 million metric tons in 1964/5 to 72 million metric tons in 1965/6 and 74·2 in 1966/7 added the last straw. Paddock's *Famine 1975!* became a best seller.

## INFLUENCES ON AGRICULTURAL PERFORMANCE

With this background review of the nature and magnitude of the 'stagnation' period in agriculture, we may now examine what the economists thought were its causes. The literature on this question is so voluminous that we will have to pick and choose from the lengthy array of explanations. They could be perhaps divided into three major groups: institutional and organisational (land tenure, credit, marketing, extension, education); economic (incentives-disincentives, prices, taxes, subsidies); technological (seed, fertilisers, irrigation).

Obviously, these are not mutually exclusive categories and their interactions should not be ignored. The classification has been suggested merely for the convenience of discussion. Further, in view of the limitation of space, we shall confine ourselves only to a few items under each category.

First, however, it would be helpful to indicate the line of reasoning which will be adopted. If one or more of the above factors explained the stagnation, it would be reasonable to expect that their full or partial removal would explain the success of the post-1965 period. More specifically, it should be possible to identify the inhibiting factors which were present during the stagnation period but were absent during the success period. Discussions of this issue usually bundle together several factors, institutional, economic, technological, to explain stagnation or growth. At that level of generalisation, no one can disagree with such propositions but, at the same time, little can be said about the critical factor which makes all the difference between stagnation and growth. Of course the ideal procedure would be some sort of regression analysis, but factors like land reforms do not lend themselves to the kind of quantification needed for the purpose. Under these circumstances, the sort of 'yes-yes', 'no-no' analysis suggested above would be more appropriate, especially as we are dealing with both stagnation and growth situations spanning a period of twenty years.

Another profitable line of inquiry would be a cross-country analysis, for rates of growth between countries have varied a great deal. For example, growth rates of agricultural production in Taiwan and North Korea have been markedly higher than those in India and Burma. They have been higher in Mexico than in several other countries in Latin America. We could perhaps identify the factor(s) which may explain such glaring differences. Such analysis has indeed been made and much has been made of the findings. But, as we shall see, their scientific validity becomes doubtful on closer scrutiny. Quite often, a single factor, such as land reform or fertiliser/product price ratio, is picked up as an explanatory factor. Such partial analysis is not only questionable, it breaks down when extended beyond the restricted sample. But more importantly, the essential distinction between reproducible and non-reproducible factors like soil, climate and, to some extent, even the natural advantage in the creation of irrigation facilities, must be made. This distinction is critical to the question of the transferability of successful experiences.

One observation, however, can be made on the basis of cross-country analysis. The phenomenon of the low rate of growth in agricultural production till 1964 was pervasive throughout Southeast Asia. After 1965, when the breakthrough began to occur, it was shared by most of the countries. In both periods, these countries were pursuing diverse economic and political policies, and yet, in all these countries, there were almost identical results; stagnation before 1964, and breakthrough thereafter. If economic policies matter, the near identical outcome of diverse policies is difficult to explain. Further, as we shall see, in none of these countries is there clear evidence of deliberate reversal of policies which could explain the rather sudden emergence of success after a long spell of failure. It is obvious that the only thing which happened was the discovery of the high yielding varieties and their commercial application, an event in which economists and economic policies played a marginal role. Indeed when the curtain rose, many economists were caught rehearsing their gloomy roles. The fact that they are now reappearing with another set of gloomy roles[3] is a different story.

Let us now quickly review the Indian scene and try to identify the factor(s) which could explain the sudden change in the agricultural situation. We shall determine whether what has happened deserves the name of 'green revolution', and whether a confident full-stop, a sarcastic question mark or a hesitant 'can't say' sign would be most appropriate after the phrase.

This should be a simple task: nothing really noteworthy happened either in policy, or in fact, to most of the explanatory factors mentioned above. Some observers, mainly from abroad, have no doubt suggested that the disastrous droughts of 1965 and 1966 taught a lesson to Indians who belatedly realised the sheer folly of 'neglecting' agriculture. It is difficult to see where they found the evidence of this purging. The 'neglect' story was perhaps no more than a professional racket which yielded a crop of best selling scientific fictions and rewarding international travel. My position, which I have repeatedly stated, is that I can see no evidence of deliberate neglect in the past,[4] or of any sudden later awakening.

*Land Reform*

The one remarkable piece of land reform—the abolition of intermediary tenures—was enacted during the early 1950s, and even the critics concede that it was almost fully implemented. The pre-

[3] For example, a review of Lester Brown's book *Seeds of Change* in the *Times of India* was captioned 'Seeds of Disaster'.

[4] The Pearson Report (see Pearson 1969) is amongst the few authorities which support this view.

dominantly *zamindari* states like Uttar Pradesh, Bihar and West Bengal did not reveal any impact of this reform on their agricultural production during that time. Tenancy laws and laws imposing ceilings on holdings were also enacted during the 1950s in most of the States, but were so indifferently implemented that it is difficult to assess the extent to which they benefited or hindered agricultural growth. The relevant point is that nothing new happened in this sphere immediately preceding 1965 which could explain the sudden emergence of the green revolution. If anything, there was a distinct sign of waning interest in land reform and a frozen cynicism about it. Of course, no one openly said anything against land reforms; prudence suggested the less said the better. It was only after the green revolution which is alleged to have made the rich richer and the poor poorer that there was a rekindling of interest in land reforms. With agriculture stagnant, there was little attraction in land grabbing, either for the rich or the poor. Thus land reform, far from being a catalytic agent of change, assumed relevance only in the aftermath of growth. Incidentally, it has been in the States like Punjab, Andhra Pradesh and Tamil Nadu, States not renowned for their zest for land reforms, that the green revolution appears to have taken hold.

### *Other Institutional and Organisational Factors*

Much the same story can be told regarding community development, farmers' education and the rest of the institutional and organisational factors. Even in regard to the credit policy, the Reserve Bank and the Planning Commission stuck to their slogan 'co-operation has failed, co-operation must succeed'. They must have been taken by surprise when one of the major grounds for imposition of social control on commercial banks, and for their subsequent nationalisation, was stated to be the latter's neglect of agricultural finance. The multi-agency idea was mooted by the Ministry of Food and Agriculture when it was convinced of the utter incapability of the co-operatives to cope with the credit requirements of the new technology. The reluctant Reserve Bank acquiesced, but has now adopted the ingenious strategy that 'where co-operatives have failed, commercial banks should rush in, revive the co-operatives and retire'. Under the new policy commercial banks will now finance the primary societies as a transitional measure, in some selected districts where co-operative Central Banks are weak. In these districts, commercial banks will be precluded from financing farmers who are not members of the co-operative societies; a blatant example of compulsory co-operation.

The Intensive Agricultural District Programme (IADP) should, however, be mentioned as a major pre-1965 policy decision. Even though it has been criticised for accentuating regional disparities, it was a step in the right direction. It was a pioneering effort in the direction of intensive agricultural development and by the time the high yielding varieties arrived, the IADP districts had built up an impressive organisational infrastructure. While the program's achievements should not be belittled, it should be noted that it was launched early in the 1960s while the green revolution was not heard of until the high yielding varieties arrived in 1965. As the Fourth Report of the Expert Committee on Assessment and Evaluation (India, Ministry of Food and Agriculture 1969) states, the IADP has played the role of the pathfinder rather than that of a pacesetter. It could not quieten the *Famine 1975!* chorus.

*Price Levels and Incentives*

The gravamen of the charge against India's agricultural policy was its urban bias which deliberately encouraged low foodgrain prices in the interest of urban consumers.[5] This issue has been dealt with at length elsewhere (Dantwala 1967) and only a few key points are highlighted here. Presumably, PL-480 imports were the instrument with which foodgrain prices were kept low. Wheat was by far the major component of these imports, yet wheat is the foodgrain in which the green revolution is most evident. In spite of marginal imports of PL-480 cotton and soyabean, prices of commercial crops like cotton and oilseeds have remained consistently high while price ratios between foodgrain and non-foodgrain crops were highly favourable to the latter. Yet there is no sign of a breakthrough in their production. Indeed the productivity index of groundnuts has been declining. None of this need be interpreted as representing the view that farmers in India or elsewhere are insensitive to price incentives or that prices do not matter for production. The fault, if any, lay with the economists' analysis and not with the farmers' response.

First, the analysis of price response of individual or substitutable crops to relative prices was erroneously applied to aggregate production of the agricultural sector. Inadequate note was taken of the situation that prevailed in input markets and of the production and supply bottlenecks. With these constraints, high product prices would have resulted in higher input prices rather than higher production. Futhermore, in the highly inflationary situation which then existed, with domestic prices soaring high above international prices (witness the 1966 devaluation) higher prices were no remedy for boosting aggregate agricultural or even foodgrain production. In the context of a traditional agriculture, the potential for increasing production

5 With a few exceptions, most foreign economists have blamed India's policy preference for low farm product prices and cheap food for the failure of Indian agriculture. See, amongst others, Schultz 1964 and Mason 1966.

through price policy is extremely limited. It can, however, play a more positive role by maintaining the profitability of technological transformation (see Mellor 1968 and 1969).

Cross-country comparisons of fertiliser/farm product price ratios are even less tenable. Lipton (1968 : 102, quoting Edward Mason) laments: 'Fertilizer prices relative to farm prices have been higher in India than in Pakistan.' So what? Pakistan belatedly realised that the policy of heavily subsidising fertiliser prices was not very productive. Fertiliser prices have been much higher in Taiwan than in Pakistan, and yet all through the 1950s and the early 1960s, agricultural growth rates in Taiwan surpassed those in Pakistan. Those who talk of urban or consumer bias in India's agricultural policy, of low food prices and high fertiliser prices, would do well to examine the soaring Consumer Price Indexes and the recurring hikes in wage and salary awards.

For several years, domestic prices for almost all agricultural commodities have been far above international prices.[6] So, when Lipton tries to establish a link between the 'big farmer' and the 'urban bias' of the planners (1968 : 104), he strains his readers' indulgence. I would concede the big farmer bias in India's agricultural policy; but by the same token, it cannot be pro-consumer or even pro-urban. Lipton suggests that 'the true class struggle is between the food buyer and the food grower', but this should not be equated with rural/urban, because, as pointed out later, rural households have to buy 30-50 per cent of their food requirements.[7] And why does the big farmer tolerate cheap food? 'In exchange for subsidised fertilizers and seeds, and for a truce on land reform,' says Lipton (1970 : 147). However food is not cheap, fertilisers are not only not subsidised, they bear excise duty, and the big farmer lobby succeeds in obtaining higher procurement prices than those recommended by the Agricultural Prices Commission. But then Lipton would like to question the 'origins of planners and politicians'. Presumably, only the Visiting Expert, as he calls himself, has an immaculate origin.

Incidentally, is not high price advocacy by India's well wishers a syndrome of big farmer bias? As Mellor (1968) explains, for a (low income) cultivator who sells only 10 per cent of what he produces, a 10 per cent relative increase in agricultural price increases his real income by only 1 per cent. For a (big) cultivator who sells 70 per cent of what he produces, real income rises by 7 per cent with a 10 per cent relative rise in prices. Conversely,

[6] See Appendix Table I.

[7] For a confusion of a similar nature in Lipton's discussion of the Intensive Agricultural Area Programme, see the review of his book *Crises in Indian Planning*, by Srinivasan 1970.

> for a low income consumer (rural or urban) who spends 70 per cent of his income on food, a 10 per cent increase in food prices will represent a 7 per cent decline in real income. For a high income urban consumer, spending only 20 per cent of his income on food, a 10 per cent rise in food prices provides only 2 per cent decline in income.

Of course, a high income consumer spends a larger sum on food than does a low income consumer.

Advocates of high agricultural prices should study the profile of poverty and consumer expenditure. For 1967/8, Minhas (1970 : 102) estimates that 210 million persons in rural India were below the poverty line defined in terms of per capita annual consumption of less than Rs240. About 154 million persons had per capita annual consumption of less than Rs200. A majority of these would be net purchasers of food. Thus genuine rural bias should be in favour of low rather than high food prices. In Andhra Pradesh, the rice belt of India, rural households taken together are required to purchase 53·5 per cent of their foodgrains consumption from the market. Even the percentage of food purchased to total consumption of producer households is as large as 34·7 (India, Agricultural Price Commission 1970). The coincidence of (relatively) higher foodgrain prices and production breakthrough should not be ignored, for it is undeniable that the high prices of foodgrains considerably improved the profitability of high yielding varieties.

Equally helpful was the establishment of the Agricultural Prices Commission which recommended statutory minimum support prices, at which the Government accepted the responsibility of buying the entire available supplies of foodgrains. In fact, the Government offered to buy all market supplies at procurement prices which were much higher than support prices. Further, as mentioned above, the Chief Minister of States and the Government of India yielded to pressure from States and agreed to higher prices than those recommended by the Prices Commission. Foreign experts, and critics of urban bias, can now at least take satisfaction from the fact that the policymakers in India were more inclined to accept their advice on incentive prices rather than that of their own experts and of the Commission expressly established to study the situation and to make recommendations.

The cost to the Central Exchequer of the unwillingness to contain prices will amount to Rs1060 million per year from 1970 onwards in lieu of higher dearness allowances. With state government employees waiting in the lobby, the cost to the government may well amount to Rs2000 million per year, implying either more deficit financing or a cut in development expenditure. The fact is that policy is trying to keep pace with prices and not the other way around. Therefore, the only credit that price policy can take is that it is doing nothing effective to check their rise, and in fact, is making poorer millions of urban and rural poor. Except perhaps for wheat, the prevailing prices owe nothing to policy as such. But if this is what is meant by learning the lessons from the disastrous droughts, the lesson can be said to have been learned.

There is, however, one snag. Commercial crops are not responding to prices which are relatively higher than those of foodgrains. The key question to ask is: what makes for the sustained profitability of cultivation? High prices in the second round, reduce profitability as a result of higher costs of cultivation. Technological innovation without altering input/product price ratios, improves the input/output ratio in physical terms and hence the profitability of the enterprise. And is there anything like an upper limit to incentive prices, beyond which they become a disincentive to investment and exports? Does the fact of positive response of production to prices mean that the key to higher production is just higher prices? To the protest—who says so?—all that one can say is that none of the critics of low price policy has cared to indicate the limitations of incentive price prescriptions.

### *The Impact of Agricultural Policy*

The foregoing review indicates that during the years immediately preceding the green revolution, agricultural policies did not change in any significant way that could explain its sudden emergence. If the earlier policies were in any way responsible for the failure of agriculture, they at least did not impede its subsequent success. As a matter of fact, in the eyes of some critics, some of the policies in operation in the later period such as compulsory procurement below market prices and zonal restriction on movement of foodgrains, constituted a severe disincentive, yet they apparently did not prove an obstacle to the breakthrough. It can, of course, be argued that in their absence, the success would have been greater.

As every one knows, the key factor behind the green revolution was the new technology with the high yielding varieties at its core. Once again, it may be noted that several countries in Asia and Latin America, each following diverse economic policies, almost simultaneously experienced the breakthrough.

All this casts serious doubts on the relevance of the economic analysis of the time and its ability to explain the process of stagnation or growth. It would surely be silly to deduce from this that economic policies after all do not matter, but it may be appropriate to say that the policies which appeared to matter perhaps did not. Stated another way, it could be suggested that in the absence of a technological breakthrough, the many otherwise highly relevant policies of land reform, economic incentives and institutional changes, jointly and severally could not have lifted agriculture above its stagnation threshold. Land reforms and other institutional changes were necessary and highly desirable. There are many reasons why they were not successfully carried out, but a thought occurs that perhaps the very stagnation of agriculture inhibited their success. When yields were miserably low, it was difficult to get enthusiastic about lowering the ceiling on holdings or drastically reducing the rents. In India, extremes apart, if there is injustice or exploitation, it is by and large, of the poor by the less poor. It is true, even after the abolition of the intermediary tenures from over 170 million acres of land, there are several pockets of absentee landlordism. Yet the bulk of the Indian agricultural sector is poor and depressed even by India's urban standards. Ninety eight per cent of Household Operational Units are in the size group of less than 12 hectares, irrespective of the quality of the land. A Reserve Bank Survey in 1961 indicated that out of the twelve districts studied, the value of gross product of the big (top 10 per cent) cultivators exceeded Rs4000 in only two districts (Jullundur—Rs4052 and Akola—Rs6450). It was below Rs2000 in seven out of twelve districts (Dantwala 1961). Under such circumstances, how much scope was there for divesting or expropriation of the 'rich'? Now that the possibilities of making a fortune in agriculture have opened up, land reform can play a vigilant corrective role.

When yields were in the neighbourhood of 1 ton per hectare, it was a mockery to talk of incentives through higher prices. But if new technology is to be adopted, it is essential to assure the innovating farmers that their efforts to adopt it will not be frustrated by unremunerative prices.

It may be argued that the discovery of high yielding varieties was not a gift from heaven, and that without appropriate policies they would not have seen the light of the day. Whose policies? Probably of the Rockefeller and the Ford Foundations whose generous grants made it possible for the scientists in Mexico and Philippines to evolve the high yielding varieties. The revolution was a reward of scientific research and not of the wisdom of economists; few even dreamt of it and many even do not believe it. It is our professional privilege and duty to question its occurrence or its consequences—Palace Revolt, Pandora's Box, Seeds of Disaster, dualism—but let us not get into a stance that nothing good can ever happen in this country. As for the policy makers, it was indeed fortunate that around 1965, India had a Minister for Agriculture who, in spite of his many detractors, pinned his faith on the new strategy and the scientists behind it.[8] Mercifully

[8] During 1968/9, weather conditions were less favourable than those in the preceding year (India, *Annual Report of the Ministry of Food and Agriculture,*

his successors have as yet done little to harm it by their irrelevant radicalism, except by their congenital inefficiencies and red-tape; but let us keep our fingers crossed.

### THE NATURE OF RECENT AGRICULTURAL CHANGE

We have assumed throughout that a revolutionary change has occurred, or rather is occurring, in agriculture not only in India but simultaneously over several developing countries. If this assumption itself is wrong, much of the earlier thesis regarding the role of agricultural policy must fall through. Though in the brief span of this chapter it is not possible to present a full case justifying the assumption, a few points may be submitted for consideration.

It is legitimate to judge the occurrence of the revolution by reference to the growth in production. It is generally conceded that as far as wheat is concerned, both in India and Pakistan, and in Mexico, what happened is truly revolutionary.[9] Success in maize, *bajra* and *jowar* though not as spectacular and widespread, is also encouraging. But production of rice in India—39·8 million metric tons in 1968/9 as against 39·3 millions in 1964/5—presents an unrelieved spectacle of stagnation. It is highly relevant to note in this context that between 1960-4 and 1968, in (then) West Pakistan, there was an increase of 75 per cent in rice production as against 15 per cent in (then) East Pakistan, 11 per cent in India and 13 per cent in Asia as a whole.

Once again, policies have nothing to do with such glaring disparate performance. The Annual Report (1969) of the Department of Agricultural Economics of the International Rice Research Institute examined this problem in depth and observed that the difference resulted mainly from environmental conditions under which the two crops, wheat and rice, are grown, and not from cultivation problems. Under favourable conditions (e.g. for rice in Pakistan) the potential of the new rice varieties equals or exceeds that of high yielding wheat. The authors of the report contend that 'the major obstacle to the spread of the new rice varieties has not been poor farm management, lack of farmers' knowledge, nor inadequate extension,' (IRRI 1969: 2, 4) nor, one may add, low prices, unsatisfactory land tenure or inadequate

---

*1969/70*). In spite of this, rice production improved by 2·2 million metric tons, when that of other *kharif* cereals declined by 3·8 million metric tons. As a matter of fact, as the Economic Survey of Asia and the Far East for 1968 points out, unprecedented drought conditions affected the region as a whole after the middle of the 1960s (United Nations 1968: 128).

9 Wheat production in India has gone up from 12 million metric tons in 1964/5 to about 20 million in 1969/70.

institutional credit. The latter group of factors could not have affected one cereal (wheat) favourably, and another (rice) in a contrary direction, or the same commodity (rice) favourably in (then) West Pakistan and the other way in (then) East Pakistan.

Analyses of rice crop area and production in South and South-east Asia indicates the critical relevance of physical environmental conditions. The double cropped irrigated area which constitutes 10 per cent of the total rice area, produces 25 per cent of total crop. Fifty per cent of the rice area is rain fed and another 20 per cent upland; these together produce only 52 per cent of rice production. For the new varieties of rice to succeed, adequate water supply and good water control are essential. Wheat cultivation in India and Pakistan and rice in the latter, meet these requirements.

It is true that an explanation does not undo the fact of the disappointing performance of rice. New varieties suited to monsoon conditions are being evolved, but if even these fail, rice cultivation will have to shift to the *rabi* season, or to drier regions, with adequate irrigation. Thus, it may as well be conceded that judged by the production performance alone, the green revolution cannot yet be said to have spread to rice cultivation, at any rate in India. Which other test, equally relevant and convincing, can we suggest to establish the pervasiveness of its occurrence?

A notable feature of the Indian agricultural scene since 1965 has been the marked spurt in the use of modern inputs. The spread in the adoption of high yielding varieties has surpassed earlier international records. The Pearson Report (1969) calls it 'one of the authentic marvels of our time'. The area planted to high yielding cereals in Asia in the 1964/5 crop year was estimated at 80 hectares; by 1968/9, almost 14 million hectares were covered. In the entire history of agricultural progress anywhere in the world, no innovation has been adopted with such speed. 'Farmers in these societies who are planting the new seeds are achieving in matters of years, progress that took decades in other countries' (Brown 1970: 41). The agricultural transformation taking place in India has been described as 'one of the most amazing agricultural stories of all time (Streeten and Lipton 1968: 132). One may wisely discount such hyperboles. But there is no reason to question the testimony of the International Rice Research Institute (1968: 336) which says: 'The initial thrust of the new technology has been far more rapid than any previous historical experience. In many areas, the level of adoption has exceeded even the most optimistic predictions.' In Japan, it took almost two decades for the new dwarf varieties to catch on nationally. In India, from a modest beginning in 1965, more than 9 million hectares have been sown to the high yielding varieties in 1968/9. These days, the mention of plan targets may evoke amusement. Even so, it may be mentioned that the target of 10 million hectares under high yielding varieties of rice by 1973/4 would constitute 15 per cent of the entire rice crop area of South and Southeast Asia. Even the 2·6 million hectares covered in India in 1968/9 constituted 56 per cent of the area planted to high yielding varieties in South and Southeast Asia in that year.

The spurt in fertiliser consumption has been equally remarkable. As has been observed: 'The jump in fertiliser usage in India in recent years has been one of the biggest and fastest known anywhere in the world' (Rockefeller Foundation 1969: 14). Let us, however, simply record the facts. The annual increase in fertiliser (NPK) consumption between 1960/1 and 1965/6 was at the rate of 90,000 metric tons per annum; between 1965/6 and 1969/70, the rate went up to 300,000 metric tons. Whether this can be interpreted as a sign of agricultural transformation or just a big farmer boom may be left to the dialecticians. But what has been happening since 1965 is certainly more than an ordinary acceleration.[10]

The same story can be told about the post-1965 acceleration of the spread of farm mechanisation; but it need not be recounted here. Attention may, however, be invited to the remarkable upsurge in minor irrigation. Between 1966 and 1969, 175,000 tubewells were reported to have been dug. Their total number at the end of the Third Five-Year Plan was only 80,000. Between 1961 and 1965, the number of energised pump sets/tubewells increased from 192,814 to 512,925, with an annual average increase of 64,000. By March 1970, the number of energised pump sets had gone up to 1,348,842; and the annual rate of increase since 1966 went up threefold at more than two hundred thousand. Currently, more than 250,000 pump sets are being energised each year.[11] And, as one associated with the Agricultural Finance Corporation, I may add that the farmers' demand is far in excess of the State Electricity Board's material and financial resources available to meet it.

The question as to whether this remakable upsurge in the use of modern inputs should be characterised as transformation of traditional agriculture or merely an emergence of dualism should be lifted above a mere semantic debate. K. N. Raj (1969) has very ably discussed this

[10] Note should be taken of the liberalisation of fertiliser policy in 1965 under which fertiliser plants were permitted to be set up in the private sector, and fertiliser price and distribution constraints were relaxed. This was a welcome departure, but its impact on domestic production has been limited. In 1968/9, domestic production of nitrogenous fertilisers accounted for only 45 per cent of total consumption, as against 40 per cent in 1965/6.

[11] The data on energisation of pump sets/tubewells can be easily checked and, therefore, is quite reliable.

issue in a scholarly article in the first issue of the Journal of Development Planning. After a meticulous scrutiny of production performance in Mexico, Taiwan and India, he observes that 'the result is that the availability of modern technology leads very often to dualism in the agricultural sector (as in Mexico) rather than to transformation of traditional agriculture' (1969: 37). More importantly, Raj contests the view that 'the institutional framework of agrarian economies is not a serious obstacle to growth and that what is needed is concentration on specific questions relating to the mechanics of the technological change required in each case and the supply of the necessary inputs' (1969: 15). Raj has tried to show that what has happened, in Mexico, is an 'enclave' development and in Taiwan, a continuation of the trends established in an earlier period of its history—in fact under the Japanese colonial rule. Regarding India also, Raj would contest the transformation thesis. As his contribution would therefore appear to counter the entire theme of this chapter, it merits respectful attention. Before discussing some of the basic issues, one or two preliminary points may be dispensed with.

Raj's empirical data pertain to a period which begins with 1950, and even earlier for Taiwan, and ends for Mexico at 1959, for Taiwan at 1963 and for India at 1964/5. We are more concerned with the phenomenon which emerged as a result of the adoption of the newly discovered high yielding varieties in the middle 1960s. Though the high yielding varieties of wheat were commercially applied in Mexico somewhat earlier, they and those of rice came to India and Southeast Asia only by 1965. Our argument is that the technological revolution in the tropics and sub-tropics really began after the discovery and application of the dwarf varieties with their high yielding, quick maturing, photo-non-sensitive characteristics.

### *Transformation vs Dualism*

Coming to the basic issues, we shall first discuss the transformation versus dualism hypothesis and then the relative roles of institutions and technology in accelerating growth in agriculture. It is quite true that agricultural growth which has taken place in Mexico and India in the post-war period and even after the discovery of the high yielding varieties has been confined to a few regions and has not been a national phenomenon.[12] Our contention is that the transformation of traditional agriculture is a qualitative technical phenomenon and its authenticity is not conditional upon a given geographical coverage. That it has not permeated the entire country cannot detract from the

[12] Raj concedes that 'Taiwan is clearly a more genuine case of transformation of traditional agriculture than Mexico's' (1969: 22).

fact of its occurrence as testified by the qualitative change in the input-mix in agricultural production leading to the shift in the production function. Even so, it has occurred not on a few experimental farms, but literally over millions of farms covering large regions such as North and Pacific North in Mexico, and Punjab and Tamil Nadu in India. And within the regions where it has occurred, it has not been confined to privileged pockets.

As the Programme Evaluation Organisation sample survey has shown in the rabi season of 1968/9, high yielding wheat covered 92·3 per cent of cropped area in the package districts and 66·5 per cent in all areas (Mukherjee 1970). These regions are bigger than several independent nation States. Had Punjab or Tamil Nadu or Andhra Pradesh been independent nation States, the phenomenon would have been indisputably considered as agricultural transformation. But simply because they are constituent units of a continental country, their more rapid growth compared with other States, is seen as leading to dualism. Even so, if the phenomenon is inherently incapable of spreading to wider areas, its characterisation as transformation will perhaps need qualification. No innovation, however beneficial, can become suddenly universally applicable. The right question to ask is whether it is inherently incapable of spreading. The discovery of life-saving drugs can rightly be regarded as a revolutionary accomplishment, but currently it is perhaps guilty of generating 'dualism' inasmuch as it saves the lives of those who can afford the treatment and mockingly bypasses the millions of the poor. Michael Lipton, alluding to urban bias in Indian Planning (1969: 108), says: 'What of the big rise in spending on animal husbandry in the Fourth Five-Year Plan? Meat and dairy products are likelier consumed by towns people than by villagers, and by rich people, than poor.'

The spread of the high yielding varieties is certainly constrained by environmental factors. Research hereafter may succeed in evolving varieties congenial to all major environmental situations but, at present, their application will necessarily be confined to areas of assured water supply. Even here, if water control is poor in these areas the results will be below par. This is an inherent limitation but it is not as big as is generally believed. Here it becomes necessary to distinguish between cultivated area and cultivating households. The area under irrigation is 20 per cent of the net cultivated area, but as B. Sen (1970), drawing on the National Sample Survey (NSS) data, has pointed out, '22·7 million farms, consisting of 45 per cent of the total, have irrigation facilities'. More importantly, out of these, as many as 10·5 million farms are in the size group 0·4-2·0 hectares, and another 3·5 million in the size group under 0·4. As much as 63 per cent of the farm area of the former and 89 per cent of the latter is irrigated. These two groups constitute 61 per cent of all farm households with irrigation facilities. As Sen (1970) puts it: 'Credit for this goes to the inherited pattern of the distribution of irrigated acreage among farms.'

There are, no doubt, other limitations to the spread of the high yielding varieties technology. For example, doubt has been expressed about the capacity of small farmers and tenants to benefit from the rather expensive high yielding variety cultivation. I shall discuss this issue a little later but assuming that this class of cultivators is likely to be inhibited by institutional factors such as non-availability of credit or insecurity of tenure one should agree with Raj (1969) when he states 'this is where the need to make institutional changes in agriculture comes in.'

To sum up, I contend that what has happened in Sinaloa and Sonora in Mexico, and is happening in Punjab and Pakistan is, in every sense, a transformation of backward agriculture. Our effort now should be to extend it to as wide an area as is inherently capable of utilising it. Even then, vast tracts of agriculture in India and elsewhere will not come within the reach of the high yielding variety transformation. If, as a result, disparities are widened, it will at least be as a consequence of a forward thrust. Those who will be pushed ahead will not necessarily be a privileged few. There are millions of big as well as medium and small farmers and even tenants who, but for the high yielding varieties, would have remained poor for a long time. If this is to be viewed as dualism, it is at least partially benevolent and must be preferred to a more egalitarian stagnation. The new technology has reduced the threshold of non-viability to something like 1 hectare irrigated. This itself is a great contribution to the softening of dualism. Besides, in the Indian context, one of the most rewarding egalitarian devices is cheaper food; and only through the new technology can food prices be reduced without reducing farmers' incomes.

### *The Roles of Institutional Reform and Technology*

As to the role of institutional reform (land tenure, etc.), at the outset I should like to avoid a possible misunderstanding. As an ardent believer in egalitarianism, I advocate and support all institutional changes which have the potential for improving the social and economic conditions of the under-privileged.[13] The points I wish to make are: firstly, the technological barrier has for long rendered the

[13] I dislike making avowals of faith and do not feel the necessity of proving my *bona-fides*. If in spite of this, I do so, it is only because of my anxiety to avoid being knocked off before I am fully heard.

much needed institutional changes practically sterile and ineffective,[14] and secondly, where growth has occurred, institutional reform as a factor has a very low pay-off and technology a very high one (United Nations 1968: 127).

Taiwan is the oft-cited illustration of the contribution of land reforms to agricultural growth. But as Raj himself has shown, the growth rate in Taiwan's agriculture under the 'highly oppressive' land tenure system was slightly higher than that achieved in the post-land reform period. 'If 1952 is regarded as the year ending the post-War reconstruction period, the growth rate for the post-War period then became 3·22 per cent, which is slightly *lower* than the pre-War rate of 3·32 per cent of 1906-1939 (Ho 1966: 32). Raj's other quotation regarding the use of police force under the Japanese rule to extract higher yields is so telling that we shall not press the point. In fact, the contribution of high yielding variety technology is well brought out by comparing the growth rate for the period 1951-5 to 1956-60 with that from 1956-60 to 1961-5. Both of these periods were covered by land reforms, but only the latter, to some extent, by the high yielding variety technology. The growth rate in the yield of brown rice in the earlier period was 2·96 per cent as against 3·26 for the latter.[15] Growth

[14] One of the best known students on land reform, Doreen Warriner (1969: 374, 379, 382) writes: 'In India, the question of the relation of reform to development is highly problematical, because, although it seems wrong and unwise that so much legislation should have been enacted without being implemented, yet in the present political situation, the ceiling legislation cannot be implemented, *and in the present economic situation it is difficult to believe that the tenancy conditions can be improved*' (my italics). She adds: 'Agrarian reforms really do liberate . . . But no conclusion emerges that agrarian reform is necessary to development. It is unfortunately customary to prove that reform is a condition of development by using Humpty-Dumpty definitions' (p. 374). 'It can only be regarded as such if all other things needed for growth will [can!] not be undertaken without it.'

[15] Raj (1969: 25) has observed that 'the rate of growth of output of rice has been much lower than that of these other agricultural commodities, in fact, only about 3 per cent per annum'. Though true for output, the following table for yield (land productivity) may be kept in mind:

GROWTH-RATES (YIELD): QUINQUENNIAL AVERAGES

| *Crop* | *1951–5 to 1955–60* | *1956–60 to 1961–5* | *1951–5 to 1961–5*[a] |
|---|---|---|---|
| Brown rice | 2·96 | 3·26 | 3·11 |
| Sugarcane | 3·33 | 0·24 | 1·79 |
| Sweet potatoes | 5·24 | 0·61 | 2·91 |
| Peanuts | 3·45 | 2·94 | 4·15 |

[a] For 1951–5 to 1961–5, annual averages, compound.
*Source:* Compiled by Myers in Shand (1969: 34, 48).

rates for other crops, sugarcane, sweet potatoes and peanuts, which were not touched by the high yielding variety technology, however show a sharp decline in the latter period.

In relation to the relative roles of technology and institutional factors, Barker and Mangahas (1970), using linear probability functions for adoption of high yielding varieties on specified farm type in Central Luzon (Philippines), found that

> the effect of owner operatorship is generally positive, but rather small. From the viewpoint of land reform, it appears that transfer from share tenancy to owner operatorship *per se* is of lesser importance than such other aspects of the reform programme as irrigation and extension system to the tenant. Farm size was found to have a very small effect, supporting the contention that no minimum farm size is necessary for the new varieties to be economically acceptable.

Incidentally, all independent variables showed higher values under favourable environmental factors such as irrigation and water control compared to dry regions.

Nearer home, the evidence regarding the inhibitory influence of tenancy and small size of holding, factors wholly or partially amenable to land reforms, is mixed. Some field investigations have revealed that the technology has bypassed the small farmer. Chowdhury's meticulous study, however, shows that 'the participation by small farmers in the programme is less than that by the large farms, but the *effective participation* of the former in terms of the proportion of the cultivated area put under HYV crops is no less satisfactory than that of the large farm holders' (1970: A-91). Chowdhury also finds that the efficiency (as judged by net return per additional rupee spent) in the cultivation of high yielding varieties does not improve with the increase in farm size. In fact, in one of the two districts studied (though not in the other), the net return per rupee spent was highest on relatively small farms with 1-2 hectares of land. In terms of total gains, however, it is obvious that larger farms will gain more, unless there is a marked decrease in the efficiency of resource use with the increase in size. Mukherjee (1970: A-15) also found that for wheat in Punjab, all size groups of farmers participated in the high yielding varieties program; though in Tamil Nadu, for paddy, there was positive relationship between the size of holding and the proportion of participants. Nonetheless, for the lowest size group (below 1 hectare) the proportion of participants was as high as 70·3 per cent. It should also be noted, as pointed out by B. Sen (1970), that

> without reference to the size-group, the statement that inequality is bound to increase between the irrigated farmer and the non-irrigated farmer is somewhat misleading. Surely, a small farmer with

2 or 3 acres of irrigated land has now got the prospect of raising income to the level of a farmer with perhaps 10 to 15 acres of dry land.

The probable disadvantage of tenancy for the adoption of the high yielding variety technology is more relevant to the issue of land reform. Here again, the evidence is not conclusive either way. The enquiry conducted by the Department of Economics of the University of Bombay into tenancy reform in Maharashtra and Gujarat indicates that the tenant-cultivators were distinctly better adopters of improved practices compared with owner-operators. We were, however, unable to separate the performance of the former group on owned and leased holdings. The findings of the Programme Evaluation Organisation are, however, more conclusive and striking. Both in Tamil Nadu for paddy and in Punjab for wheat, judged by the percentage of area under high yielding varieties, the performance of tenant-cultivators has been consistently better than that of the owner-cultivator. Furthermore, the tenants were able to obtain even better yields than the owners. Mukherjee concludes: 'What little evidence has been produced does show that no significant difference in the rate of adoption has been observed between the owners and the tenants and also among the different size-of-holdings groups.' (1970: A-22).

The findings of Sen and Mukherjee must not, however, generate a feeling of complacency regarding the desirability of institutional changes. One may not share Martin Abel's (1970: A5) gloomy forebodings (perhaps a Paddock contamination) that the 'green revolution' may after all turn out to be 'merely a small palace revolt . . . [and that] . . . India could easily end up as an importer of 10 million tonnes of cereal grains by 1980,' but one should not lightly brush aside Wolf Ladejinsky's warning (1969: 160) that 'the share-croppers are, if anything, worse off now than before because as ownership of improved land is prized very highly there is mounting determination among owners not to permit the tenants to share in the rights of the land they cultivate. Their preference is to be rid of them.'

One's ideas about relevant radicalism in land reforms and other spheres may differ from those who have to win elections, but there is undoubtedly a lot of exploitation still in the agrarian structure and now that the green revolution has vastly improved the profit potential of Indian agriculture, the economic feasibility of transfer of gains to the disadvantaged groups in agriculture has improved *pari passu.* It has also augmented the temptation for the big farmer to entrench himself more securely and for the wealthy and the influential to enter the field. This must not be allowed to happen at the cost of the vulnerable sections in agriculture. A preventive land reform is, therefore, even more urgent than a positive one. What is needed is selective action rather than a generalised, largely ineffective legislation. Keeping in view that limited and none-too-efficient administrative-organisational resources of the State Governments, the former approach would have a better pay-off.

It would be prudent to recognise that even if all the desirable land reforms and other institutional changes were fully implemented, their impact on traditional agriculture would be rather limited. It is difficult to see how the small farms in dry regions, for example, would be made economically viable through land redistribution. Because of the low productivity, the ceiling on holdings in these regions cannot be drastically reduced. It is common knowledge that the tenancy problem in these regions is not acute.

Because some academicians, known for their progressive views, recognise the limited potential of institutional changes, they have suggested practical solutions which appear politically non-revolutionary. Minhas (1970: 113), for example, has suggested an integrated program of compulsory consolidation of land holdings and compulsory public works. After examining the impact of an 'imagined but radical' land redistribution policy, he is constrained to observe that, by itself,

> it would not be able to solve the problem of abject poverty in 1970. The size of the cake is small and the claimants are far too many. One may, however, envisage an even more drastic land redistribution policy for solving the problem of rural poverty. Nonetheless, this kind of arithmetic of redistribution needs to be tempered with a sense of realism. Realistic radicalism and political demagogy must be separated from each other.

It may be emphasised that this structural reorganisation of the agrarian economy will at best remove 'the inabilities besetting the small farmers and enhance their absorptive capacity for modern inputs and technology.'

My contention, in sum, is that chances of effective implementation of institutional reform, in the absence of marked technological breakthrough, are meagre. One must seriously challenge Dandekar's assertion (1970: 1233) that 'there is no doubt that the Communist strategy (of reorganisation of the agrarian structure) can break through the vicious circle of poverty and rescue an overpopulated agricultural country from the conditions of overpopulation'. All that it can achieve is to cover up poverty and exploitation of the peasantry under a facade of institutions which can be called socialistic simply because they are so labelled. The only abiding solution to poverty in poor underdeveloped economies, is to improve the profitability of human endeavour and ensure that its gains are evenly distributed.

Even in a stagnant agriculture, one finds a tiny privileged sector which can extract affluence from general poverty. This is economically and ethically indefensible and must be eliminated through agrarian reforms. At the other extreme, there is a highly vulnerable class of socially and economically disadvantaged persons which needs protection, relief and rehabilitation. These two extremes should be the immediate concern of agrarian reform. In a stagnant agriculture, agrarian reform by itself, whether communist or Maoist, cannot solve either the problem of poverty or overpopulation (unemployment). However, once some such innovation as a 'miracle seed' shatters the technological barrier to rewarding human endeavour, economic and institutional policies acquire a grave responsibility of ensuring that the gains of technology are not monopolised by the privileged and the powerful.

## CONCLUSION

If an impression has been created that the green revolution has solved the basic problems of Indian agriculture, or will soon enough do so, it needs to be effaced. At best, it will solve the food problem, bring higher incomes to many farmers and through its forward and backward linkages give fillip to general economic development. Apart from the unresolved problem of dry regions, agronomic research has not yet evolved corresponding high yielding varieties for India's major commercial crops like cotton, oilseeds and jute. Though the area devoted to these crops is relatively small compared to foodgrains, a breakthrough in their productivity is crucial to the stability and growth of the industrial and export sector.

Even during the bleakest period of the two consecutive droughts of 1966 and 1967, I had not despaired of India's capacity to feed her growing population although, of course, this does not mean that I approved of the fast rate of increase in population. In fact, when everybody was talking about food self-sufficiency as the most important problem in Indian agriculture, I drew attention to the massive additions to the labour force in agriculture as the most explosive and perhaps the most intractable problem of Indian agriculture in the 'seventies. I quote a somewhat lengthy passage from an article I published in 1968 (Dantwala 1968: 241-2):

> The consequences of a massive addition of 60 to 87 million persons to the agricultural work-force within the short span of 15 to 20 years can indeed be very serious . . . The dark spot of the future (however) is not food, but the farmer himself. In 1961, 132 million workers—or about 70% of the total work-force—were engaged in agriculture. Based on the population projections currently in use, and assuming that the ratio of agricultural work-force to the total as well as the participation rates remain unchanged, the work-force in agriculture will grow to 192 to 219 million by

TABLE 13.1

PRICES OF SELECTED AGRICULTURAL COMMODITIES, 1962–9

| Year[a] | *Wheat (Rs/100 kilograms)* | | *Groundnut kernels*[d] *(Rs/100 kilograms)* | | *Cotton*[d] *(Rs/100 kilograms)* | |
|---|---|---|---|---|---|---|
| | *India (Moga, Punjab)* | *United States (Export Price, No. 1 Hard Winter, Gulf Ports)* | *India (Bombay–Karad Bold)* | *Nigerian (at European Ports)* | *India (Bombay–Cambodia B–1 inch)* | *United States*[c] *(Middling 1 inch)* |
| 1962 | 46 | 30 | 89 | 81 | 372[b] | 352 |
| 1963 | 46 | 31 | 112 | 87 | 372[b] | 347 |
| 1964 | 56 | 31 | 125 | 98 | 372[b] | 324 |
| 1965 | 60 | 28 | 184 | 111 | 389[b] | 309 |
| 1966 | 72 | 32 | 185 | 138 | 408[b] | 368 |
| 1967 | 90 | 36 | 141 | 123 | 500 | 413 |
| 1968 | 85 | 47 | 191 | 153 | 547 | 375 |
| 1969 | n.a. | n.a. | 221 | 165 | 586 | 360 |
| | | | (Up to Aug. 1970) | | | (Up to March 1970) |

[a] Year: wheat—January–December, groundnut—November–October, cotton—India, September–August, U.S.A., August–July.

[b] Official ceiling prices.

[c] Average of fifteen markets.

[d] For groundnut and cotton, read 1962/3 for 1962 and so on.

*Source:* India, Agricultural Prices Commission (1968).

1976 and 1981 respectively. The assumption of a constant ratio of agricultural to non-agricultural work-force may appear unreasonable, because with rapid economic development, the ratio is generally expected to decline. It did not, however, so decline between 1951 and 1961. As it is, there is lot of unemployment in the non-agricultural sector of the economy, and even on the assumption of a constant ratio, the non-agricultural sector will have to absorb an additional 26 million people in the years between 1961 and 1976 and 38 million between 1961 and 1985. If the growth of employment in this sector over the last 16 years since 1951 is any indication, fulfilment of the above requirements of additional employment will itself be a formidable task.

The green revolution has given us a breathing time. Its greatest contribution it that it may help to relieve the tension and distraction caused by chronic food storages, and bring back the economist and the planners to the more abiding issues of Indian Planning.

14

# An Interim Judgment

*R. T. Shand*

THE new foodgrain technology is now commonly expected to have a decisive influence upon the balance between rates of growth of food production and population in Asia. Some economists go further and assign it a leading role in overall economic development of countries in the region. It now receives primary emphasis in many national development policies and is given high priority in advice from international experts on development.

It is only logical to expect therefore that its impact should be subjected to continued systematic evaluation to provide a basis for formulating realistic development policies to make maximum use of its benefits. Yet to date remarkably few such appraisals have been undertaken in the breadth and depth appropriate; indeed the studies in this volume comprise a significant proportion of the major fieldwork which has been carried out on the topic.

Assessments to date have typically lacked a sound empirical base, and this has led to widely differing expectations concerning the nature and spread of the impact. Overall, optimism has prevailed. For some it has provided a cautious 'breathing space' argument, but at the extreme it has led others to conclude that the problem of balance between food supply and population has been overcome and that subsequent rural prosperity will serve to stimulate industry and perhaps to shift some of the economies into the much sought-after phase of self-sustaining growth.

The temptation to embrace this extreme optimism is all the greater in the face of the problems and of the limited success of other growth policies followed in the region. The view is also attractive because early success with the new technology has been bought relatively cheaply in terms of public resources. As Dantwala observes, progress so far has been primarily a scientific accomplishment.

Moreover, many of the claims made for the new technology have arisen from unjustified assumptions concerning the significance of indicators currently used to measure the level of its adoption and its impact.

One popular indicator is the area sown to high yielding varieties. It is, for example, commonly assumed that large areas under these varieties necessarily imply substantial additions to total output. For wheat, this has been the case. Kaneda reports that Pakistan production in 1969/70 was 67 per cent greater than in 1966/7. In India, wheat output rose 64 per cent between 1964/5 and 1969/70, a result all the more remarkable since, in the two drought years of 1965/6 and 1966/7, no progress was made. For rice, the relationship in Pakistan was similarly positive. In India it was not. Despite the reportedly large areas sown to the high yielding varieties (estimated at 4 million hectares in 1969/70) there was no evidence that total Indian rice output had been significantly increased as a result (Lockwood, Mukherjee and Shand 1971: 111-23).

A second indicator sometimes used is the number, or growth in numbers, of those using high yielding varieties. In practice, however, the use of high-yielding varieties does not necessarily mean a heavy or a lasting commitment to the new technology in terms of the area under crop. For wheat in India, the Programme Evaluation Organisation (PEO) survey data generally did show that adopters devoted a large proportion of irrigated wheat area to the varieties. However, in some irrigated wheat areas without assured water, individual growers were found to have reduced their areas under the varieties in later years compared with their initial commitments. Again in India for rice, the same PEO surveys showed the proportion of rice producers in the survey sample who grew high yielding varieties in areas with conditions considered to be most suitable for them rose between 1967/8 and 1969/70 from 16 to 32 per cent, which might suggest a major increase in individual commitment to them. However, amongst these adopters, the proportion of the rice area devoted to the new varieties remained almost static at around one quarter, so that within the communities surveyed, the proportion of the total area under the high yielding rice varieties expanded only from 4 to 8 per cent (Lockwood, Mukherjee and Shand 1971: 18-30).

A third indicator is total output and in recent years there has been a tendency for some observers to assume an unqualified association between increases in total foodgrain output and the spread of the new technology, thus ignoring or playing down the influence of other factors. In Pakistan, it is likely that the new input package was mainly responsible for production increases of both wheat and rice, but it was also significant (Chapter 9) that both before and after the introduction of the new varieties, there was considerable investment in new and improved irrigation, which enabled intensification of production in areas already under cultivation and probably encouraged new areas to be brought into use. The advent of the new varieties undoubtedly enhanced and accelerated the profitability of such investment, but it appears very likely that these new facilities would still have been profitable and that a substantial rise in yields and output would have occurred even without the new varieties. In India, where wheat production rose 67 per cent in the five years to 1969/70 and yields rose 32 per cent, the total area under the crop also rose by a notable 24 per cent. It is by no means clear how much of the area increase can be attributed to the profitability of the high yielding varieties. Other factors such as climate and relative market price levels can also have a considerable influence.

Realistic appraisals of the new technology require much clearer appreciation of the limitations of these partial indicators and a more circumspect employment of them. There is a danger that their uncritical use will encourage complacency and will delay recognition of the limitations of the new technology and obscure the need to investigate the constraints to more sustained and widespread progress.

This final chapter reviews early experience of the new technology: its adoption, its direct impact on the farming community and its indirect influence or secondary impact on the wider economy. It also discusses future prospects and policies that could remove current or impending constraints and strengthen the direct and indirect impacts of the new technology.

## FACTORS AFFECTING ADOPTION OF THE NEW TECHNOLOGY

First we may note that the success of the new technology to date has depended primarily upon progress achieved in the development of high yielding varieties which are suited to the production and market environments of the region. By contrast, there has been relatively little, though highly significant, adaptation of these environments to meet the needs and characteristics of the new varieties.

Experience has shown that only for wheat have variety and environment been well suited over a major proportion of the cropped area. Mexican varieties and local adaptations of them were shown to be suited extensively to the Punjab of Pakistan and India and they have gained market acceptance there. Kaneda showed that the high yielding varieties covered roughly one half of total Pakistan wheat area in 1969/70. In India, more than one half of the total irrigated wheat area in resource-favoured villages surveyed by the PEO was devoted to the new varieties (Lockwood, Mukherjee and Shand 1971: 13). It is significant that these were also the areas where irrigation facilities were already well developed and where, in addition, exploitation of underground water resources by means of tubewells led to a considerable improvement in these supplies. Since the new technology was introduced, large-scale adaptation of growing conditions has been virtually confined to these areas of India and Pakistan where water availability and reliability have been improved. In India, PEO survey data also strongly suggest that adoption of high yielding varieties both in terms of numbers of growers and, amongst adopters, in terms of the proportion of wheat area sown to the varieties was markedly lower in areas with only canal water than in areas where canal water was supplemented by tubewells. Indeed in some of the former areas, in Rajasthan and Bihar, a decline in participation was recorded from 1968/9 to 1969/70.

For rice, the major initial problems have been associated with realising the high yielding potential of newly developed varieties under the widely varying conditions of production and market preference within the region. Earlier chapters documented the unsuitability of initial varietal releases on waterlogged soils and in areas which experienced deep and variable flood conditions such as the major river floodplains of Thailand and Burma. Other problems were disease susceptibility, especially under conditions of high humidity, and widespread market dissatisfaction with grain appearance, flavour and milling qualities, dissatisfaction frequently expressed in a lower price offered for high yielding varieties relative to traditional varieties. The early successes with rice variety releases were registered in Pakistan and parts of Luzon in the Philippines which offered a high output response with a good degree of water control and where the market did not penalise the new varieties unduly.

Philippines data for adoption on a country-wide basis are not particularly reliable but do suggest that use of IRRI varieties was widespread, with a considerable overall proportion of rice area devoted to them. Certainly in the resource-favoured areas studied by Barker and Cordova, the proportion under these varieties was high. However national output trends before the 1972 floods strongly suggest that output response to them was not as great as had been expected and that hopes of the impact on crop yields have been too high. This may well be due to the use of the high yielding varieties in many areas with levels of other inputs (water, fertilisers, etc.) far short of those required to gain full advantage of the high yield potential of the varieties.

Sajogyo and Collier's data suggest varietal adaptation problems for Indonesia too. In the three areas with reputedly favourable conditions for adoption of high yielding varieties, only one showed a significant proportion of rice producers wholly committed to the new varieties (41 and 29 per cent in wet and dry seasons of 1971 respectively). Admittedly, in all three areas the proportion of farmers growing some high yielding varieties with other varieties was higher, but without data on area commitment to these varieties, no firm judgment can be given as to overall involvement. Indications are that it was almost certainly minor in two of the three areas.

*Profitability*

Available direct evidence of the profitability of the new technology on the farm is limited but does not tend to support the expected: that substantial net returns can be achieved from the expenditures only if all the necessary inputs in the package are available in full measure.

Conditions conducive to high profits have undoubtedly existed in the Punjab of India and Pakistan. In the latter, Chapter 9 shows that government policy played an important role in assuring profits by subsidising inputs, and facilitating their supply by arranging favourable foreign exchange rates. Both public and private investment had been heavy before the introduction of the high yielding varieties, particularly in developing water supplies which enabled greater cropping intensity. Price support schemes assisted by ensuring an attractive market for increased output.

Evidence from elsewhere in the region affords other insights concerning the returns from the new technology. Barker and Cordova's results suggest an important correlation between profitability and water availability under three quite different situations. In three areas, Calamba, Cabuyao and Biñan, cash expenditures on inputs per hectare rose substantially with the adoption of the new technology, so that variable costs almost doubled in three years. There, however, the similarities ended. In Calamba, a 21 per cent increase in net returns was recorded over the three years to 1969 (in terms of 1966 prices), in Cabuyao the increase was 96 per cent, while in Biñan there was no increase.

The record for Biñan farmers is clearly an important one to explore. As Barker and Cordova show, conditions there are more typical of resource situations in the Philippines than they are in the other two, particularly availability of irrigation water (for which Biñan was worst of the three). Although a substantial yield increase of 30 per cent or more was achieved, it was accompanied by higher input expenditures and no net income gain accrued to the farmer. Barker and Cordova suggest that input levels may have been too high, but without assured water, producers would always face the risks of climatic variation which could reduce or eliminate profits from the new technology. Such farmers may be forced to plant the high yielding varieties simply to avoid a loss which might arise if local varieties previously grown suffered a decline in price. Given the added risk arising from higher cash costs, the new situation could hardly be regarded as a major improvement for them.

At the other end of the spectrum, productivity and yields were already high in Calamba; it is notable that the assured supplies of water already allowed 98 per cent of the area to be doubled cropped. Before the introduction of the new technology average variable costs per hectare were 40 per cent higher in Calamba than in Biñan, but yields were more than 50 per cent higher. The additional expenditures per hectare incurred with the new technology were very similar in both areas but because Calamba was apparently more favourably placed for resources, especially water, the absolute response in yield and gross return was substantially greater there. This result is suggestive of the impact of the new technology in areas already well served with water supplies.

In the third area, Cabuyao, the new technology was introduced while water supplies were being increased and improved, as was the case recently in the Punjab. Barker and Cordova show that between 1954/5 and 1966/7 the proportion of land double cropped rose from 16 to 58 per cent and this may have been extended further after 1967. The increase in yield here was the highest of all three areas between 1966 and 1969 (91 per cent) and so were the increases in net returns per hectare at current prices (117 per cent). This may broadly indicate the magnitude of response possible at the farm level with irrigation schemes specifically designed and timed to complement the dissemination of high yielding varieties.

The group of farms studied by Castillo and colleagues was initially operating at a lower level of productivity than the Biñan groups surveyed by Barker and Cordova. Average yields were around 1·2 tons per hectare compared with 1·98 tons for Biñan farmers (Table 6.1). By 1965 however, the former's average per annum (two seasons) had been raised to around 1·8 tons per hectare which was roughly similar to the Biñan average in 1966/7. As the respective chapters show, both groups of farmers proceeded to adopt high yielding varieties and associated inputs, and both recorded substantial yield increases in the following years. What is significant for Castillo's group is that, even with the widespread adoption of the new technology by survey farmers, only 41 per cent assessed their incomes as being greater in 1969 than at the beginning of the project in 1963, while 31 per cent considered their incomes were lower. On average, the change was probably close to zero which would broadly tend to corroborate the experience of the Biñan farmers.

The Indonesian data on profitability are difficult to interpret but, as the survey population was drawn from areas thought to be most likely to adopt the new technology, i.e. from resource-favoured areas, they provide some basis for the conclusion that profitability has been highly variable and thus far has not been sufficiently positive to have increased net farm incomes.

### THE WIDER IMPACT OF THE NEW TECHNOLOGY

Disappointment has been expressed in recent years over the lack of stimulus to industry from the spread of the new technology through backward and forward linkages with agriculture. The lack of such secondary impact has been the more acutely felt because of the exaggerated hopes placed on the new technology as an activating factor for economic growth. However the preceding chapters suggest reasons for this.

It is true that there has been a notable stimulus, through backward linkages, to a few industries such as fertilisers and chemicals, and to a lesser extent to engineering and machinery industries associated particularly with tubewell construction. This however has been limited in varying degrees by a water constraint which has confined the levels of profitable fertiliser application, and by the above mentioned problems of adaptation which high yielding varieties released early encountered.

Amongst forward linkages, greater activity has been encouraged in trade, commerce and processing industries associated with increased foodgrain output. However, this stimulus, like the demand for purchased inputs, has been concentrated geographically and most of the benefits to processing at least have probably been felt locally.

For a number of reasons much less can probably be expected immediately of the other, potentially more important, forward linkage, the expenditures from higher net farm incomes resulting from the new technology. First it has been suggested above that, so far, only a very limited number of producers have benefited. Second, there is, in addition, a most uneven distribution of land ownership amongst producers who have benefited, which has meant a further concentration of income benefits. Thus, while it appears that the new technology is largely neutral to farm size, in absolute terms the large landowners have benefited most. Third, the pattern of income disposal of the large landowner is less likely to lead to a significant stimulus to demand for consumer goods than to further investment, perhaps in the agricultural sector itself, e.g. in the purchase of additional land or in money lending for agricultural purposes.

The only major industrial stimulus likely to have registered on a national scale so far has been in the fertiliser and chemical industries and, much less, in the engineering industries. This is a narrower impact than has been hoped for. This conclusion does not deny that progress has been made. It does however require the qualification of the belief that countries engaged in spreading the new technology are at present necessarily entering a period of sustained economic growth based on a close interrelationship between agriculture and industry.

### FUTURE PROSPECTS

Prospects for the future spread of the new technology depend upon two related developments. The first is the further adaptation of high yielding crop varieties to the production and market environments of the region. The second is the adaptation of these environments, particularly conditions of production, to the high yielding varieties.

We have noted, during the initial years of the new technology, a heavy emphasis upon the first of these; the numbers of varieties released are now multiplying rapidly, but in these early years success depended on the adaptability of a very few introduced over a wide range of agro-climatic conditions, sometimes without adequate preliminary field trials. It is hardly surprising that the results have been mixed, ranging from general success with wheat to failure with some of the hybrid crops. The early experience has, however, been valuable for plant breeders and agronomists in pointing out the nature of field problems which beset these varieties.

There is still considerable scope for the development of varieties better able to respond to existing production and market opportunities and such work is in progress. The spectacular breakthrough for wheat is yet to be matched for rice, and there is no assurance that this can be achieved with a single new variety or even with a small number. A considerable number, each specifically adapted, will probably be needed, so that the breakthrough may well be more gradual. Considerable progress has been made in this direction already. Welsch and Tongpan discussed an imminent advance with a high yielding 'floating' rice which could flourish under deep water conditions. Such varieties would have particular value in the floodplains of Thailand and Burma. Similarly cross breeding is already improving the grain quality and palatability of high yielding rice varieties and should enable them to compete successfully on the market with favoured local and export quality varieties. Breeding programs may also succeed in developing new high yielding strains of glutinous rice, for which there is a distinct market. For wheat too, new varieties have already been released which outyield the Mexican importations and are superior in grain quality.

How long particular plant breeding objectives will take to achieve is a matter for speculation. Experience shows the period required is often considerable. Iinuma describes the lengthy process by which a fruitful synthesis of foreign and domestic agricultural science took place in Japan and Hayami shows that Japanese research in crop production (pure line and cross breeding) took two decades from 1904/5 to fulfil its potential. The transfer (by local testing, adaptation and breeding for specific environmental conditions) of Japanese rice technology to other countries through the introduction of ponlai rice also took place slowly; Carr and Myers suggest about fourteen years. It is true that modern scientific communication between countries has become faster and more sophisticated, and that techniques of plant breeding and testing have greatly improved. On the other hand, the task ahead of today's plant breeders is more substantial. In all, we still may count ourselves fortunate if we see the major plant breeding objectives achieved within fifteen years.

While it is reasonable to expect foodgrain production potential to improve further with the development of new high yielding varieties, the relative scale of their contributions, and the frequency of their appearance, is likely to diminish. Moreover, however high their production potential, its realisation will depend upon the availability of complementary inputs. Amongst these, water is the most critical, and it is this that is most likely to constrain the effective use of new high yielding varieties as they are developed for a widening range of conditions.

Irrigation statistics show this clearly. Welsch and Tongpan suggest that only one quarter of the total rice area planted in Thailand is irrigated and, of this, little is controlled. Around 5 per cent of the rice area was under short-straw high yielding variety in 1969/70 and although RD1 and RD3 have recently been released, only 10 per cent of the crop area can effectively make use of them, because of the water supply constraint.

Silcock (1969: 125-6) gave a similar overall figure for the total Thai rice area irrigated in the late 1960s, and commented that in the northern valleys the proportion double cropped was less than 2·5 per cent while elsewhere it had not exceeded 0·3 per cent. Most important, storage irrigation is still not a feature of the Central Plain. It is served by dams which are essentially diversionary, designed to mitigate floods and disaster. As Silcock notes, whilst the ultimate objective of policy is to provide a reliable water supply for two crops per year over a large part of the Central Plain, the main storage currently available, the Chainat Dam, cannot achieve this, and other major storage dams are required in the northern mountains to impound the water of the Chao Phaya's tributaries.

In Chapter 4, Lockwood and Mukherjee detail the faults of the Indian irrigation system; in the *kharif* season irrigation is largely protective (a drought insurance); in the *rabi* season irrigation is limited, major systems utilise gravity flow without field control of timing or quantity, and without provision for drainage to prevent waterlogging and flooding. Under these conditions, fertiliser is wasted and pests and diseases are encouraged.

Whitcombe has made a detailed indictment of the effects of major canal construction. This shows how misleading existing statistics of water availability can be under conditions of low and declining soil productivity. The large areas of saline/alkaline land, with problems of depleted soil nitrogen, deterioration of mechanical soil qualities, and erosion, have seriously reduced soil fertility and thus the capacity of existing irrigated land to respond to the High Yielding Varieties Programme. Other points are equally disturbing. Existing irrigation systems, where viable, cannot supply the increased water required by the high yielding varieties; modification is thus necessary to meet new needs (e.g. through contraction of command areas or, where suitable, using groundwater reserves). Whitcombe also gives good reasons why utilisation of underground water is not necessarily an answer to the inadequacies of the existing canal systems. Finally, she draws attention to the damage created in the name of rural development by the infrastructure. Road and rail embankments have produced waterlogging problems; canal irrigation has given rise to lateral seepage and loss of water; flush irrigation has exceeded downward water percolation and has also led to waterlogging.

In broader vein again, Mukherjee and Lockwood point out that in 1966/7, only 20 per cent of the 120 million hectares under foodgrains in India were irrigated, and only 17 per cent were double cropped. Dantwala gives the double cropped proportion of irrigated rice area at only 10 per cent. Thus a very sizable proportion of the 24 million irrigated hectares was under wheat. Such figures do not indicate the effectiveness of irrigation. Crawford (1969: 80-1) pointed out that, while irrigated crop area was reported to have expanded by 60 per cent (14 million hectares) during the fifteen years of the first Five Year Plans, actual achievement was substantially less. In some instances the figures for irrigation referred to total areas commanded by major canals or supplied by tubewells, and not to areas actually watered. Furthermore the systems were not designed to provide water for intensive cropping but to supplement monsoon water and insure against drought, which partly explains the small proportion of area which is double cropped.

In the Philippines, Hooley and Ruttan (1969: 223-9) showed that, whereas around 25 per cent of total cultivated land was irrigated at the turn of this century, the proportion had fallen to only 8·2 per cent in 1960. Admittedly the irrigated area increased, from 320,000 to 621,000 hectares, but this did not prevent an overall decline in resource productivity in Philippines agriculture for the period. Barker and Cordova found that in one of their three survey areas the irrigation system had deteriorated to the point where in 1967 only 64 per cent of land was double cropped as against 95 per cent in 1955.

Richter shows that in Burma only a little more than 10 per cent of net sown area was double cropped in 1966/7, and a similar proportion was irrigated. However, nine tenths of the multiple cropping occurred, not on irrigated land, but on river islands and banks. Double cropping cannot readily be expanded without heavy investment, since most irrigation schemes are main season gravity flow systems without provision for storage or effective control. Most of this was sown to rice; 13 per cent of gross sown area was irrigated. Kaneda (quoting Mohammad) makes the point that most perennial canals in the Indus Basin afforded only a low cropping intensity in the summer season and it was the non-perennial canals which raised the proportion of gross area to be irrigated to about 70 per cent in that season. The proportion irrigated was lower in the winter season. Thus tubewells clearly could raise the intensity of land use further in this area.

These illustrations are given in aggregative terms, and do not consider the effectiveness and timeliness of water supplies. Further close examination and evaluation of the meanings and significance of 'irrigation' as they are accepted in the region is needed. This could show a spectrum of irrigation systems in terms of effectiveness, ranging from dryland farming with minor seasonal supplementation at one end of the spectrum, through schemes giving mainly drainage and drought protection or drainage and flood protection, with irrigation taking a minor role, to full water control for one season and comparatively rarely, to full control for two seasons. For exploitation of the new technology, the water constraint is already apparent in the shortage of land adequately served by water even for one season: for two seasons, the water constraint has been removed in very few areas indeed. Production is heavily dependent upon the effectiveness of the monsoon rains in summer, and upon accumulated storage for the winter seasons. Therefore, given the climatic variability of the region, the gains already achieved from the new technology cannot be consistently relied upon from year to year, particularly in areas without supplementary sources such as tubewells.

The development of drought resistant higher yielding foodgrain varieties would be of enormous value and would greatly extend benefits obtained from the recently developed water sensitive varieties. Although little has been achieved in this direction so far, research is obviously warranted. Success, however, is unpredictable, and cannot be incorporated in our analysis, which can deal only with water sensitive technology at present being developed.

Within the known technological framework, it is in respect of the water input that agricultural development policy and planning must take major and generally new initiatives if the growth of agricultural output is to be sustained at the high rates now considered desirable by most countries. Some countries have neglected this factor and have relied largely upon interaction between the new varieties and existing facilities. Some, like India, have had a continuing program of facility expansion, but one which fails to meet the exacting field requirements of the new technology. The nearest approach to provision of the full requirements has probably been in Malaysia where land development programs for rice production have deliberately introduced high yielding varieties under double crop conditions with controlled and assured water. India and Pakistan have benefited from fortuitous opportunities to exploit underground water reserves in the Punjab which provided an essential element of conditions for production. Reserves such as these are available in few regions.

Policy now must be more far-reaching than previously, with long term programs of water resource development which can be intergrated with the existing and anticipated potential of high yielding foodgrain varieties so as to extend production benefits progressively to increasing numbers of producers. This should be combined with far more detailed research at the farm level to determine production costs, net returns to farmers and the availability of inputs at the time needed. Only then can policy measures designed to bring income benefits to an increasing number of producers be more than hit or miss affairs.

The historical precedents for such action are convincing. Hayami shows (Chapter 2) that the proportion of total sugar cane and paddy area that was irrigated in Taiwan rose from 56 per cent in 1915 to 78 per cent in 1935, a major feat in view of the 40 per cent increase in area sown to the two crops over the period. Even more significantly, the ratio of double cropped to total cropped area remained at or above two thirds during this whole period. Myers (1969: 42) reported that in Taiwan, total expenditures on irrigation works rose from 28·5 million Taiwan dollars between 1911 and 1920 to 213 million dollars in the next decade and were still high in the 1930s at a total of 118 million dollars. Most of the expenditures in the 1920s were of a capital nature while in the 1930s they were principally on maintenance work. Myers quotes Rada and Lee's assessment that 'more than 60 per cent of the increase in the output of paddy for the 1922-38 period could be attributed to the increase in irrigated area'. Quite clearly, the expansion of the area under effective irrigation played a crucial role in the spread of *ponlai* rice in these two decades. Hayami attributes the lag in the development of high yielding varieties in Korea to a constraint of inadequate water control brought about by a delay in investment in irrigation facilities. Finally, Hayami says that in Japan 'even at the beginning of the Meiji Restoration almost all paddy fields were irrigated', which suggests that by that time favourable conditions had already been created for the rapid dissemination of improved varieties which took place later in the Meiji period. Crawcour (1969: 6-8) has pointed out that at the time of the Restoration, double cropping was already the rule in the commercialised provinces of southern Honshu and that before that, flood control, irrigation and land reclamation had been undertaken on a considerable scale by shogunates or local administration. The further work carried out after the Restoration was designed to improve drainage facilities, which suggests that the supply of water to the fields was already effective and dependable in major producing areas.

Long-term programs of irrigation development pose major policy problems. They require heavy investment, extensive technical and economic assessments of water resources, and complex comparisons of the relative merits of new large-scale irrigation schemes with others which would upgrade or rehabilitate existing systems. Long-standing legal/political disputes have to be settled (e.g. States' rights to watersheds in India) and, at the farm level, such programs challenge existing patterns of land use and ownership.

Such programs also open up serious questions with regard to patterns of international trade in foodgrains, particularly for the rice market, which has historically been sensitive to small variations in supply. Currently a rather pessimistic view is taken for the future of this market and for those who supply it, because of the increase in marketed supplies expected to follow increases in foodgrain productivity in importing countries. In this author's view, it is simply not possible to make firm projections for these markets. Trends will depend on imponderables such as the future time sequence of new varietal development; on the resources allocated to water development and on the effectiveness of design and implementation of irrigation schemes; on policies relating to self-sufficiency in foodgrain production in importing countries; and on the incidence of changes in comparative advantage in production amongst producers. There do not appear to be convincing *a priori* reasons for concluding that the terms of trade will necessarily decline for exporters in the long run.

While individual irrigation schemes have to be thoroughly investigated technically (to avoid the unfavourable long-term effects catalogued by Whitcombe) and economically, it does appear that, in combination with the existing and further development of high yielding varieties (and assuming availability of other inputs in the package—fertiliser, plant protection materials, credit, etc.), they provide the best hope for sustaining a high rate of growth of foodgrain production. This strategy should provide opportunities to spread the new technology and to increase incomes of larger numbers of the farming community, though as we argue below, the strategy for development does not end there.

Finally, we turn to the secondary impact of the new technology—to the linkages between agriculture and the rest of the economy, and thus to the likelihood of more significant intersectoral relationships in the future. Prospects will, of course, depend principally upon the nature of the development program involving the high yielding varieties. However other factors, some of which have received scant attention until now, will also be influential.

If agricultural development programs continue to rely principally upon improving high yielding varieties for existing irrigated areas, relationships are unlikely to develop or diversify much further. Varietal breakthroughs will undoubtedly extend the new technology into new areas, but because of the limitations of water availability, foodgrain production will tend to reach a plateau and, for the majority who are without assured and adequate water supplies, income gains will be neither large nor regular. In short, the pattern will be similar to that which followed the successful introduction of the wheat varieties.

The alternative strategy, combining new varietal development with provision of an assured water supply over an expanding area, offers firmer grounds for optimism concerning industry linkages. Such a strategy would have backward linkages encouraging a relatively steady and continuous growth of demand for inputs. In forward linkages, a more dependable and marked growth rate of output would offer similar advantages to trade and to processing industries. Above all, the strategy would strengthen linkages between consumer goods industries and the farm sector, through an increasingly widely distributed growth of income amongst foodgrain producers. Just how quickly this would develop and how great the impact would be on overall economic growth would depend on a number of other factors.

One would be the size of the development budget for agriculture and in particular for the provision of irrigation, for this would determine the numbers of producers affected over time. Another would be the initial standard of living of the producers who benefit from the development program. In rural sectors which have already experienced severe population pressure on resources and where living standards approach minimum subsistence levels, a large proportion of the short-run increase in foodgrain output will be consumed on the farm; where indebtedness is a feature, much will also be channelled into meeting outstanding obligations. The gain to farmers in disposable cash income will be low and the additional demand forthcoming for market goods will be meagre. Conversely, the advantages of implementing such a strategy whilst farm living standards are fairly high and before indebtedness becomes a problem are clear. Increases in income are translated more quickly and more completely into consumption expenditure. In this sense the timing of the agricultural development effort in Japan and Taiwan was propitious. It minimised this source of linkage weakness. This has probably not been a major influence thus far with the new technology, because the limited numbers who have benefited with substantially increased disposable income have been located in resource-favoured areas and have already been earning relatively high incomes. It is when (and if) the new technology is extended more widely into less favoured areas, that the linkage weakness may show up.

A third factor is the distribution of land ownership amongst producers of crops affected by the new technology. If, as was argued above, the distribution is very uneven, the impact of higher incomes upon the demand for consumer goods is likely to be much less, since a considerable proportion of it will accrue to a few wealthy landowners. How much less will depend upon the terms of tenancy agreements; i.e. upon the proportion of increased net profits which accrue to the two parties. This may be the more important in highly populated countries and areas such as India and Java, when added to the factor of low initial living standards. The combination of the latter two factors is likely to inhibit greatly the emergence over time of linkages through consumer goods demand. Thus, apart from the persuasive argument for redistribution of land ownership on the grounds of social justice, a powerful economic argument may well lie in the obstructive influence of concentration in ownership and income on general economic progress. One might even question whether close forward linkages could be developed in the forseeable future without effective redistributive measures.

A fourth factor is the nature of consumer demand which growth in disposable rural cash incomes will engender. Even beyond the point where additional output is consumed on the farm, the pattern of expenditures will continue to be dominated over a significant range of income levels by products produced within the rural sector itself. Consumers will be concerned with improving the quality of their diet and this should lead to a more diversified local market for crop and animal products, spreading the income gains of foodgrain producers to other rural producers. Since these others can be expected to have expenditure patterns similar to those of the foodgrain producers, however, the spread effects of higher incomes will remain mostly within the sector.

The pattern will vary from country to country but it is clear that such diversification in demand will require increased production in a number of food industries not initially affected by the new technology. Some of this may well be satisfied from irrigated areas for which new production possibilities are opened up. It is also clear, however, that unless there is a significant production response of such products, much of the advantage of higher incomes will be dissipated in rising prices for these commodities. Logically then research programs should anticipate this demand and include such products so as to avoid supply bottlenecks, price inflation and the absorption of the valuable additional purchasing power.

Apart from food, demand for better clothing, housing and household goods will feature in the expanding budget of rural low-income earners. Once again rural products, this time non-food, will provide the raw materials for many of these items (wool, cotton, timber, etc.).

Provided that all these conditions are met, much of the secondary impact of rising incomes of foodgrain producers will initially be on the rural sector itself, in the form of increasing demand both for food and for non-food products. Such demand will give rise initially to industrial activity associated with the processing, transport and marketing of rural produce. To the extent that these industries are dispersed amongst rural villages and towns, the secondary impact, at least in relation to disposal of higher net cash incomes, will be felt first in the countryside and may well have a significant local employment impact. It seems unlikely to give rise immediately to a demand for a wide range of new consumer goods.

Secondary industry stands to benefit most initially from the increased rural demand for producer goods, but without an accompanying and growing rural demand for consumer goods, the functional relationship between agricultural and non-agricultural industries may remain weak.

In the rural areas, processing, transport and marketing firms are usually small-scale and locally based and are seldom accurately represented in industrial statistics. This provides one reason why increased activity amongst these firms is often not recorded statistically. Many proprietors of such firms are also agricultural producers so that the use of agricultural income to finance expansion of non-agricultural activities does not show as an intersectoral transfer. Furthermore, because they are rurally based, labour required in the expansion of these non-agricultural activities is recruited locally and does not require any significant locational shift in population. For these reasons, the initial growth of small non-agricultural rural industry is likely to be centred spatially within the rural sector itself and will tend to be obscured in published statistics.

In conclusion, several policy matters deserve emphasis. First, simple reliance upon new high yielding varieties alone for a continuous long term increase in foodgrain output does not appear justified, though there is undoubtedly much scope for the productive adaptation of new varieties to existing environmental and market conditions in the region. For the new technology to have a marked and widespread impact on foodgrain production and producers, programs of varietal research and extension must be simultaneously supported by major investment in irrigation facilities; otherwise output will level off on a plateau. Such a policy also seems necessary if agricultural development is to initiate substantial expansion in other sectors of the economy. There are, moreover, other important factors which can reduce the impact of progress in foodgrain production on other industries in the short run and which have serious policy implications. It is also suggested that initially, the impact of such progress is likely to be felt primarily within the rural sector itself, and special policy initiatives will be necessary to stimulate production in rural products other than foodgrains in order to meet rising demand.

Much of the foregoing argument is necessarily speculative. Far too little is known about historical experience in these policy areas. Further historical research in this field, difficult though it is, could prove most valuable. Additionally, much value could come from country-based evaluation studies. It is all too clear from the studies contained in this volume that too little is as yet known about the spread of the new technology, about its impact upon rural producers, and about past and likely future secondary effects on the rest of the economy. This is all the more lamentable in view of the hopes that rest upon the new technology. The consequences of the success or failure of the new technology are too critical for the present imperfect state of our knowledge about it to be acceptable. Major monitoring and evaluation studies on a continuing basis (e.g. as made by the Programme Evaluation Organisation of the Indian Planning Commission) by central governments seem necessary.

# References

*Chapter 1*


Furushima, T. (1949). *Nippon Nogyo Gijyutsushi [The History of Japanese Agricultural Technology]*, vol. 2, Tokyo.

Iinuma, J. (1969a). *Meiji Zenki no Nogyo Kyoiku [Agricultural Education in the First Half of the* Meiji *Era]*, Kyoto.

——— (1969b). 'The Meiji System: the revolution of rice cultivation technology in Japan', *Agricultural History*, 43 (2): 289-96.

Saito, Y. (1968). *Nippon Nogakushi [The History of Japanese Agricultural Science]*, Tokyo.

Susuda, R. (1970). 'Meiji noho keiseiki ni okeru nogakusha to rono no koryu [Intercourse between agronomists and experienced farmers in the development of the Meiji System]', *Journal of Rural Community Studies*, **31**.

Tsukuba, T., (1959). *Nippon Nogyo Gijyutsushi [The History of Japanese Agricultural Technology]*, Tokyo.

Yasuda, S. (ed.) (1952). *Meiji iko ni okeru nogyo gijyutsu no hattatsu [The Development of Agricultural Technology since the Meiji Era]*, Tokyo.


*Chapter 2*


Barker, Randolph and Quintana, E. V. (1969). 'Studies of returns and costs for local and high yielding rice varieties', *Philippine Economic Journal*, **7**: 145-61.

Clark, Colin (1967). Review of J. I. Nakamura (1966) in *Journal of Agricultural Economics*, **18**: 428-30.

Hayami, Y. (1964). 'Demand for fertilizer in the course of Japanese agricultural development', *Journal of Farm Economics*, **46**: 766-79.

——— (1968). 'On the growth of Japanese agriculture: a review article', *Rural Economic Problems*, **4**: 79-88.

——— (1971). 'Elements of induced innovation: a historical perspective for the green revolution', *Explorations in Economic History*, **8**: 445-72.

——— (1972). 'Rice policy in Japan's economic development', *American Journal of Agricultural Economics*, **54**: 19-31.

——— and Ruttan, V. W. (1970a). 'Factor prices and technical change in agricultural development: the United States and Japan, 1880-1960', *Journal of Political Economy,* **78**: 115-41.

——— and ——— (1970b). 'Korean rice, Taiwan rice and Japanese agricultural stagnation: an economic consequence of colonialism', *Quarterly Journal of Economics,* **84**: 562-89.

——— and ——— (1971). *Agricultural Development: an International Perspective,* Illinois.

——— and Yamada, S. (1968). 'Technological progress in agriculture', in L. Kleine and K. Ohkawa (eds.), *Economic Growth: the Japanese Experience Since the Meiji Era,* Illinois.

——— and ——— (1969). 'Agricultural productivity at the beginning of industrialisation', in K. Ohkawa, B. F. Johnston and H. Kaneda (eds.), *Agriculture and Economic Development: Japan's Experience,* Tokyo.

Ishibashi, Y. (1961). *Teikoku Nokai Kome Seisanshi Chosa Shynsei [Compilation of Rice Production Cost Surveys by the Imperial Agriculture Society],* The National Research Institute of Agriculture, Tokyo.

Iso, E. (1964). *Horaimai Danwa [Discourse on the* Ponlai *Rice],* Udokukai.

Kawano, S. (1941). *Taiwan Beikoku Keizairon [A Treatise on the Rice Economy of Taiwan],* Tokyo.

Kayo, N. (ed.) (1958). *Nihon Nogyo Kiso Toke [Basic Agricultural Statistics of Japan],* Tokyo.

Kobayakawa, K. (ed.) (1959). *Chosen Nogyo Hattatsushi: Seisakuhen [History of Korean Agriculture: Policy Volume],* Tokyo.

Korean Government-General (a), *Chosen Tochikairyo Jigyo Yoran [Summary Report of Korean Land Improvement Projects],* Seoul.

——— (b), *Nogyo Tokeihyo [Agricultural Statistics],* Seoul.

Nakamura, J. I. (1966). *Agriculture and Economic Development of Japan 1873-1922,* Princeton.

——— (1968). 'The Nakamura versus LTES estimates of the growth rate of agricultural production', *Keisai Kenkyu,* **19**: 358-62.

——— (1969). Incentives, productivity gap and agricultural development in Japan, Taiwan and Korea. Columbia University, New York, mimeo.

Nihon Nogyo Hattatsushi Chosakai [Research Committee for the History of Agricultural Development] (1956). *Nihon Nogyo Hattatsushi [History of Japanese Agricultural Development],* vol. 9, Chuokoronsha, Tokyo.

Ogura, T. (ed.) (1963). *Agricultural Development in Modern Japan,* Tokyo.

Rosovsky, H. (1968). 'Rumbles in the ricefields: Professor Nakamura *vs* the Official Statistics, *Journal of Asian Studies*, **27**: 347-60.

Taiwan Government-General *Taiwan Nogyo Nenpo [Yearbook of Taiwan Agriculture]*, Taipei.

—— Bureau of Colonial Development (1928). *Shuyonosanbutsu Keizai Chosa [Economic Survey of Major Agricultural Products]*, Taipei.

Tobata, S. (1956). *Nihon Nogyo Hattatsushi [History of Japanese Agricultural Development]*. Nihon Nogyo Hattatsushi Chosakai [Research Committee for the history of Agricultural Development], Tokyo.

—— and Ohkawa, K. (1935). *Chosen Beikoku Keizairon [A Treatise on the Rice Economy of Korea]*, Tokyo.

Yanaihara, T. (1929). *Teikokushugika no Taiwan [Taiwan under Imperialism]*, Tokyo.

*Chapter 3*

Babcock, J. M. (1962). 'Adoption of hybrid corn: a comment', *Rural Sociology*, **27**: 332-8.

Bank of Taiwan (1958). *Jih-chu shih-tai Tai-wan ching-chi shih [The Economic History of Taiwan during the Period of Japanese Rule]*, vol. 1, Taiwan.

Civil Affairs Bureau (1946). *Tai-wan sheng wu-shih-i nien lai t'ung-chi ti-yai [A Statistical Summary of Taiwan Province for the Past Fifty-one Years]*, Taipei.

Eigichi, I. (1925). 'Suito naichishu [The Japanese rice seed]', *Taiwan Nojiho*, **165**: 1-61.

Food and Agricultural Organisation of the United Nations (1966). *Rice—Grain of Life*, FAO, Rome.

Griliches, Z. (1958). 'Research costs and social returns: hybrid corn and related innovations', *The Journal of Political Economy*, **66**: 419-31.

—— (1962). 'Profitability versus interaction: another false dichotomy', *Rural Sociology*, **27**: 327-30.

Joint Commission on Rural Reconstruction, Rural Economics Division (1966). *Taiwan Agricultural Statistics, 1901-65*, Taipei.

Katsuji, I. (1920). 'Inyu naichibei naichishusui saibai no shorei [Promoting the cultivation of the Japanese type paddy rice seed]', *Taiwan Nojiho*, **165**: 1-20.

Myers, R. H. and Ching A. (1964). 'Agricultural development in Taiwan under Japanese colonial rule', *The Journal of Asian Studies*, **23** (4): 555-70.

Rinji Taiwan kȳukun chōsa dainibu [Second section of the Commission for the Investigation of Old Customs] (1905). *Chosa Keizai shinyo hokoku [A Report of Economic Investigation Materials]*, vol. 1, Tokyo.

Saiwasa, U. (1923). *Taiwan beikoku menkan [The Taiwan Rice Yearbook]*, Taipei.

Schultz, T. W. (1968). *Economic Growth and Agriculture*, New York.

Sheng, W. (1959). 'Jih-chu shih-tai T'ai-wan mi-keng nung-yeh chih chi-shu K'ai-fa [The technical development of rice farming in Taiwan under Japanese rule]', *T'ai-wan ching-chi shi-ch'i-chu [Taiwan Economic History]*, **7**, Bank of Taiwan, Taipei.

Taiwan Sotokufu (1897-1942). *Taiwan Sotokufu Tokeisho [Statistical Yearbook of the Governor-General of Taiwan]*, 46 vols., Taipei.

Taiwan sotokufu shokusan kyoku (1927). *Juyo nosanbutsu keizai chosa, suito, Taisho juyonen dai niki saku [An Economic Survey of Main Farm Products: Paddy Rice]*, Taipei.

—— (1938). *Taiwan no nogyo [Taiwan Agriculture]*, Taipei.

—— (1940). *Noka rodo chosa: suito jisaku noka [An Investigation of Farm Labour: Rice Farm Cultivators]*, Taipei.

*Chapter 4*

Ishikawa, S. (1969). 'Direction of technological change in agricultural production in the ECAFE region during the 1970s', paper submitted to the Fifth U.N. Inter-regional Seminar on Development Planning, Bangkok, mimeo.

Johnston, B. F. and Cownie, John (1969). 'The seed-fertiliser revolution and labour force absorption', *American Economic Review*, **59** (4): 569-82.

Ladejinsky, Wolf (1969a). 'The green revolution in Bihar: the Kosi area, a field trip', *Economic & Political Weekly*, 4 (39): 147-62.

—— (1969b). 'The green revolution in Punjab: a field trip', *Economic & Political Weekly*, 4 (26): 73-82.

—— (1970). 'The green revolution in India', *Statesman* (New Delhi), September.

Lockwood, Brian, Mukherjee, P. K., and Shand, R. T. (1971). *The High Yielding Varieties Programme in India Part 1*, Program Evaluation Organisation, Planning Commission and Department of Economics, and Research School of Pacific Studies, Australian National University, Canberra.

Programme Evaluation Organisation, Planning Commission (1967-70). *Reports on the Evaluation of the High Yielding Varieties Programme*, New Delhi.

*Chapter 5*

Vink, G. J. (1941). *The Basis of Indonesian Farm Business,* Wageningen.

*Chapter 6*

Bernal, E. A. (1967). 'The role of landlords in Philippine agricultural development', Unpublished M.S. thesis, College of Agriculture, University of the Philippines, Laguna.

Liao, S. H. (1968). 'Factors affecting productivity and adoption of improved farm practices in rice farms', Unpublished M.S. thesis, College of Agriculture, University of the Philippines, Laguna.

Von Oppenfeld, H., et al. (1957). *Farm Management, Land Use and Tenancy in the Philippines,* Central Experiment Station Bulletin 1, College of Agriculture, University of the Philippines, Laguna.

*Chapter 7*

*Bangkok Bank Monthly Review* (1971). Bangkok.

Barker, R. (1971). 'The evolutionary nature of the new rice technology', paper presented to the Conference on Agriculture and Economic Development, Tokyo, 1971, to be published in Food Research Institute Studies, Stanford.

Chuchart, C. and Tongpan, S. (1965). 'The determination and analysis of policies to support and stabilize agricultural prices and incomes of the Thai farmers (with special reference to the rice premium)', unpublished paper, Ministry of National Development and SEATO, Bangkok.

Corden, W. M. (1967). 'The exchange rate system and the taxation of trade', in T. H. Silcock (ed.), *Thailand: Social and Economic Studies in Development,* Canberra.

——— and Richter, H. V. (1967). 'Trade and the balance of payments' in T. H. Silcock (ed.), *Thailand: Social and Economic Studies in Development,* Canberra.

Dasananda, S. (1968). 'Rice cultivation in Thailand', paper presented to the Technical Staff Meeting, the International Rice Research Institute, Los Banos, Philippines (mimeo).

Hsieh, S. C. and Ruttan, V. W. (1967). 'Environmental, Technological and Institutional Factors in the growth of Rice Production: Philippines, Thailand and Taiwan; *Food Research Institute Studies,* **7** (3): 307-41.

Ingram, J. C. (1971). *Economic Change in Thailand 1850-1970,* Stanford.

Israngkura, Vanrob (1967). *A Study on Rice Production and Consumption in Thailand,* Ministry of Agriculture, Bangkok.

Jackson, B. R. (n.d.). 'Thailand rice breeding program', (mimeo).

———, Panichapat, W. and Awakul, S. (1969). 'Breeding, performance and characteristics of dwarf, photoperiod non-sensitive rice varieties for Thailand' *Thai Journal of Agricultural Science,* **2**: 83-92.

Kulthongkham, S. and Ong, S. (1964). *Rice Economy of Thailand.* Ministry of Agriculture, Division of Agricultural Economics, Bangkok.

Netherlands Engineering Consultants (NEDECO) (1969). 'Report of the Farm Management Survey', Land Consolidation Project, Phase II, to the Royal Thai Irrigation Department, Ministry of National Development, Thailand.

Pookamana, P. and Jackson, B. R. (1970). 'Seed production and distribution', paper presented to the IRC Working Party on Rice Production and Protection, Thirteenth Session of the International Rice Commission, Teheran.

Ruttan, V. W. and Soothipan, A. (1970). 'Rice Production in Thailand: Comment', *World Crops,* **22**: 390-5.

———, Soothipan, A. and Venegas, E. C. (1966). 'Changes in Rice Growing in the Philippines and Thailand', *World Crops,* **18**: 18-33.

Silcock, T. H. (1970). *The Economic Development of Thai Agriculture,* Canberra.

Sopanaratna, P. (1968). 'High Yielding Varieties Program and Rice Production Development in Thailand', Rice Department, Ministry of Agriculture, Bangkok (mimeo).

Thailand, Department of Customs (annual). *Annual Statement of Foreign Trade of Thailand,* Bangkok.

———, Government (1971). 'National Economic Development Board Appendix B, Estimate of Rice Production', in *National Income of Thailand: 1968-1969 Edition,* Bangkok.

———, ——— (1970). *Marketing Channels and Price of Agricultural Commodities in the Upper Region of the Chao Phaya River Basin in Thailand, 1969.* Ministry of Agriculture, Division of Agricultural Economics, Bangkok (in Thai).

———, Ministry of Agriculture, Division of Agricultural Economics (1967). *Agricultural Statistics of Thailand.* Bangkok.

———, ———, Rice Department (1962). *Annual Report on 1962 Rice Production in Thailand. University of Agriculture,* Bangkok.

———, ———, ——— (1970). 'Impact of the High Yielding Varieties of Rice on Thailand', paper presented to the Tenth FAO

Regional Conference for Asia and the Far East, Canberra. Ministry of Agriculture, Bangkok.

Trescott, P. B. (1968). 'Rice production in Thailand: Comment', *World Crops,* **20**: 49-56.

Usher, D. (1967). 'The Thai rice trade' in T. H. Silcock (ed.), *Thailand: Social and Economic Studies in Development,* Canberra.

Welsch, Delane E. (1971a). 'Agricultural problems in Thailand: some policy alternatives', *Bangkok Bank Monthly Review,* **12** (3): 89-104.

—— (1971b). 'Some speculations on the long run future of rice in Thailand', background paper prepared for the Annual Conference of the Agricultural Economics Society of Thailand, Bangkok (mimeo).

—— and Sprague, E. W. (1970). 'Technical and economic constraints on grain production in South-East Asia', in SEADAG, *Agricultural Revolution in South-East Asia: Vol. 1; Impact on Grain Production and Trade,* New York.

—— and Tongpan, Sopin (1971). 'Rice in Thailand', in *Proceedings of the Rice Policy Conference,* International Rice Research Institute, Los Banos, The Philippines.

Yantasast, A., Prechachat, C. and Jackson, B. R. (1970a). 'Breeding dwarf varieties of rice for tolerance to deep water', *Thai Journal of Agricultural Science,* **3**: 119-33.

——, —— and —— (1970b). 'Deep Water and Floating Rice', paper presented to IRC Working Party on Rice Production and Protection, Thirteenth Session of the International Rice Commission, Teheran (mimeo).

*Chapter 8*

Barker, R. (1969). 'Economic aspects of high yielding varieties of rice, with special reference to national rice policies', F.A.O., *Monthly Bulletin of Agricultural Economics and Statistics,* **18** (6): 1-9.

British Broadcasting Corporation. *Summary of World Broadcasts, Part 3, The Far East,* London, daily.

Burma, Government of the Union (annual 1952-65). *Economic Survey.*

—— (1951-65). *Quarterly Bulletin of Statistics.*

—— (1965, 1967). *Statistical Year Book.*

—— (annual 1964/5 to 1971/2). *Report to the People by the Union of Burma Revolutionary Council on the Revolutionary Government's Budget Estimates.*

—— (1967). *Statistical Pocket Book.*

—— *Forward,* fortnightly, 1963 to date.

Burma Socialist Programme Party (1971). *Guidelines for the Framing and Implementation of the National Plan for the Economic Development of the Union of Burma*, Rangoon (in Burmese).
——— (1971). *Report on Guidelines for the Framing and Implementation of the National Plan for the Economic Development of the Union of Burma*, Rangoon (in Burmese).
*Far Eastern Economic Review*, weekly.
*The Guardian*, Rangoon, daily, 1969 to date.
Knappen, Tippetts, Abbett and Associates (1953). *Comprehensive Report on Economic and Engineering Survey of Burma for Burma Economic Council*, vol. 1, London.
Nuttonson, M. Y. (1963). *The Physical Environment and Agriculture of Burma*, Washington.
Richter, H. V. (1969). 'The Union of Burma', in R. T. Shand (ed.), *Agricultural Development in Asia*, Canberra.
——— and Edwards, C. T. (1972). 'Recent economic developments in Thailand', in R. H. and E. C. Chapman (eds.), *Studies of Contemporary Thailand*, Canberra.
Silcock, T. H. (1970). *The Economic Development of Thai Agriculture*, Canberra.
Thailand, Ministry of Agriculture (1967). *Agricultural Statistics of Thailand*, Bangkok.
Whitcombe, Elizabeth. (1972). *Agrarian Conditions in Northern India*, vol. 1, Berkeley.
*Working People's Daily*, Rangoon, daily 1969 to date.

*Chapter 9*

Aresvik, Oddvar (1969). 'Fertiliser sale in January 1969', Report to the Minister of Agriculture, Government of Pakistan, Islamabad (mimeo).
Bose, S. R. (1969). 'East-West contrast in Pakistan's agricultural development', paper presented at the International Economic Association Conference on the Economic Development of South Asia, Kandy (mimeo).
Eckert, Jerry B. (1970). 'The beginning of change in West Pakistan's agriculture' (mimeo).
Falcon, W. P. (1967). 'Agricultural and industrial interrelationships in West Pakistan', *Journal of Farm Economics*, **49** (4): 1,139-54.
——— and Gotsch, Carl H. (1968). 'Lessons in agricultural development —Pakistan', in Gustav F. Papanek (ed.), *Development Policy —Theory and Practice*, Cambridge.

Ghaffar, Mohammad and Clark, Edwin H., II (1968). *Installation of Private Tubewells in West Pakistan* 1964-1967, Pakistan Institute of Development Economics, Research Report No. 71, Karachi (mimeo).

Gotsch, Carl H. (1968). 'Regional agricultural growth: the case of West Pakistan', *Asian Survey,* **8** (3): 188-205.

Hirschman, Albert O. (1958). *The Strategy of Economic Development,* New Haven.

Ishikawa, Shigeru (1967). *Economic Development in Asian Perspective,* Tokyo.

Johnston, Bruce F. and Cownie, John (1969). 'The seed-fertiliser revolution and labour force absorption', *American Economic Review,* **54** (4): 569-82.

Kaneda, Hiromitsu (1969). 'Economic implications of the "green revolution" and the strategy of agricultural development in West Pakistan', *Pakistan Development Review,* **9** (2): 111-43.

—— and Ghaffar, Mohammad (1970). 'Output effects of tubewells on the agriculture of the Punjab: some empirical results', *Pakistan Development Review,* **10** (1): 68-87.

Lewis, Stephen R., Jr (1968). 'Effects of trade policy on domestic relative prices: Pakistan 1951-64', *American Economic Review,* **58** (1): 60-78.

Mohammad, Ghulam (1964a). 'Waterlogging and salinity in the Indus Plain', *Pakistan Development Review,* 4 (3): 358-403.

—— (1964b). 'Some strategic problems in agricultural development in Pakistan', *Pakistan Development Review,* 4 (2): 223-60.

—— (1965). 'Private tubewell development and cropping patterns in West Pakistan', *Pakistan Development Review,* **5** (1): 1-53.

Pakistan, Central Statistical Office (1966). *Pakistan Statistical Yearbook, 1964,* Karachi.

——, Economic Adviser to the Government (1969). *Pakistan Economic Survey, 1968-69,* Islamabad.

United States Agency for International Development (1968). *Statistical Fact Book (1968),* Washington.

*Chapter 10*

Hume, A. O. (1879). Agricultural Reform in India. Madras

Whitcombe, Elizabeth (1971). *Agrarian Conditions in Northern India. Volume I: The United Provinces under British Rule 1860-1900,* Berkeley.

*Chapter 11*

Castillo, G. T. (1964). *Man with Many Faces. The Researcher In a Program of Planned Change*, A/D/C Paper No. 1, New York.
——— (1970). 'The dynamics of "coordination" and team approach in a rural development project', *Solidarity*, 5 (7): 41-64.
——— Dimaano, C. M., Calleja, J. C. and Parcon, S. F. (1964). 'A development project in action: a progress report on a Philippine care', *Asian Studies*, 2 (1): 37-66.
de Guzman A. M. and Dimaano C. M. (1968). 'Coralan rice farmers' response to change in cropping pattern. A case study', *The Philippine Economic Journal*, 7 (2): 224-41.
——— *et al.* (1969). *Work Roles of Barrio Extension Workers in a Cooperative Rival Development Project: An Analysis.* Farm and Home Development Office, College of Agriculture, University of the Philippines (mimeo).
Dimaano, C. M. and de Guzman, A. M. (1969). *Farmers' Association for Commodity Exchange and Services of Santa Maria.* Farm and Home Development Office, College of Agriculture, University of the Philippines.
Pahud, Soledad L. (1969). 'Some factors associated with the adoption of recommended rice practices', unpublished M.S. Thesis, College of Agriculture, University of the Philippines.

*Chapter 12*

Bardhan, Pranab (1970). 'Green revolution and agricultural labourers', *Economic and Political Weekly*, Special Number 29-31.
Bose, S. R. and Clark, B. H. (1969). 'Some basic considerations on agricultural mechanisation in West Pakistan', *Pakistan Development Review*, 9 (3).
Falcon, N. P. (forthcoming). *The Green Revolution: Generations of Problems.*
Fischer, Louis (1944). *A Week with Gandhi*, Bombay.
Fleming, Donald (1970). 'Big science under fire', *Atlantic Monthly*.
Mommsen, Theodore (1958). *The History of Rome*, New York.

*Chapter 13*

Abel, Martin (1970). 'Agriculture in India in the 1970's', *Economic and Political Weekly*, 5 (13): A5-14.
Barker, R. and Mangahas, M. (1970). 'Environmental and other factors influencing the performances of HYV of wheat and rice.' Paper submitted to the XIV International Conference of Agricultural Economists, Minsk, USSR.

Blyn, George (1966). *Agricultural Trends in India, 1891-1947: Output, Availability and Productivity,* Philadelphia.

Brown, Lester (1965). *Increasing World Food Output: Problems and Prospects,* U.S. Department of Agriculture, Foreign Agriculture Economic Report No. 25, Washington.

——— (1970). *Seeds of Change,* New York.

Chowdhury, B. K. (1970). 'Disparity in income in the context of HYV', *Economic and Political Weekly,* **5** (39): A90-6.

Dandekar, V. M. (1970). 'Agricultural Growth with Social Justice in Over-populated Countries.' *Economic and Political Weekly,* **5** (Special Number): 29-31.

Dantwala, M. L. (1961). 'Agrarian structure in twelve districts', *The Economic Weekly,* **13** (Special Number): 27-9.

——— (1967). 'Incentives and disincentives in Indian agriculture', *Indian Journal of Agricultural Economics,* **22** (2).

——— (1968). 'Rejuvenating sick agriculture: Indian experience', *Netherlands Journal of Agricultural Science,* **16** (4): 234-42.

Ho, Yhi-Min (1966). *Agricultural Development in Taiwan, 1903-1960,* Nashville.

India, Agricultural Prices Commission (1968). *Annual Report 1967-8,* Ministry of Food and Agriculture, New Delhi.

———, ———, (1970). *Report on Price Policy for Kharif Cereals, 1969-70 Season,* National Sample Survey, 18th Round, New Delhi.

———, Ministry of Food and Agriculture, Economic and Statistical Adviser (1966). *Growth Rates in Agriculture 1949-50 to 1964-65,* New Delhi.

———, ———, Expert Committee on Assessment and Evaluation (1969). 'Modernising Indian Agriculture. *Report on the Intensive Agricultural District Programme (1960-68),* vol. 1, New Delhi.

International Rice Research Institute (IRRI) (1968, 1969). *Annual Report,* Manila.

Ladejinsky, Wolf (1969). 'Green revolution in Bihar-Kosi area: a field trip.' *Economic and Political Weekly,* **4** (39): 147-62.

Lipton, Michael (1968). 'Agriculture: urban bias and rural planning', in Streeten, Paul and Lipton, Michael (eds.) *Crisis of Indian Planning,* London.

——— (1970). 'India's agricultural performance, achievements, distortions and ideologies', *Asian and African Studies,* **6** (Special Number): 127-48.

Mason, Edward (1966). *Economic Development in India and Pakistan,* Centre for International Affairs, Harvard University, Cambridge.

Mellor, John (1968). 'The functions of agricultural prices in economic development', *Indian Journal of Agricultural Economics,* **23** (1): 23-37.

——— (1969). *The Economics of Agricultural Development,* Bombay.

Minhas, B. S. (1970). 'Rural poverty, land redistribution and development strategy', *Indian Economic Review,* New Series, **5** (1): 97-128.

Mukherjee, P. K. (1970). 'The HYV programme: variables that matter', *Economic and Political Weekly,* **5** (13): A15-22.

Paddock, William and Paul (1968). *Famine 1975!*, London.

Pearson, Lester (Chairman) (1969). *Partners in Development,* London.

Raj, K. N. (1969). 'Some Questions Concerning Growth, Transformation and Planning of Agriculture in Developing Countries.' *Journal of Development Planning,* **1**: 15-38.

Rockefeller Foundation (1969). *A Special Report from the Rockefeller Foundation,* New York.

Schultz, T. W. (1964). *Economic Crises in World Agriculture,* Ann Arbor.

Sen, B. (1970). 'Opportunities in green revolution', *Economic and Political Weekly,* **5** (13): A33-40.

Sen, S. R. (1967). *Growth and Instability in Indian Agriculture,* Agricultural Situation in India XXI, 10, Ministry of Food and Agriculture, Government of India.

Shand, R. T. (ed.) (1969). *Agricultural Development in Asia,* California.

Srinivasan, T. N. (1970). Review of Michael Lipton, 'Crises in Indian planning', *Journal of Economic Literature,* **8** (2): 455-8.

Streeten, Paul and Lipton, Michael (eds.) (1968). *Crisis in Indian Planning,* London.

Warriner, D. (1969). *Land Reform in Principle and Practice,* London.

United Nations (1968). *Economic Survey of Asia and the Far East,* United Nations, Economic Commission for Asia and the Far East, Bangkok.

United States, Agency for International Development (1970). *War on Hunger,* Washington.

*Chapter 14*

Lockwood, Brian, Mukherjee, P. K. and Shand, R. T. (1971) *The High Yielding Varieties Programme in India,* Canberra.

Shand, R. T. (ed.) (1969). *Agricultural Development in Asia,* Canberra.

# Index

# Index


Abe, Kameji, 14
Abel, Martin, 278
Agrarian reform, 79, 159, 189, 235, 240, 242, 244, 256, 263-4, 279-80; and the Green Revolution, 245-55; Asia, 235-58; Burma, 159-60; India, 238-40, (Minimum Reform Program) 240-2, 243-4, 244-5, 245-55, 263-4, 275-80; Japan, 235, 237-8, 241, 243; Pakistan, 238; Philippines, 110-11, 118, 220, 277; Taiwan, 237-8, 241, 243-4, 276; U.S.S.R., 237; *see also* India, Land tenure
Agricultural and Rural Development Corporation (ARDC) (Burma), 150-1
Agricultural development, strategies, 182
Agricultural policy, *see* countries
Agricultural Prices Commission (India), 266, 268
Agricultural Research Institute (India), 54
Agro-Economic Survey (Java), 81-107 *passim*
Alkalinity, 188, 189, 192, 197
Alternaria blight, 199
Amritsar District (India), 69
Andhra Pradesh (India), 63, 66-8, 69, 247
Aquifers, 196-7
Areas planted: as indicators, 283; food grains, 53, 56, 157; industrial crops, 156-7; rice, 15, 17, 20, 33, 71, 129-34, 148-9, 157, 159, 173, 283; wheat, 55, 169, 173, 283
Australian National University, The, 58

Bacterial leaf blight, 138
Bakanae disease, 138
Bali, 81
*Barrios*, 108-23 *passim*, 202-34 *passim*
Bihar (India), 62, 63, 66-8, 73, 247, 248
'Bimas', 80-1, 93, 106
Biñan Municipality (Philippines), 108-23 *passim*
Blast disease, 138
Britain, rice technology in Japan, 5, 12
Bundelkhand (India), 189
Bureau of Agricultural Extension (Philippines), 202
Burma: agrarian reform, 159-60; agricultural zones, 145-50; Department of Agriculture, 150-1; government planning, 159-60; industrial crops, 156-7; irrigation, 144, 147-8, 149, 156, 157, 292; land tenure, 159; land use, 145-6, 292; population, 147; price policies, 159-60; research and extension, 150-2; rice (area) 144, 148-9, 157, (exports) 144, (fertilisers) 153, 155, 157, (field crops) 156-9, (high yield strains) 152-3, (production) 147, 148, (returns) 155, (value of seed) 158, (yields) 153; separation from India, 52; subsistence crops, 150; water control, 144, 154

Cabuyao Municipality (Philippines), 108-23 *passim*
Calamba Municipality (Philippines), 108-23 *passim*
Canals, 190; and crop patterns, 190; and the hydrological cycle, 191; development, 190, 192; and water distribution, 193-4
Canal Department (India), 185, 194
Capital: India, 76-7; Pakistan, 162; Philippines, 218-20
*Carabao* (tractor), 232
Celebes, 81
*Changwats*, 129
Chao Phya delta (Thailand), 129
Chen-Cheng, General, 244
Chiba Prefecture (Japan), 4
Chin Special Division (Burma), 144-60 *passim*
College of Agriculture (Philippines), 230
Combine harvesters, *See* Harvesters, combine
Co-operatives, 74
Corn, hybrid, 39

Costs of production: Japan, Korea, Taiwan, 20-1, 25; Philippines, 112-14, 120-2, 286-7; Taiwan, 46-7, 48, 49
Cotton, 159
Council of Agricultural Research (India), 54
Crawford, Sir John, 58
Credit: India, 55, 57-8, 74-5, 248, 264; Philippines, 118-20, 220-1, 231-2; Taiwan, 48-9
Cropping patterns: Japan, Korea, Taiwan, 22, 23; Philippines, 110

Dandekar, V. M., 279
Deforestation, 190
Diesel engines, 175, 176-8, 180
Diseases, *see* Bacterial leaf blight, Bakanae disease, Blast disease, Gall midge, Tungro
Doab (India), 183, 185, 187, 191
Drainage, *see* Water resources and control

Education, aspirations (Philippines), 223-5
Elasticity of demand, 38
Engineering industry, Pakistan, 165-6, 175-9

Farmer associations: India, 76; Japan, 14; Philippines, 230, 231; Taiwan, 31, 32; Thailand, 140
Farmers, as inventors: progressive (Philippines), 217-18, 219, 220-2, 229; *rono* (Japan), 2, 3, 6-7, 12-13, 14, 16
Farmers' Association for Commodity and Exchange Services, (FACES) (Philippines), 231
Farm practices: Japan, Korea, Taiwan, 2, 3, 9-24, 31, 32, 41, 49, 276; Pakistan, 164-5; Philippines, 203-7, 216-17, 275-80; Thailand, 133-4; *see also* Technology, new
Farm sizes: celing, 237-8; India, 64-9; Java, 85, 87-9; Pakistan, 180-1, 237-8; Philippines, 110-11, 205
Fertilisers: and soil fertility, 192; as a factor, 25, 26; Burma, 153, 155, 157; consumption, 57, 70, 80-1, 97-9, 272; dosage, 70, 71, 74, 115, 186; expenditure, 48-9; India, 55, 57, 70, 71, 74, 272; Japan, 9, 16-17, 20, 22, 26; Java, 80, 97-9; Korea, 9, 20, 22, 26-7; nitrogenous, 70, 71, 97-9, 114-15; Pakistan, 173-5, 181; Philippines, 113, 114-17, 203, 204; phosphatic, 98-9; price, 141-2, 266; product price ratios, 266; recommendations, 192-3; Taiwan, 9, 16, 17, 20, 22, 26-7, 46; *see also* Inputs, packages
Fesca, Max, 1, 5
Fischer, Louis, 258
Five Year Plans, *see* India
Flooding, 63, 75, 144, 154; *see* also Water resources and control
Food and Agriculture Organisation (FAO) Rice Hybridisation Project, 54
Foodgrains, international trade prospects, 294; *see also* Areas planted, names of grains, Yields
Ford Foundation, 54
Fukuoka Agricultural School (Japan), 6, 7

Gall midge, 138
Gaya District (India), 69
Ghandi, Mahatma, 258
Germany, rice technology in Japan, 2, 5, 12
Government, role in agriculture, 286; Burma, 159-60; India, 72-3; Japan, 3-6; Philippines, 225-8
Green Revolution, 9, 73, 107, 124, 172-5, 192, 202, 231, 235, 245-55, 255-7, 268, 280, 281
Groundnuts, 158
Ground water, *see* Water resources and control
Grow More Food Campaign (India), 53
Gujarat (India), 241

Harvesters, combine, 180
Haryana (India), 62, 65-9 *passim*, 73
Hayashi, Enri, 3, 7
Herbicides: India, 71; Pakistan, 181; Philippines, 116, 117, 203, 204
High Yielding Varieties: as indicators, 283; rice, 9, 16, 19, 24, 27; *see also* India
High Yielding Varieties Programme (India), 51, 52, 56, 58, 62, 70, 72, 73, 74, 76, 77, 78, 79, 186, 194, 195, 196, 197, 291
Hokkaido University, 4
Hume, A. O., 183, 191
Hydrological surveys, 195, 196

Incomes and wages, 288; India, 78-9, 248, 249, 253; Japan and Taiwan, 125; Java, 92; Philippines, 208, 209-11, 222-3, 234; Thailand, 142
India: agrarian reform, 52, 78-9, 238-55, 263-4, 269, 275-80; agricultural policy, 72-3, 263, 265, 268-70, 286; areas planted, 53, 55, 71; capital investment, 76-7; co-operatives, 74-5; credit, 55, 57-8, 74-5, 248, 264; distribution of seed, 174; double cropping, 29, 77; electricity supply, 57; farmers' associations, 76; farm practices, 51, 52, 271-2; farm sizes, 64-9, 239-40, 242, 277; fertilisers, 55, 57, 70, 71, 74, 272; Five Year Plans, 53, 54, 56, 77, 274; high yielding variey adoption and packages, 55-72, 74, 235; incomes, 253; irrigation, 57, 59, 63, 69, 75-6, 185, 192, 248, 272, 274-5, 277-8, 284-5, 289-90; labour, 76-7, 78, 252-5, 265-8, 280-1; land tenure, 52, 79, 237-9, 240, 241, 249-52, 277-8; mechanisation, 77, 252-5, (pumps) 76, (tractors) 255, (tubewells) 65, 76, 272; millet, sugarcane, sorghum, maize, 69, 73-4, 270; population growth, 52; prices, 260-3, 265-8, 281; productivity, 52, 55, 260-1, 273; research and extension, 54-5, 187-8, 194-6, 197, 280; returns, 286; Rice States, 69-72; taxation, 78; water resources and control, 63, 75-6, 187-8, 194-6, 197, 280; welfare, 78-9, 248, 249, 257; Wheat States, 65-9; yields, 53-4, 261, 276-8, (rice) 260-1; *see also* Royal Commission on Agriculture
Indonesia, *see* Java
Indus Basin, 171, 172, 173
Industrialisation and mechanisation: India, 175, 254-5; Japan, 10; Pakistan, 165-6, 180-1, 182; Philippines, 232-3; Thailand, 141; U.S.A., 180
Inputs: 'leading', 174; packages, (India) 55-72, (Java) 86-7, 93-4, 106, (Philippines) 110-11, 114-18, 203-7, 235, 285; *see also* Capital, Labour, Land, Water resources and control
Insecticides, *see* Pesticides
Intensive Agricultural Area Programs (India), 186
Intensive Agricultural Development Program (India), 186
Intensive Agricultural District Programs (IADP) (India), 54, 55, 77, 264-5
International Bank for Reconstruction and Development, 171
International Centre for Corn and Wheat Improvement (CIMMYT), 9, 54
International Rice Research Institute (IRRI), 9, 55, 138, 139, 270, 271
International Wheat Research Centre (Mexico), 154
Irrawaddy Division (Burma), 158
Irrigation: and double cropping, 290-3; and linkages, 295; and high yielding varieties, 192; availability by countries, 290-3; Burma, 144, 147-8, 149, 154, 157, 292; canal, 187, 290-1; characteristics, 290-1; development programs, 294; effectiveness, 291-3; future policy, 292-3; India, 57, 59, 63, 69, 75-6, 185, 192, 248, 272, 274-5, 277-8, 289-90; Japan, Korea, Taiwan, 16-22, 23, 24, 42, 293 294; Pakistan, 171, 173-4, 182; Philippines, 108, 123, 205, 284-5, 291-2; Thailand, 129-34, 290; *see also* Water resources and control and countries by name

Japan: agrarian reform, 237-8; agricultural schools, 4, 5; *see also* names of schools; agricultural societies, 14; farm practices, 2, 3; fertilisers, 9, 16-17, 20, 22, 26; irrigation, 16-22; Japanese plough, 1-2, 8, 31; tenure, 15, 237; rice technology (imported) 1-8, 12, (exported) 9-12, 18-27; rice production (Meiji era) 1-8, 9, 11-19, 294, (Meiji system) 7-8, (Modern) 9-10, (role of government) 3-5, (role of local authorities) 6, (*rono* system) 2, 3, 6-7, 12-13, 14, 16, (Tokugawa period) 13, 16-18; Rice Production Development Program, 19, 22; Rice Riot, 18-19; Samurai outlook, 4-5; water resources and control, 18, 22; yields, 10-12, 17, 19, 20, 22, 28-9
Java: population, 80; rice production (Agro-Economic Survey) 18-105 *passim,* (farmers) 82, 86, (farm sizes) 85, 87-9, (fertilisers and pesticides) 80, 97-9, (incomes) 92, (labour) 76-7, 95-7, (land tenure) 85, 90-1, 107, (Paddy Centra-Scheme) 80, (package

Java (Cont.)
programs) 86-7, 93-4, 106, 285, (regions) 84-5, (returns) 101-3, 105, 287-8, (yields) 99-101, 102, 104
Jute, 159

*Kabupaten*, 82-5
*Kaing*, 147, 148, 149
Kerala (India), 63, 66-8, 241
Kerner, Oscar, 5
*Khaira*, 200
*Kharif* (monsoon) crops, 57, 58, 60-1, 63, 64, 66-7, 69, 70, 71, 75, 77, 171, 172, 190, 194, 199
Komaba Agricultural School (Japan), 4, 5, 6, 7, 12
Korea: Japanese rice technology imported, 9-12, 19, 20-1, 22, 26-7; water resources, 293
Kosi District farmers (India), 248-9
Labour: India, 76, 77, 78, 252-5, 265-8, 280-1; Java, 76-7, 95-7; Pakistan, 169, 252; Philippines, 113, 117-18; Taiwan, 41-7

Ladejinsky, Wolf, 278
Laguna (Philippines), 108-23 *passim*
Land availability: Japan, 11-12; Taiwan, 25; *see also* Agrarian reform
Land reform, *see* Agrarian reform
Land Reform Program (Philippines), 208
Land tenure: Burma, 159; India, 52, 79, 237-9, 240, 241, 249-52, 277-8; Japan, 25, 237-8; Java, 85, 90-1, 107; Philippines, 110-11, 118-22, 208-9, 219, 277; Taiwan, 25, 46, 48, 49, 237-8, 276; *see also* Agrarian reform
Liao, S. H., 108
Lipton, Michael, 266

Mabitac municipality (Philippines), 202-34 *passim*
Maharashtra (India), 65-9 *passim*, 241
Maize, 157, 159; hybrid, 55, 73
Mandalay Division (Burma), 153
Marketing: Philippines, 232; Thailand, 134-7; *see also* Prices
Maruo Jujiro, 14
Mechanisation, *see* Industrialisation
Meerut Division (India), 185, 187, 196
Mexico: agricultural performance, 273, 275; rates of return on hybrid corn, 39; wheat varieties, 55, 56, 59-62, 65-7, 68, 69, 72, 79
Millet, hybrid, 55, 73, 199
Milling of rice, 134-5
Mintras, B. S., 279
Mixed cropping, 200
Momsen, Theodore, 244
Money-lenders, 231; *see also* Credit
Monsoons, *see Kharif*
Mortgage laws (India), 189
Mukherjee, P. K., 277, 278
Multiple Cropping Programme (India), 56, 73, 76
Mysore (India), 66-8

Nakamura, Naozo, 2
National Agricultural Experiment Station (Japan), 7, 13; branches, 18; South Korea branch, 24
'New Strategy', 55-6, 57, 72, 73, 79
Nishigahara Experiment Farm for Staple Cereals and Vegetables (Japan), *see* National Agricultural Experiment Station
North-western Provinces (India), 183, 185

Orissa (India), 66-8
Oudh (India), 183, 184
Output, as an indicator, 283; *see also* Yields
Ownership, *see* Land tenure

Packages, *see* Inputs
Paddock, Paul, 259
Paddy, *see* Rice
Paddy Centra-Scheme (Java), 80
Pakistan: areas planted, 169, 173, 283; capital, 162; engineering industry and mechanisation, 165-6, 253-4, (combine harvester) 180, (diesel engine) 175, 176, 180, (pumps) 175, 180, (strainers) 175, 176, (tractors) 166, 167, (tubewells) 170-2, 174, 175-9; farm practices, 164-5; fertilisers, 173-5, 181; ground water, 170-1; irrigation, 171, 173-4, 182; pesticides and weedicides, 181; Plans, 168-70; population, 161-2; price supports, 170; productivity, 162-5, 168-70, 173; returns, 286; size of farms, 180-1, 237-8; water resources and controls, 70-2; yields, 169, 173
Palay, *see* Rice
*Panchayats*, 257
Pearson Report, 271

Pegu Division (Burma), 158
Pesticides: Java, 97-9; Pakistan, 181; Philippines, 116, 117, 203
Philippines, rice production: credit, 118-20, 220-1, 231-2; costs and returns, 112-14, 120-2, 286-7; diffusion of high yielding varieties, 111-12, 285; farmers (attitudes to farming) 215-18, (attitudes to investment) 218-20, (community patterns) 228-30, (expectations of government) 225-7, (material conditions) 211-13, (political participation) 227-8, (progressive) 217-18, 219, 220-2, 229, (responsibilities and aspirations) 222-5, (status perception) 213-14; farm practices and inputs, 203-7, 275-80, (chemicals and fertilisers) 113, 114-17, 203, 204, (irrigation) 108, 123, 205, 291-2, (labour) 113, 117-18; farm sizes, 110-11, 205; incomes, 208, 209-11, 222-3, 234; land tenure, 110-11, 118-22, 208-9, 219, 277; marketing, 232; taxation, 228; tractors, 232-3; yields, 110, 112-14, 122, 203, 205, 206; *see also* College of Agriculture
Photosensitivity, 19
Planning Commission (India), 51
Population growth: Burma, 147; India, 189; Java, 80; Pakistan, 161-2
Prices and price policy: Burma, 159-60; India, 265-8; Japan, 10-11; Pakistan, 167, 170; Taiwan, 32; Thailand, 135-6
Production function, 24-7, 172
Profits, *see* Returns
Programme Evaluation Organization (PEO) (India) 51, 58, 69, 72, 274, 278, 283, 298
Pulses 159
Pumps, 76, 108, 141, 175, 180
Punjab (India), 52, 62, 65-71 *passim*, 73, 74, 175, 179, 242, 247, 252-3, 254, 284
Purnea (India), 248-9

*Rabi* crops, 57, 58, 60-1, 63, 64, 66-7, 68, 69, 70, 71, 72, 73, 75, 77, 172, 190, 193, 194, 197, 199, 271
Raj, K. N., 272-3
Rajasthan (India), 62, 65-9 *passim*
Rents, *see* Land tenure
Research and extension: Burma, 150-2; India, 269, 280; Korea, 19, 22; Philippines, 207; Taiwan, 36-7, 39-40, 41; Thailand, 137-40, 142-3
Returns, 286-8; Burma, 155; India, 286; Java, 101-3, 105, 287-8; Pakistan, 286-7; Philippines, 112-14, 119, 120-2; Taiwan, 48
Rice: as a wage good, 10; cropping pattern, 205, *see also* Mixed cropping; eating quality, 155, 289; glutinous, 111, 112; high yielding varieties, 9, 16, 20, 22, 24, 26-7, 34, 35, 48-9, 58-64, 67, 69-70, 72, 79, 86-9, 111-12, 139-42, 153-6, 173, 199-201, 203-7, 247, 270, 284-5, 289, 290, 292; improved varieties, 2, 14, 15, 17, 31; milling, 232; native, 32; upland, 49; weeding, 203-4; *see also* Areas planted, Costs, Farm practices, Fertilisers, Herbicides, Irrigation, Mechanisation, names of rice producing countries, Pesticides, Prices, Research and extension, Rice varieties, Yields
Rice Department, Thailand, 139-40
Rice varieties and strains: ADT-27, 54, 57, 63; BPI-76, BPI-76-1, 204; C-4-63, 153; C-18, 204; C-53-39 (ngwetoe), 153; chailai, 19-21, 24-5, 27, 28-50 *passim*; gampai-15, 138; ginbozu, 24; indica, 24, 55, 57, 63; IR varieties, 57, 62-3, 64, 81, 111, 112 135, 138, 139, 152, 153, 154-5, 204, 205; intan, 112; japonica, 24; kamenoo, 14, 15; kokurato, 24; leb mue nahnga, 111, 138; leuang tawng, 138; malagkit, R4, 112; National Improved varieties (bengewan, sigadis, remadja, shinta), 80-108 *passim*; peta, 138, 204; PL480, 167-8, 265; ponlai (horai), 19-21, 24-5, 27, 28-50 *passim*; RD1, RD2, RD3, 138-43 *passim*, 283; rikku no. 132, 24; shinriki, 14, 15; taichung native-1, 138; tainan, 204; Taiwanese native varieties, 28-50 *passim*; tamanishiki, 24; tjeremas, 204; yagyaw, *see* IR varieties
Royal Commission on Agriculture, 1929 (India) 52
Rural Bank (Philippines), 231
*Ryotwari* areas, 239

Saline Control and Reclamation Projects (SCARP) (Pakistan), 170
Salinity, 185, 188, 189, 191, 192, 197

Salt Tax, 185
Samurai, 4, 7
Sanrijuka Sheep Station (Japan), 4
Santa Maria Municipality (Philippines), 202-34 *passim*
Sapporo Agricultural School (Japan), 4, 7
Seed, rice: classes, 140
Seed Corporation (India), 57
Seeds: bed preparation, 42; government distribution (Burma), 152, 158-9, (India) 174; *see also* High yielding varieties
Sen, B., 277, 278
Social returns of rice production (Taiwan), 36-41
Sorghum, hybrid, 55, 73
Stamp, Sir William, 185, 194
Strainers, 175, 176
Stewart, A. B., 185
Sugarcane, 33, 34, 73-4, 159
Sumatra, 81
Surplus foodgrain, 38, 124

Taiwan: agrarian reform, 237-8, 276; agricultural associations, 31-2, 33; agricultural stations, 31-2; credit, 48-9; fertilisers, 9, 16-17, 20, 22, 26-7, 166; foodgrain production, 28, 273; irrigation, 22, 23, 24, 42, 293; land tenure, 25, 251-8; rice, (exports to Japan) 30, (Japanese technology imported) 9-12, 19-20, 28-31, (management impact of new technology) 41-6, (ponlai) 28-50 *passim*, (social impact of new technology) 36-41, (Taiwan technology in India) 62, (varieties) 31; water resources and control, 18, 22; yields, 10-12, 17, 19, 20, 22, 28-9, 30, 33, 49, 273
Tamil Nadu (India), 51, 66-8, 70-1, 73
Tarai (India), 186, 196
Taxation and agriculture: India, 78, 228; Japan, 19; Pakistan, 167; Philippines, 228; 'rice premium' (Thailand), 136-7, 141
Technology and techniques, *see* Farm practices, Technology, new
Technology, new, 25, 27, 142-3; adoption of, 285; and agricultural transformation, 273-5; and dualism, 273-5; and equity, 257; and farm size, 275, 277; and physical environment, 271, 274-5; and policies, 142, 144; and *rabi* season, 271; and tenancy, 275, 277, 278; and welfare, 256-7; areas sown, 283; effects (on farm incomes) 122, (on industry) 288-9, (on net farm incomes) 288; employment, 249-52; environmental adaptation, 284-5; evaluation of, 282; factors affecting, 284-9; farm size, 247, 248, 256; for rice, 285; for wheat 284; future prospects, 289-98; grower numbers, 283; income polarisation, 247, 249; indicators of progress, 283-4; inflation, 250, (in India) 271, (in Japan) 271, (in Philippines) 285; irrigation, 284-5; land prices, 250-1; linkages, 288-9; long-term policy, 298; mechanisation, 252-5; output changes, 283-4; participation, 246, 247; profitability, 250, 286-8, (and double cropping) 286-7, (and government policy) 286, (and location) 286, (and water availability) 286-7, 290; regional income balance, 248; rents, 250-1; secondary impact, 295-7, (and consumer demand patterns) 296-7, (and land ownership) 296, (and rural industry) 297; tenancy, 249-52; tenure, 251; *see also* Agrarian reform, Farm practices
Tenancy, *see* Land tenure
Thailand, rice production: areas planted, 129-34, (classification) 129-31; dissemination and adoption of new varieties, 139-42; exports, 124-8, 136-7; farm practices, 133-4; floating rice, 133, 138-9, 142; glutinous rice, 133-4; irrigation, 129-34, 290; marketing, 134-7; milling, 134-5; pricing, 135-6; research, 137-40, 142-3; rice premium, 136-7, 141; storage, 135; variety breeding and selection, 137-40; yields, 126
Tokyo University, 3, 4, 7
Tractors, 141, 166, 167, 232-3, 255
'Trickle-down' effect, 256
Tsukiji experimental station (Japan), 3-4
Tubewells, 76, 170-2, 174, 175-9, 185, 194-8, 272, 284-5; and the water table, 196; clustering, 196-7; tubewell industry, 175-8, 182
Tungro (yellow-orange leaf virus), 138, 199

Union Department of Agriculture (India), 55
United Provinces (India), 183, 184
United States: mechanisation, 180; rates of return on hybrid corn, 39; rice technology in Japan, 2, 3-8
University of the Philippines, 202
Uttar Pradesh (India), 62, 65-9 *passim*, 73, 247

Vink, G. J., 96

Wages, *see* Income and wages
Water and Power Development Authority (Pakistan), 170
Waterlogging, *see* Water resources and control
Water resources and control: Burma, 144, 154; India, 63, 75-6, 187-8, 194-6, 280; Japan, Taiwan, Korea, 18, 22; Pakistan, 170-2; *see also* Irrigation
Welfare: India, 78-9, 248, 249; Philippines, 222-5; Taiwan, 36-41
West Bengal (India), 63, 66-8
Wheat, *see* Areas planted, Fertilisers, Herbicides, India, Irrigation, Mechanisation, Mexico, Pakistan, Pesticides, Prices, Research and extension, Yields
Work-animal inputs, Taiwan, 43, 44-5, 46, 47, 48

Yields: foodgrains (Burma) 158, (India) 53, 56, 276-8, (Taiwan) 29, 30; rice, 285-6, (Burma) 153, (India) 260-1, (Japan, Korea, Taiwan) 3, 10-12, 17, 19, 20, 22, 28-9, 30, 33, 49, 273, (Java) 99-101, 102, 104, (Pakistan) 169, 173, (Philippines) 110, 112-14, 122, 203, 205, 206, (Taiwan) 29, 30, (Thailand) 126, 129-32; wheat, 250, 284-5, (India) 53-4, 261, 276-8, (Pakistan) 169, 173, 174
Yokoi, Tokiyoshi, 7

*Zamindari* areas, 238-9, 243-4, 269

Text set in 10 point Baskerville face, one point leaded,
and printed on 85 gsm Matt Valecote at
The Griffin Press, Adelaide, South Australia